Data Mining and Multi-agent Integration

Longbing Cao
Editor

Data Mining and Multi-agent Integration

Editor
Longbing Cao
Faculty of Engineering and Information Technology
University of Technology, Sydney
Broadway, NSW 2007
Australia
lbcao@it.uts.edu.au

ISBN 978-1-4899-8440-1 ISBN 978-1-4419-0522-2 (eBook)
DOI 10.1007/978-1-4419-0522-2
Springer Dordrecht Heidelberg London New York

Printed on acid-free paper

Springer is part of Springer Science+Business Media (www.springer.com)

Preface

Data Mining and Multi-agent Integration aims to reflect state-of-the-art research and development of *agent mining interaction and integration* (for short, agent mining).

The book was motivated by increasing interest and work in the agents data mining, and vice versa. The interaction and integration comes about from the intrinsic challenges faced by agent technology and data mining respectively; for instance, multi-agent systems face the problem of enhancing agent learning capability, and avoiding the uncertainty of self-organization and intelligence emergence. Data mining, if integrated into agent systems, can greatly enhance the learning skills of agents, and assist agents with predication of future states, thus initiating follow-up action or intervention. The data mining community is now struggling with mining distributed, interactive and heterogeneous data sources. Agents can be used to manage such data sources for data access, monitoring, integration, and pattern merging from the infrastructure, gateway, message passing and pattern delivery perspectives. These two examples illustrate the potential of agent mining in handling challenges in respective communities.

There is an excellent opportunity to create innovative, dual agent mining interaction and integration technology, tools and systems which will deliver results in one new technology. For example, if an open complex agent system is powered with actionable knowledge discovery capabilities, it then has the potential to deal with very complex problem solving with super-intelligent information processing, knowledge discovery, collective intelligence emergence, and actionable decision-making skills in complex environments. Currently, systems of this magnitude are not possible without the integration of agents and data mining.

This book, as the first in this area, does not intend to cover the field of agent mining. Rather, it features the latest methodological, technical and practical progress on promoting the successful use of agent mining. In 22 chapters, the book reflects state-of-the-art agent mining research and development. The book is divided into three parts. Part I provides an introduction to agents and data mining integration. Part II addresses data mining-driven agents, and Part III focuses on agent-driven data mining.

Part I has three introductory chapters. Chapter One presents a comprehensive introduction to interaction and integration of agents and data mining, which covers driving forces, disciplinary frameworks, agent-driven distributed data mining, data mining driven agents, mutual issues in agent mining, applications and case studies, trends and directions, and agent-mining community development. Chapter Two presents a brief overview of agent mining interaction, and two case studies. Chapter Three provides a survey on agent-based distributed data mining.

Part II has nine chapters outlining the latest progress and techniques for data mining driven agents. Chapter Four explores agent behavior by particle swarm optimization based web usage clustering. Chapter Five enhances agent learning through mining temporal patterns of agent behavior. Chapter Six summarizes the use of Web and structure mining to form an agent system which detects user interaction information. Chapter Seven presents an e-commerce-oriented distributed recommender system with peer profiling and selection strategy. Chapter Eight uses multi-class classification to build a multi-agent-based intrusion detection system. Chapter Nine proposes genetic algorithms and regular expressions to automatically learn about software entities. Chapter 10 proposes a words weight vectors driven network module for establishing and maintaining a knowledge network. Chapter 11 proposes goal mining of query logs for commonsense knowledge to equip intelligent agents. Chapter 12 proposes an agent-based interactive diagnostic workbench with diagnostic rules.

In Part III, we present 10 chapters on agents-driven data mining. Chapter 13 proposes an extensible multi-agent data mining system powered by association rule and classification. Chapter 14 proposes an anytime multi-agent approach to on-line unsupervised learning, which handles continuous agglomerative hierarchical clustering of streaming data. Chapter 15 proposes an agent system utilizing a divide and conquer approach and data dependent schemes for clustering large data. Chapter 16 proposes a multi-ant colony and a multi-objective clustering algorithm by combining the results of all colonies. Chapter 17 proposes an interactive environment for psychometrics diagnostics, where agents supervise the users actions and data mining for searching of potentially interesting information. Chapter 18 uses static agent to execute mine firewall policy rules, and mobile agent to exploit optimized rules to detect eventual anomalies. Chapter 19 uses game-theory modeling for competitive knowledge extraction, hierarchical knowledge mining, and Dempster-Shafer result combination. Chapter 20 discusses a normative multi-agent enriched data mining architecture and ontology frameworks. Chapter 21 presents static and dynamic agent societies responsible for group formation, to execute a data mining classification process. Chapter 22 describes an agent based video contents identification scheme using a watermark based filtering technique.

This book is directed to students, researchers, engineers and practitioners in both the agent and data mining areas, who are interested in the marriage of agents with data mining, or are experiencing challenges in one area that could be managed by incorporating the other technology. The book will also be of interest to students and researchers in many related areas such as machine learning, artificial intelligence, intelligent systems, knowledge engineering, human-computer interaction, intelligent information processing, decision support systems, knowledge management, organizational computing, social computing, complex systems, and soft computing.

We would like to convey our appreciation to all contributors, including the accepted chapters' authors, and many other participants who submitted work that has not been included in this book. Our special thanks go to Ms. Melissa Fearon and Ms. Valerie Schofield from Springer US for their kind support and efforts in bringing the book to fruition. In addition, we also appreciate our reviewers, and Ms. Ziye Zuo and Mr. Yong Yang's assistance in co-ordinating the book chapter collection and formating the book.

Longbing Cao
University of Technology Sydney, Australia
March 2009

information processing, decision making systems, knowledge management, cognitive [illegible] computing and complex systems, [illegible] computing.

We would like to [illegible] authors [illegible] and [illegible] other participants [illegible]. Special thanks go to Ms. Melissa Fearon and Ms. [illegible] for their support and efforts [illegible] Zhu [illegible] and [illegible] the book.

[illegible]

Foreword

Integration of advanced technologies is a natural progression in the development of science. Such integration may result in a new technology whose "power" can be metaphorically described by the equality $2+2=5$, or even more. The integration of agents and data mining is such an emergent technology. The potential benefits of integrating agent and data mining were recognized when multi-agent systems (MAS) was at its infancy stage and when data mining has already reached close to their maturity. Since that time research has demonstrated the value of integration.

These advantages can be evaluated from many viewpoints and perspectives. Indeed, enrichment of MAS with data mining capabilities can significantly enhance intelligence of multi-agent systems by enabling them to adapt to unpredictable changes of environment, and by providing them with a property to on-line improve agents' collaboration utility through learning. The latter is especially important for novel classes of applications, in particular, for mobile, ubiquitous and peer-to-peer networking systems where high level of autonomy of agents pursuing their local goals has to be harmonized with the global system goal. Distributed data mining and learning technologies can provide such applications with powerful means for on-line coordination and agreement of local and global system objectives. An example is given by so-called agent network self-configuration task. Indeed, a specific feature of mobile, ubiquitous and peer-to-peer (P2P) computing systems is that they operate in dynamic environment. E.g., mobile devices may move and freely enter into and exit from a mobile network changing the set of network nodes and their communication topology and thus changing the set of application agents existing in the network and the availability of services as well. Consistent and coherent operation of such networks is organized using a "virtualization" idea implemented in terms of overlay computing. In such dynamic environment, an important task is self-configuration of overlay networks to support the optimality of agent network operation at the communication, infrastructural and application layers. This task cannot be effectively solved without intensive use of distributed data mining and P2P machine learning technologies. The list of applications in which multi-agent systems, data mining and machine learning have been combined to great effect is growing rapidly. This book contains carefully selected examples.

From the data mining and machine learning perspective, multi-agent systems leads to new architectures and to novel styles of software development. Modern data mining and machine learning applications lay down new requirements for enabling information and software technologies that are used at design and implementation phases. In particular, these, often challenging, requirements are caused by the distributed nature and heterogeneity of data sources that are selected dynamically by a data mining application and depend on the current state of the environment. Data sources may be massive and may continue to generate formidable volumes of data. It is well recognized that, in such applications, the most promising technology is agent-based. It is capable of supporting multi-strategy data mining and of achieving scalability in the presence of massive quantities of distributed data. Agent technology can also support intelligent interaction with users. Experience shows that no other currently existing information technology can compare with the power of MAS technology for data mining problems in this class. However, this class of problems is not rare and includes modern real life data mining challenges.

Recently, researchers are intensively using agent technology in many novel data mining applications. Among them are: web mining, text mining, and brain computing. On the other side, agent-based applications use data mining and machine learning for coordination strategy mining, for interaction protocol selection and mining. A sign of the maturity of the integration of agent and data mining technologies is the development of several software tools specifically targeted for joint use of agent and data mining, for example, Agent Academy.

We acknowledge the important role of the FP5 European project KDNet and, in particular, the AgentLink ALAD Special Interesting Group (SIG) organized within the framework of this Project, in consolidating the data mining and agent communities. ALAD SIG was a pioneer task force intended to achieve this consolidation leading to joint research. Another dedicated SIG is the Agent-Mining Interaction and Integration SIG [1], which actively links leading researchers in this emerging field, and organizes regular events for promoting and encouraging research networking in both agent and data mining communities. These task forces have put ahead many other analogous initiatives including special sessions of the international conferences, dedicated international workshops, journal special issues, etc.

As an explicit aftereffect of the AMII SIG initiative, this book on Data Mining and Multiagent Integration, edited by Longbing Cao, presents the first collection of the latest research outcomes on the topic. A single volume can not present a complete coverage of the current state-of-the-art in the perspectives of integrating agents and data mining, however it covers all major research topics of agent mining research, and presents many novel ideas, applications and other solutions that can be useful for the researchers and industry interested in research on and practical use of novel information technologies. This book will be of interest to new scientists working in agent and data mining integration and interaction, and will further promote the consolidation of this crucial international scientific community.

Prof. Vladimir Gorodetsky,
St. Petersburg, SPIIRAS

[1] AMII-SIG: www.agentmining.org

Contents

Part I
Introduction to Agents and Data Mining Interaction

Chapter 1
Introduction to Agent Mining Interaction and Integration

Longbing Cao

Abstract In recent years, more and more researchers have been involved in research on both agent technology and data mining. A clear disciplinary effort has been activated toward removing the boundary between them, that is the interaction and integration between agent technology and data mining. We refer this to *agent mining* as a new area. The marriage of agents and data mining is driven by challenges faced by both communities, and the need of developing more advanced intelligence, information processing and systems. This chapter presents an overall picture of agent mining from the perspective of positioning it as an emerging area. We summarize the main driving forces, complementary essence, disciplinary framework, applications, case studies, and trends and directions, as well as brief observation on agent-driven data mining, data mining-driven agents, and mutual issues in agent mining. Arguably, we draw the following conclusions: (1) agent mining emerges as a new area in the scientific family, (2) both agent technology and data mining can greatly benefit from agent mining, (3) it is very promising to result in additional advancement in intelligent information processing and systems. However, as a new open area, there are many issues waiting for research and development from theoretical, technological and practical perspectives.

1.1 Introduction

Autonomous agent and multi-agent systems (AAMAS, refer to here as *agents*) [44] and knowledge discovery from data (KDD, or otherwise known as *data mining*)[10] have emerged and developed separately in the last twenty years. Both areas are currently very active.

Longbing Cao
Faculty of Engineering and Information Technology, University of Technology Sydney, Australia, e-mail: lbcao@it.uts.edu.au

L. Cao (ed.), Data Mining and Multi-agent Integration, DOI: 10.1007/978-1-4419-0522-2_1,

Agents primarily focus on issues from many aspects, from theoretical, methodological, and experimental to practical issues in developing agent-based computing and agent-oriented intelligent systems, which are a powerful technology for autonomous intelligent system analysis, design and implementation. The major topics of interest consist of research on individual agents, multi-agent systems (MAS), methodology and techniques, tools and applications. The agent technology contributes to many diverse domains such as software engineering, user interfaces, e-commerce, information retrieval, robotics, computer games, education and training, ubiquitous computing, and social simulation.

Currently, agent studies have been spread from programming to organizational and societal factors to study agents and agent-based systems. The research on agents has far exceeded the original community scope of artificial intelligence and software. Researchers from many other areas have started to discuss, develop, wrap and use the concept of agents, covering almost all aspects of the social sciences such as law, business, organizational, behavior sciences, finance and economics, tourism, not to mention the extensive family of natural science and technology. The benefits from agents are expected to be very comprehensive and diverse, from academic disciplines, to the sciences, the social sciences and the humanities.

Similarly, *data mining* originally focused on knowledge discovery in databases, but it has experienced a migration from data-centered pattern discovery, to knowledge discovery, actionable knowledge discovery, and currently to domain-oriented decision delivery [11]. Data mining and its tools is becoming a ubiquitous information processing field and tools, involving techniques and researchers from many areas such as statistics, information retrieval, machine learning, artificial intelligence, pattern recognition, and database technologies. Data mining is increasingly widely tested in varying applications and domains, for instance, web mining and services, text mining, telecommunications, retail, governmental service, fraud, security, business intelligence studies.

Besides the emphasis of in-depth data intelligence, recent efforts in data mining cover many additional areas and domain problems. Data mining researchers recognize the need to involve the environment, human intelligence, domain intelligence, organizational intelligence, and social intelligence in the mining process, models, the findings and deliverables. This will trigger another wave of migration from the discovery of knowledge to the delivery of deep knowledge-based problem-solving systems and services.

The above analysis of trends and directions of both areas shows that these two independent research streams have been created and originally evolved with separate aims and objectives. They used to target individual methodologies and techniques to cope with domain-specific problems and challenges in respective areas. However, both are concerned with many similar aspects and factors, such as human roles, user-system interaction, dynamic modeling, domain factors, organizational and social factors. In fact, both areas contribute to the advancement of intelligence, and intelligent information processing, services and systems. In fact, they need each other, as evidenced by typical topics of agent-based data mining in the middle 1990s.

Consequently, we see a clear trend of the interaction and integration between agents and data mining. Its development has reached the level of a new and promising area, and is moving towards becoming a first-class citizen in the science and technology family [12, 5, 6]. This edited book, as the first one on this exciting topic, once again evidences the strong need and potential of agent-mining interaction and integration (*agent mining* for short).

This chapter presents an overall picture of this emerging field, data mining and multi-agent integration. We first analyze the respective and common challenges in agents and data mining areas. These challenges motivate and drive the need and emergence of agent mining. A scientific framework and theoretical underpinnings are presented, which illustrate the synergy methods and foundations of agents and data mining. Further, we briefly summarize the research on three major directions in agent mining, namely agent-driven distributed data mining, data mining-driven agents, and mutual issues in agent mining. Applications and open issues are then discussed. Finally, we discuss the development of agent mining community. Information provided here can benefit new researchers, and enable them to quickly step into this field.

1.2 Driving forces of agent mining interaction and integration

The emergence of agent mining results from the following driving forces:

- the critical challenges in agents and data mining respectively,
- the critical common challenges troubling agents and data mining
- the complementary essence of agents and data mining in dealing with their challenges, and
- the great add-on potential resulting from the interaction and integration of agents and data mining.

Agents and data mining are facing critical challenges from respective areas. Many of these challenges can be tackled by involving advances in other areas. Fig. 1.1 illustrates these challenges. In this section, we specify both individual and mutual challenges in agent and mining disciplines that may be complemented by the interaction with the other disciplines.

1.2.1 Challenges in agent disciplines

As addressed in some retrospective publications, traditional agent technology has been challenged in many aspects such as developing organizational and social intelligence. In the following analysis, we concern ourselves with the challenges that may benefit from the involvement of data mining. We explain this from the follow-

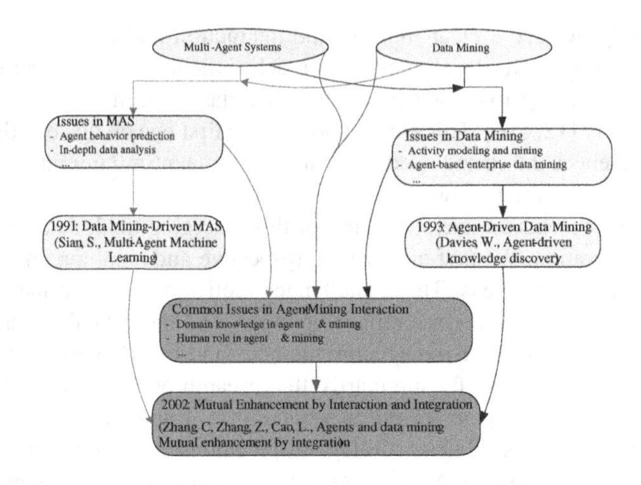

Fig. 1.1: Challenges in agents and data mining

ing aspects: agent awareness, agent learning, agent actionability, agent distributed processing, agent in-depth services, and agent constraint processing.

- *Agent awareness* Agent awareness refers to the capability of an agent to recognize internal and/or external environment change, and analyze situation change. In contrast to normal sensing and perception as conducted in reactive agents, here agent awareness specifically refers to situation analysis and environment modeling driven by agent learning and discovery. Agents with such a capability should self-recognize, compare and reason the changes taking place in the environment. To this end, it is necessary for agents to accumulate learning capability.
- *Agent learning* In open multi-agent organizations, interaction widely exists between agent and environment, and between an agent and the other agent(s). Agents are expected to learn from other agents, their environment, and from the interaction and dynamics. In addition, agents may be expected to learn from users and interaction with humans. To foster such learning capability, agents need to be fed with learning and reasoning algorithms that can support them to discover, reason or simulate interesting information from interactive and situational data. Learning capability is widely recognized to be significant for enhancing agent intelligence. On the basis of the varying objectives, agent learning has been paid unprecedented attention. Multiple forms of agent learning capability are being studied. Agent learning may be conducted in terms of agent architectures such as cognitive learning, deductive learning, distributed learning, and cooperative learning. With respect to learning objectives, agent learning may also be classified into procedural learning, action learning, rule and pattern learning, and decision-making learning. From the learning process aspect, agent learning can be categorized into reinforcement learning, discovery learning, single-trial learning, reasoning learning, and random learning. The implementation of agent learn-

ing presents either a passive or active manner. In a broad sense, learning can be in a supervised, unsupervised or hybrid manner.

- *Agent actionability* Agent actionability refers to the capability of an agent to take actions to its advantage on the basis of the knowledge obtained through in-depth analysis, reasoning and discovery. Unlike general action taken by agents, we are specifically interested in actions for recommendation, servicing, searching, discovery, conflict resolution, etc. with great benefits but low costs. To this end, agents need to balance benefits and costs, and maximize their return while minimizing the risk before taking an action or a sequence of actions.
- *Agent distributed processing* In middle to large scale multi-agent systems, agents need to deal with distributed processing tasks such as learning from agents across multiple organizations, applications or data, conducting decentralized coordination, cooperation and negotiation among agents crossing resources, and implementing information gathering, dispatching and transport among agents located in distributed applications. To tackle the above tasks in distributed conditions, agents need to make decisions after analyzing and utilizing relevant information from multiple sources. Information analysis and utilization is not a trivial job. Agents may need to develop capabilities such as data analysis and discovery, procedural learning, goal adjustment, and information fusion.
- *Agent in-depth services* Agents are often developed for providing varied services, for instance, network services such as web recommender systems, mobile agents for information searching and passing, and user services such as for user interaction and user modeling. Smart service providing relies on in-depth analysis of the service request-related data and information, as well as service historical data and service performance, in order to deeply understand service data and select the best service solutions. However, the agent community often does not work on such kinds of capabilities.
- *Agent constraint processing* Open complex agent systems often involve many types of constraints from many aspects, for instance, temporal and spatial constraints, or execution constraints from organizational aspects. Such constraints form conditions in improving agent capabilities such as learning, adaptation, actionability, and services. There is a need to understand such constraints, and to involve and best treat such constraints in an agent system and solution generation.

1.2.2 Challenges in data mining disciplines

Data mining faces many challenges when it is deployed to real world problem-solving, in particular, in handling complex data and applications. We list here a few aspects that can be improved by agent technology. These include enterprise data mining infrastructure, involving domain and human intelligence, supporting parallel and distributed mining, data fusion and preparation, adaptive learning, and interactive mining.

- *Enterprise data mining infrastructure* The development of data mining systems supporting real-world enterprise applications are challenging. The challenge may arise from many aspects, for instance, integrating or mining multiple data sources, accessing distributed applications, interacting with varying business users, and communicating with multiple applications. In particular, it has been a grand challenge and a longstanding issue to build up a distributed, flexible, adaptive and efficient platform supporting interactive mining in real-world data.
- *Involving domain and human intelligence* Another grand challenge of existing data mining methodologies and techniques are the roles and involvement of domain intelligence and human intelligence in data mining. With respect to domain intelligence, how to involve, represent, link and confirm to components such as domain knowledge, prior knowledge, business process, and business logics in data mining systems is a research problem. Regarding human intelligence, we need to distinguish the role of humans in specific applications, and further build up system support to model human behavior, interact with humans, bridge the communication gap between data mining systems and humans, and most importantly incorporate human knowledge and supervision into the system.
- *Supporting parallel and distributed mining* One of the major efforts of data mining research is to enhance the performance of data mining algorithms. This is usually conducted through designing efficient data structures and computational methods to reduce computational complexities. In many cases, computational performance can be greatly improved through developing parallel algorithms. In other cases, distributed computing is necessary such as dealing with distributed data sources or applications, or peer-to-peer computing is required. However, how to design effective and efficient parallel and distributed algorithms is an issue.
- *Data fusion and preparation* In the real world, data is getting more and more complex, in particular, sparse and heterogeneous data distributed in multiple places. To access and fuse such data needs intelligent techniques and methods. On the other hand, today's data preparation research is facing new challenges such as processing high frequency time series data stream, unbalanced data distribution, rare but significant evidence extraction from dispersed data sets, linking multiple data sources, accessing dynamic data. Such situations expect new data preparation techniques.
- *Adaptive learning* In general, data mining algorithms are predefined to scan data sets. In real-world cases, it is expected that data mining models and algorithms can adapt to dynamic situations in changing data based on their self-learning and self-organizing capability. As a result, models and algorithms can automatically extract patterns in changing data. However, this is a very challenging area, since existing data mining methodologies and techniques are basically non-automatic and unadaptable. To enhance the automated and adaptive capability of data mining algorithms and methods, we need to search for support from external disciplines that are related to automated and adaptive intelligent techniques.

- *Interactive mining* Controversies regarding either automatic or interactive data mining have been raised in the past. A clear trend for this problem is that interaction between humans and data mining systems plays an irreplaceable role in domain-driven data mining situations. In developing interactive mining, one should study issues such as user modeling, behavior simulation, situation analysis, user interface design, user knowledge management, algorithm/model input setting by users, mining process control and monitor, outcome refinement and tuning. However, many of these tasks cannot be handled by existing data mining approaches.

1.2.3 Mutual challenges in agent and mining

As addressed in [5, 6, 7], agents can enhance data mining through involving agent intelligence in data mining systems, while an agent system can benefit from data mining via extending agents' knowledge discovery capability. Nevertheless, the agent-mining interaction symbiosis cannot be established if mutual issues are not solved. These mutual issues involve fundamental challenges hidden on both sides and particularly within the interaction and integration. Fig. 1.1 presents a view of issues in agent-mining interaction highlighting the existence of mutual issues. Mutual issues constraining agent-mining interaction and integration consist of many aspects such as architecture and infrastructure, constraint and environment, domain intelligence, human intelligence, knowledge engineering and management, and nonfunctional requirements.

- *Architecture and infrastructure* Data mining always faces a problem in how to implement a system that can support those brilliant functions and algorithms studied in academia. The design of the system architecture conducting enterprise mining applications and emerging research challenges needs to provide (1) functional support such as crossing source data management and preparation, interactive mining and the involvement of domain and human intelligence, distributed, parallel and adaptive learning, and plug-and-play of algorithms and system components, as well as (2) nonfunctional support for instance adaptability, being user and business friendly and flexibility. On the other hand, middle to large scales of agent systems are not easily built due to the essence of distribution, interaction, human and domain involvement, and openness. In fact, many challenging factors in agent and mining systems are similar or complementary.
- *Constraint and environment* Both agent and mining systems need to interact with the environment, and tackle the constraints surrounding a system. In agent communities, environment could present characters such as openness, accessibility, uncertainty, diversity, temporality, spatiality, and/or evolutionary and dynamic processes. These factors form varying constraints on agents and agent systems. Similar issues can also be found from real-world data mining, for instance, temporal and spatial data mining. The dynamic business process and logics surround-

ing data mining make the mining very domain-specific and sensitive to its environment.

- *Domain intelligence* Domain intelligence widely surrounds agent and mining systems. Both areas need to understand, define, represent, and involve the roles and components of domain intelligence. In particular, it is essential in agent-mining interaction to model domain and prior knowledge, and to involve it to enhance agent-mining intelligence and actionable capability.
- *Human intelligence* Both agent and mining need to consider the roles and components of human intelligence. Many roles may be better played by humans in agent-mining interaction. To this end, it is necessary to study the definition and major components of human intelligence, and how to involve them in agent-mining systems. For instance, mechanisms should be researched on user modeling, user and business friendly interaction interfaces, and communication languages for agent-mining system dialogue.
- *Knowledge engineering and management* To support the involvement of domain and human intelligence, proper mechanisms of knowledge engineering and management are substantially important. Tasks such as the management, representation, semantic relationships, transformation and mapping between multiple domains, and meta-data and meta-knowledge are essential for involving roles and data/knowledge intelligence in building up agent-mining symbionts.
- *Nonfunctional requirements* Nonfunctional requests are essential in real-world mining and agent systems. The agent-mining symbionts may more or less address nonfunctional requirements such as efficiency, effectiveness, actionability, user and business friendliness.

1.3 Complementary essence and interaction potential of agents and data mining

Why do the above challenges matter to both sides of agents and data mining? Why is the interaction and integration between agents and data mining important? There are both explicit and implicit reasons. Explicit reasons may include the following system complexities.

- Explicit limitations and challenges in pure agent systems, as addressed in [5, 6, 7], can be complemented by data mining, for instance, data mining driving agent learning, user modeling and information analysis.
- Explicit limitations and challenges in pure data mining systems, as discussed in [5, 6, 7], can be better serviced by agent technology, for instance, agent-based data mining infrastructure, agents for data management and preparation, agent-based service provision.
- The integration of agents and data mining has the potential to result in new strengths and advantages that cannot be delivered by any single side, for instance,

leading to more intelligent agent-mining symbiont fusing capabilities of in-depth perception, learning, adaptation, discovery, reasoning, and decision.

Implicit driving forces for including the above mutual issues are equally significant.

- Agent-mining symbionts are substantially essential for dealing with complicated intelligence phenomena and system complexities in complex intelligent systems. Simple intelligent systems and other issues that can be tackled using one side of these technologies, for instance, an agent-based data integration system, may not necessarily involve both sides.
- The emergence of intelligence in agent-mining interaction may massively strengthen the problem-solving capability of an intelligent system, which cannot be carried out by either part.
- Implicit roles need to be discovered through interdisciplinary studies, which may extensively promote either one side or the whole of an agent-mining integrative system, once the roles are disclosed and properly developed.
- New research issues, opportunities, techniques and systems may be triggered in the agent mining community.

It is arguable that agents and mining are complementary. The agent-mining interaction can enhance both sides considerably through introducing new approaches and techniques to solve those domain-specific challenges that cannot be tackled well by either methods. Some typical benefits and roles in agent and mining areas that can be achieved through agent-mining interaction.

- Enhancing agents through data mining. Agent-mining interaction was originally initiated by data mining driven agent learning in 1991 [20, 40]. Data mining has the potential to enhance agent technology through introducing and improving the learning and reasoning capability of agents. Agents can be enhanced through involving data mining in broad aspects, in particular, agent learning, agent coordination and planning, user modeling and servicing, and network servicing.
- Promoting data mining through agent. Sometime around 1993, another effort was started on agent-based data mining [21, 22, 23], namely to utilize agent technology to enhance data mining. The enhancement may be embodied in terms of varying aspects, for instance, agent-based KDD infrastructure, agent-based distributed processing, agent-based interactive data mining, and agent-based data warehouse.
- Building super intelligent symbionts. As evidenced by the agent service-based trading support system F-Trade [10], the use of agent mining can lead to more intelligent systems that can best fuse the strengths of agents in building intelligent systems as well as the beauty of data mining in processing deep knowledge.

1.4 A Disciplinary Framework of Agent and Mining Interaction and Integration

This section aims to draw a concept map of agent mining as a scientific field. We observe this from the following perspectives: evolution process and characteristics, agent-mining interaction framework, and theoretical underpinnings for agent mining.

1.4.1 Evolution process and characteristics

As an emerging research area, agent mining experiences the following evolution process, and presents the following unprecedented characteristics.

- *From one-way interaction to wo-way interaction*: The area was originally initiated by incorporating data mining into agent to enhance agent learning [20, 40]. Recently, issues in two-way interaction and integration have been broadly studied in different groups.
- *From single need-driven to mutual needs-driven*: Original research work started on the single need to integrate one into the other, whereas it it is now driven by both needs from both parties. As discussed in [12, 8], people have found many issues in each of the related communities. These issues cannot be tackled by simply developing internal techniques. Rather, techniques from other disciplines can greatly complement the problem-solving when they are combined with existing techniques and approaches. This greatly drives the development of agent-driven data mining and data mining-driven agents.
- *Intrinsic associations and utilities*: The interaction and integration between agents and data mining is also driven and connected by intrinsic overlap, associations, complementation and utilities of both parties, as discussed in [5, 6]. This drives the research on mutual issues, and the synergetic research and systems coupling both technologies, into a more advanced form.
- *Application drives*: Application request is one of the key driving forces of this new trend. In Section 1.8.1, we present some major application domains and problems that may be better handled by both agent and mining techniques.
- *Major research groups and researchers* [6] in respective communities tend to undertake both sides of research. Some of them are trying to link them together to solve problems that cannot be tackled by one of them alone, for instance, agent-based distributed learning [30, 31, 32, 25, 26], agent-based data mining infrastructure [4, 5, 26], or data mining driven agent intelligence enhancement [4, 35].
- Broad research covering *theoretical, technological and practical perspectives*: Publications and projects have involved not only technological issues, but also theoretical and practical problems. A cross-disciplinary and multi-dimensional study roadmap is clear.

We also draw an evolutionary tree of this area by combining the emergence of significant landmarks and events in the life of agent mining (see Fig. 1.4.1).

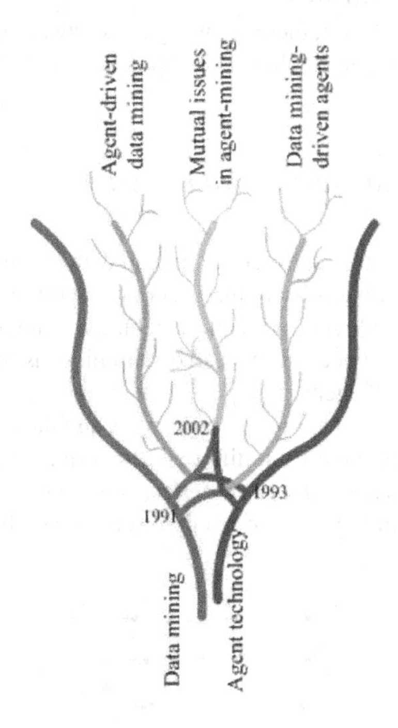

Fig. 1.2: Evolution of agent mining as a scientific area

As identified in a recent position meeting and related activities[5], there are many research topics and open issues from both sides of agent and mining interaction. In particular, issues for agent-driven data mining, and issues for mining-driven agents are attracting research interest. However, there are some mutually fundamental issues that are not receiving attention in the emerging research. These issues are significant because of their fundamental and necessary roles in establishing a symbiotic relation between agents and mining.

Through reviewing the related work in the above areas, there is a clear indication that agent-mining interaction and integration has emerged as a prominent, challenging, dynamic and exciting area. It evidences that

1. Agent-mining interaction is attracting ever-increasing attention from both agents and data mining communities,
2. The interaction and integration between agent and mining can greatly complement and strengthen each side of both communities. Some complicated challenges in either community may be effectively and efficiently tackled through agent-mining interaction,

3. Furthermore, as a newly emergent area, agent and mining interaction and integration has the potential to create new interesting symbiosis opportunities in both academic and business worlds,
4. As a new open area, however, there are many issues awaiting research and development from theoretical, technological and practical perspectives.

1.4.2 Agent-mining interaction framework

The interaction and integration between agents and data mining are comprehensive, multiple dimensional, and inter-disciplinary. As an emerging scientific field, *agent mining* studies the methodologies, principles, techniques and applications of the integration and interaction between agents and data mining, as well as the community that focuses on the study of agent mining.

On the basis of complementation between agents and data mining, agent mining fosters a synergy between them from different dimensions, for instance, *resource*, *infrastructure*, *learning*, *knowledge*, *interaction*, *interface*, *social*, *application* and *performance*. As shown in Fig. 1.3, we briefly discuss these dimensions.

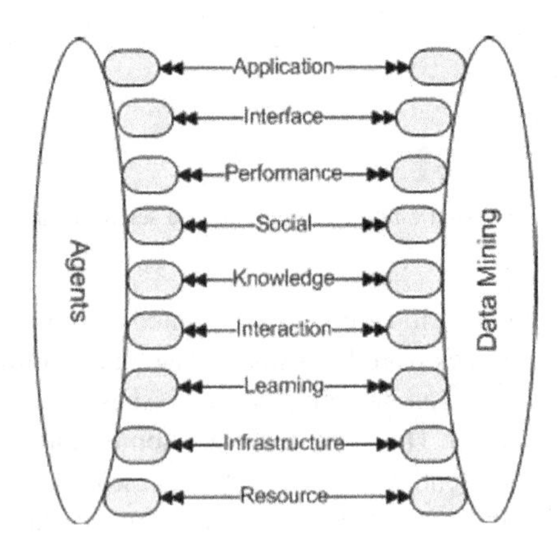

Fig. 1.3: Multi-Dimensional Agent-Mining Synergy.

- Resource layer – interaction and integration may happen on data and information levels;
- Infrastructure layer – interaction and integration may be on infrastructure, architecture and process sides;
- Knowledge layer – interaction and integration may be based on knowledge, including domain knowledge, human expert knowledge, meta-knowledge, and knowledge retrieved, extracted or discovered in resources;

- Learning layer – interaction and integration may be on learning methods, learning capabilities and performance perspectives;
- Interaction layer – interaction and integration may be on coordination, cooperation, negotiation, communication perspectives;
- Interface layer – interaction and integration may be on human-system interface, user modeling and interface design;
- Social layer – interaction and integration may be on social and organizational factors, for instance, human roles;
- Application layer – interaction and integration may be on applications and domain problems;
- Performance layer – interaction and integration may be on the performance enhancement of one side of the technologies or the coupling system.

From these dimensions, many fundamental research issues/problems in agent mining emerge. Correspondingly, we can generate a high-level research map of agent mining as a disciplinary area. Figure 1.4 shows such a framework, which consists of the following research components: *agent mining foundations*, *agent-driven data processing*, *agent-driven knowledge discovery*, *mining-driven multi-agent systems*, *agent-driven information processing*, *mutual issues in agent mining*, *agent mining systems*, *agent mining applications*, *agent mining knowledge management*, and *agent mining performance evaluation*. We briefly discuss them below.

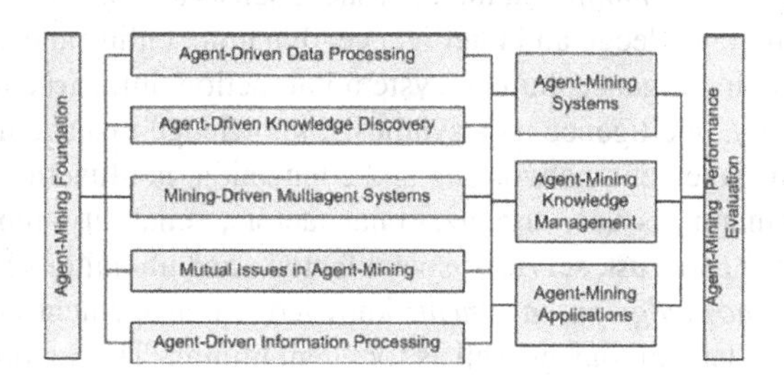

Fig. 1.4: Agent-Mining Disciplinary Framework.

- *Agent mining foundations* studies issues such as the challenges and prospects, research map and theoretical underpinnings, theoretical foundations, formal methods, and frameworks, approaches and tools;
- *Agent-driven data processing* studies issues including multi-agent data coordination, multi-agent data extraction, multi-agent data integration, multi-agent data management, multi-agent data monitoring, multi-agent data processing and preparation, multi-agent data query and multi-agent data warehousing;
- *Agent-driven knowledge discovery* studies problems like multi-agent data mining infrastructure and architecture, multi-agent data mining process modeling and management, multi-agent data mining project management, multi-agent interactive data mining infrastructure, multi-agent automated data learning, multi-agent

cloud computing, multi-agent distributed data mining, multi-agent dynamic mining, multi-agent grid computing, multi-agent interactive data mining, multi-agent online mining, multi-agent mobility mining, multi-agent multiple data source mining, multi-agent ontology mining, multi-agent parallel data mining, multi-agent peer-to-peer mining, multi-agent self-organizing mining, multi-agent text mining, multi-agent visual data mining, and multi-agent web mining;

- *Mining-driven multi-agent systems (MAS)* studies issues such as data mining-driven MAS adaptation, data mining-driven MAS behavior analysis, data mining-driven MAS communication, data mining-driven MAS coordination, data mining-driven MAS dispatching, data mining-driven MAS distributed learning, data mining-driven MAS evolution, data mining-driven MAS learning, data mining-driven MAS negotiation, data mining-driven MAS optimization, data mining-driven MAS planning, data mining-driven MAS reasoning, data mining-driven MAS recommendation, data mining-driven MAS reputation/risk/trust analysis, data mining-driven self-organized and self-learning MAS, data mining-driven user modeling and servicing, and semi-supervised MAS learning;
- *Agent-driven information processing:* multi-agent domain intelligence involvement, multi-agent human-mining cooperation, multi-agent enterprise application integration, multi-agent information gathering/retrieval, multi-agent message passing and sharing, multi-agent pattern analysis, and multi-agent service-oriented computing;
- *Mutual issues in agent mining* including issues such as actionable capability, constraints, domain knowledge and intelligence, dynamic, online and ad-hoc issues, human role and intelligence, human-system interaction, infrastructure and architecture problems, intelligence metasynthesis, knowledge management, lifecycle and process management, networking and connection, nonfunctional issues, ontology and semantic issues, organizational factors, reliability, reputation, risk, privacy, security and trust, services, social factors, and ubiquitous intelligence;
- *Agent mining knowledge management:* knowledge management is essential for both agents and data mining, as well as for agent mining. This involves the representation, management and use of ontologies, domain knowledge, human empirical knowledge, meta-data and meta-knowledge, organizational and social factors, and resources in the agent-mining symbionts. In this, formal methods and tools are necessary for modeling, representing and managing knowledge. Such techniques also need to cater for identifying and distributing knowledge, knowledge evolution in agents, and enabling knowledge use.
- *Agent mining performance evaluation* researches on methodologies, frameworks, tools and testbeds for evaluating the performance of agent mining, and performance benchmarking and metrics. Besides technical performance such as accuracy and statistical significance, business-oriented performance such as cost, benefit and risk are also important in evaluating agent mining. Other aspects such as mobility, reliability, dependability, trust, privacy and reputation, etc., are also important in agent mining.
- *Agent mining systems:* this research component studies the formation of systems, including techniques for the frameworks, modeling, design and software engi-

neering of agent-mining systems. It provides agents and data mining technologies as basic resources, thus they can perform as parts of the system. Techniques and tools for engineering and constructing agent-mining systems are important for handling various kinds of applications. A specific agent-mining system may be extended to fit into some applications based on the provided features of the system. With regards to integrated systems, applications can be constructed from the pre-defined framework of a particular problem domain or fully tailored to serve the purpose of the applications. Either way, system implementation and deployment are to be further investigated since development of the two technologies has been done in parallel. It is hard to generate dedicated platforms for agent mining systems.

- *Agent mining applications* This refers to any real-world applications and domain problems that can be better handled by agent mining technologies. Based on the need from particular applications, any issues discussed in the above topics may be engaged here. For instance, in some cases, an agent mining simulation system needs to be built for us to understand the working mechanism and potential optimization of a complex social network. In other cases, the enhancement of learning capability is the main task, and appropriate learning tools need to be used on demand.

1.4.3 Theoretical underpinnings for agent mining

Current research on agent mining mainly focuses on the application of agents in data mining or data mining in agents. No systematic work has been conducted on developing foundations for agent mining. In fact, we believe it is very important to foster research on fundamental issues such as:

- how to synergize agents and data mining.
- what methodologies are needed to synergize agents and data mining?
- what are the lifecycle, process and outcome aspects to synergizing agents and data mining?

This needs to be investigated from methodological, technical and tool perspectives. For this, we need to investigate the theoretical underpinnings for agent mining.

Obviously, to support agent mining, many disciplines related to either agents or data mining need have been involved, for instance, artificial intelligence, machine learning, and logic. We form the following multiple layers of framework for developing an agent mining discipline: (1) Theoretical foundation, (2) Fundamental technologies, and (3) Supporting techniques and tools.

From the *theoretical foundation* perspective, agent mining draws theoretical support from multiple disciplines, including mathematics, logics, information sciences, intelligence sciences, system sciences, cognitive sciences, many particular disciplines in social sciences such as business, and behavioral sciences. Mathematics and logics provide formal methods for learning, modeling and knowledge representation, reasoning, discovery, transformation and presentation. Information and intelligence sciences provide support for intelligent information processing and systems in agent mining. System sciences furnish methodologies and techniques for the understanding, modeling, simulation, deployment and system integration. Cognitive sciences incorporate principles and methods for understanding human behavior belief, and the intention and goal of human behavior, in order to involve human intelligence in the agent mining process, supporting user modeling and services, and human-centered computing. The social sciences supply the foundations for conceiving organizational and social factors and business processes surrounding domain problems using agent mining. Many areas may be involved, for instance, economics, finance, business, behavior science, choice, and social network analysis, which are important for understanding environmental impact, interaction, and requirements on deliverables.

Fundamental technologies may involve many aspects in information and technology fields, for instance, user modeling, formal methods, logics, knowledge engineering, data engineering, ontological engineering, semantic web, machine learning, artificial intelligence, software engineering, information systems, and interaction design. Specific techniques in risk management and analysis, organizational theory, sociology, psychology, economics and finance are important for incorporating organizational, social and domain factors into agent mining. Emerging areas such as organizational computing, social computing, ubiquitous computing and collective intelligence can contribute to the agent mining family as well. In fact, agent mining will involve a majority of disciplines and a body of knowledge in the modern science and technology family.

To make agent mining work, many *supporting techniques and tools* are essential. This may involve effective techniques and tools for representing, modeling, analyzing, presenting, integrating and employing agents and data mining. In particular, techniques and tools are required to support information access, fusion, processing, discovery, user interface design, visualization, dynamic and adaptive interaction, information and knowledge sharing, integration and management, and mechanisms for supporting the performance evaluation and improvement of agent mining bodies. For instance, how can we evaluate the performance of agent mining symbionts? In data mining-driven agent planning, how can we involve data mining findings to make agent planning smarter, more stable and predictable?

1.4.4 Agent mining lifecycle and process

In general, it is very hard to define or extract a generic lifecycle and process for agent mining interaction and integration, as has been done for agent system development and data mining. This is partially due to the diversity of methodologies, techniques and community interests in both agents and data mining fields. In fact, the lifecyle and process for any concrete case study or system has to be specific. However, this does not mean this has nothing to do with a general understanding of lifecyle and process for integrating agents and data mining.

If agents are mainly used to support data mining, the general lifecycle and process would fit into that of data mining and knowledge discovery, from data extraction, transformation, loading, to modeling, initial evaluation, refinement, final improvement and delivery of findings. On the other hand, if data mining is mainly used for agent system construction, a likely process is to follow the software engineering lifecycle, such as requirement analysis, system analysis, database and semantic design, architectural and detailed system design, user interface design, to system implementation, performance evaluation and refinement. In either case, the lifecycle and process need to be customized and enhanced toward catering for advantages and requests from the other end of the cycle; for instance, the data mining process needs to be supported in constructing a data mining-driven agent system.

In other cases, mutual issues are focused by involving both agents and data mining, for instance, research on the involvement of domain knowledge and intelligence into agent mining symbionts. The processes for such a case would more likely follow that usually used for handling the issue by considering the special needs from agents and data mining aspects.

However, due to the extra contributions and needs from the involvement of the other technique, it is interesting to develop a customizable lifecycle and process for agent-driven data mining and data mining driven agent systems. For instance, if data mining is used to strengthen agent intelligence, how should data mining agents be trained and tested? Extra procedures may be required for such data mining agents.

1.5 Agent-Driven Distributed Data Mining

This section particularly discusses the state-of-the-art of agent-driven knowledge discovery (otherwise known as multi-agent-driven data mining, multi-agent data mining) [12]. As discussed in the above, agent-driven knowledge discovery forms a big area for agent mining. It is actually the mostly addressed area since the proposal of the integration between agents and data mining.

1.5.1 The challenges of distributed data mining

Data mining and machine learning currently forms a mature field of artificial intelligence supported by many various approaches, algorithms and software tools. However, modern requirements in data mining and machine learning inspired by emerging applications and information technologies and the peculiarities of data sources are becoming increasingly tough. The critical features of data sources determining such requirements are as follows:

- In enterprise applications, data is distributed over many heterogeneous sources coupling in either a tight or loose manner;
- Distributed data sources associated with a business line are often complex, for instance, some is of high frequency or density, mixing static and dynamic data, mixing multiple structures of data;
- Data integration and data matching are difficult to conduct; it is not possible to store them in centralized storage and it is not feasible to process them in a centralized manner;
- In some cases, multiple sources of data are stored in parallel storage systems;
- Local data sources can be of restricted availability due to privacy, their commercial value, etc., which in many cases also prevents its centralized processing, even in a collaborative mode;
- In many cases, distributed data spread across global storage systems is often associated with time difference;
- Availability of data sources in a mobile environment depends on time;
- The infrastructure and architecture weaknesses of existing distributed data mining systems requires more flexible, intelligent and scalable support.

These and some other peculiarities require the development of new approaches and technologies of data mining to identify patterns in distributed data. Distributed data mining (DDM), in particular, Peer-to-Peer (P2P) data mining, and multi-agent technology are two responses to the above challenges.

1.5.2 The needs of agent-driven distributed data mining

The practical implementation of distributed and P2P data mining and machine learning creates many new challenges. While analyzing these challenges, [30] argues why agent technology is best able to cope with them in terms of autonomy, interaction, dynamic selection and gathering, scalability, multi-strategy and collaboration. Other reasons include privacy, mobility, time constraints (stream data which is too late to extract and then mine), and computational costs and performance requests.

- Isolation of data sources. Distributed and multiple data sources are often isolated from each other. For in-depth understanding of a business problem, it is essential to bring relevant data together through centralized integration or localized communication. From this, agent planning and collaboration, mobile agents, agent communication and negotiation can benefit.
- Mobility of source data and computational devices. Data and device mobility requires the perception and action of data mining algorithms on a mobile basis. Mobile agents can adapt to mobility very well.
- Interactive DDM. Pro-actively assisting agent is necessary to drastically limit how much the user has to supervise and interfere with running the data mining process.

- Dynamic selection of sources and data gathering. One challenge for an intelligent data mining agent acting in an open distributed environment, in which to pursue the DM tasks, for example, where the availability of data sites and their content may change at any time, is to discover and select relevant sources. In these settings, DM agents may be applied to adaptively select data sources according to given criteria such as the expected amount, type and quality at the considered source, actual network and DM server load. Agents may be used, for example, to dynamically control and manage the process of data gathering.
- Time constraints on distributed data sources. Some data distributed in different storages is dependent on time, e.g., time differences.
- Multi-strategy DDM. For some complex application settings, an appropriate combination of multiple data mining techniques may be more beneficial than the application of a particular one. DM agents may learn in, due course which of their deliberative actions to choose, depending on the type of data retrieved from different sites and the mining tasks to be pursued.
- Collaborative DDM. DM agents may operate independently on data they have gathered at local sites and then combine their respective models. Alternatively, they may agree to share potential knowledge as it is discovered, in order to benefit from the additional options of other DM agents.
- Privacy of source data. Distributed local data is not allowed to be extracted and integrated with other sources directly, due to privacy issues. A DM agent with authority to access and process the data locally can dispatch identified local patterns for further engagement with findings from other sources.
- Organizational constraint on distributed data sources. In some organizations, business logic, process and work-flow determine the order of data storage and access. This, therefore, augments the complexity of DDM. Agents located in each storage area can communicate with each other and dispatch the DDM algorithm agents instantly, once the response is over.

1.5.3 Research issues in agent driven data mining

There are many open issues in the research direction of agent driven data mining. In establishing an agent-based enterprise data mining infrastructure, one may study organization and society-oriented study system analysis and design techniques for large-scale agent systems. Correspondingly, solutions for agent service based application integration, distributed data preparation, distributed agent coordination and parallel agent computing should be considered. In many cases of data mining, people should study algorithms that can adapt to dynamic data changes, dynamic user requests. To this end, it has the potential for agents to detect and reason such changes. Automated and adaptive data mining algorithms should be studied. The following is a list of some research open issues and promising areas.

- Activity modeling and mining
- Agent-based enterprise data mining

- Agent-based data mining infrastructure
- Agent-based data warehouse
- Agent-based mining process and project management
- Agent-based distributed data mining
- Agent-based distributed learning
- Agent-based grid computing
- Agent-based human mining cooperation
- Agent-based link mining
- Agent-based multi-data source mining
- Agent-based interactive data mining
- Agent-enriched ontology mining
- Agent-based parallel data mining
- Agent-based web mining
- Agent-based text mining
- Agent-based ubiquitous data mining
- Agent knowledge management in distributed data mining
- Agent for data mining data preparation
- Agent-human-cooperated data mining
- Agent networks in distributed knowledge discovery and servicing
- Agent service-based KDD infrastructure
- Agent-supported domain knowledge involvement in KDD
- Agent system providing data mining services
- Automated data mining learning
- Autonomous learning
- Distributed agent-based data preprocessing
- Distributed learning
- Domain intelligence in agent-based data mining
- Mobile agent-based knowledge discovery
- Protocols for agent-based data mining
- Self-organizing data mining learning.

1.6 Data Mining-Driven Agents

This section discusses data mining-driven multi-agent systems [12]. In contrast to the previous section, this section emphasizes agents empowered by more informative knowledge provided by data mining.

1.6.1 The challenges of data mining-driven agents

The astonishing rates at which data is generated and collected by current applications is difficult to handle even for the most powerful of today's computer systems.

This windfall of information often requires another level of distillation to elicit the knowledge that is hidden in voluminous data repositories. Data mining can be used to extract knowledge nuggets that will constitute the building blocks of agent intelligence. Here, intelligence is defined loosely so as to encompass a wide range of implementations from fully deterministic decision trees to self-organizing communities of autonomous agents. In many ways, intelligence manifests itself as efficiency.

In rudimentary applications, agent intelligence is based on relatively simple rules, which can be easily deduced or induced, compensating for the higher development and maintenance costs. In more elaborate environments, however, where both requirements and agent behaviors need constant modification in real time, these approaches prove insufficient, since they cannot accommodate the dynamic transfer of DM results into the agents. To enable the incorporation of dynamic, complex, and re-usable rules in multi-agent applications, a systematic approach must be adopted.

Existing application data (i.e., past transactions, decisions, data logs, agent actions, etc.) can be filtered in an effort to refine the best, most successful, empirical rules and heuristics. The resulting knowledge models can be embedded into 'dummy' agents in a process equivalent to agent training. As more data is gathered, the dual process of knowledge discovery and intelligence infusion can be repeated periodically, or on demand, to further improve agent reasoning.

In data mining-driven agent systems, induction attempts to transform specific data and information into generalized knowledge models. During the induction process, new rules and correlations are produced, aimed at validating user hypotheses. Since induction is based on progressive generalizations of specific examples, it may lead to invalid conclusions. In contrast, deductive systems draw conclusions by combining a number of premises. Under the assumption that these premises are true, deductive logic is truth preserving. In MAS applications, deduction is used when business rules and agent goals are well-defined and the human expert, who constructs the knowledge base, has a fine grasp of the problem's underlying principles. Nevertheless, deduction proves inefficient in complex and versatile environments.

The coupling of the above two approaches usually leads to enhanced and more efficient reasoning systems. Indeed, this combination overcomes the limitations of both paradigms by using deduction for well-known procedures and induction for discovering previously unknown knowledge. The process of transferring DM-extracted knowledge into newly-created agents is suitable for either upgrading an existing, non-agent-based application by adding agents to it, or for improving the already operating agents of an agent-based application. We consider three distinct cases, which correspond to three types of knowledge extracted and to different data sources and mining techniques.

- Case 1. Knowledge extracted by performing DM on historical datasets which record the business logic (at a macroscopic level) of a certain application;
- Case 2. Knowledge extracted by performing DM on log files recording the behavior of the agents (at a microscopic level) in an agent-based application, and
- Case 3. Knowledge extracted by the use of evolutionary data DM techniques in agent communities.

In each case the software methodology must ensure: a) the ability to dynamically embed the extracted knowledge models into the agents, and b) the ability to repeat the above process as many times as is deemed necessary. Standard agent-oriented software engineering processes are followed, in order to specify the application ontology, the agent behaviors and agent types, the communication protocol between the agents, and their interactions.

A number of agent-based applications that cover all three cases of knowledge diffusion have been developed. Domains, that are better suited for Case 1, include the traditional data producers, such as enterprise resource planning and supply chain management systems, environmental monitoring through sensor networks, and security and surveillance systems.

A typical example of Case 2 knowledge diffusion involves the improvement of the efficiency of agents participating in e-auctions. The goal here is to create both rational and efficient agent behaviors, which, in turn, will enable reliable agent-mediated transactions. Another example is a web navigation engine, which tracks user actions in corporate sites and suggests possibly interesting sites. This framework can be extended to cover a large variety of web services and/or intranet applications.

Finally, Case 3 encompasses solutions for ecosystem modeling and for web crawling with clusters of synergetic crawler agents.

1.6.2 The needs of data mining-driven agents

Data mining-driven multi-agent systems present attractive features to create more intelligent systems as discussed below:

- The combination of autonomy (MAS) and knowledge (DM) provides adaptable systems. Knowledge discovered in data and then fed into agents can greatly enhance the self-organization and learning performance of agents.
- DM can greatly enhance the learning and knowledge processing capability of agents, through involving DM algorithms in the building-blocks of agent learning systems. As a result, agents can learn from the data and from the environment before planning and reasoning are made.
- DM can enhance the agent capability of handling uncertainty through historical event analysis and dynamic mining and active learning; through mining agent behavioral data, it is possible to reach a balance and trade-off between agent autonomy and supervised evolution; the outcomes of self-organization and emergence become much more certain, controllable and predictable.
- The rigidity and lack of exploration of deductive reasoning systems is overcome. Rules are no longer hard-coded into systems and their modification is only a matter of retraining.
- DM techniques such as association rule extraction have no equivalent in agent systems. These techniques now provide agents with the ability of learning, discovery, probing and searching.

- Real-world databases often contain missing, erroneous data and/or outliers. Through clustering, noisy logs are assimilated and become a part of a greater group, smoothing down differences, while outliers are detected and rejected. Through classification, ambiguous data records can be validated and missing data records can be estimated. Rule-based systems cannot handle such data efficiently without increasing their knowledge-base and therefore their maintenance cost.
- The presented approach favors the combination of inductive and deductive reasoning models. In some of the demonstrators presented, there were agents deploying deductive reasoning models ensuring, thereby ensuring system soundness. Nevertheless, these agents decide on data already preprocessed by inductive agents. In this way, the dynamic nature of the application domains is satisfied, while the set of deductive results (knowledge-bases of deductive agents) becomes more compressed and robust.
- Even though the patterns and rules generated through data mining cannot be defined as *sound*, there are metrics deployed to evaluate the performance of the algorithms. Total mean square error (clustering), support-confidence (association rules), classifier accuracy (classification), and classifier evaluation (genetic algorithms), among others, are employed for evaluating the knowledge models extracted. The engagement of data mining performance evaluation into agents satisfies the need of agents for knowledge quality and model evaluation, model testing and model comprehension.
- Usually DM tools are introduced to enterprises as components-off-the-shelf. These tools are used by human experts to examine their corporate or environmental databases to develop strategies and take decisions. This procedure often proves time-consuming and inefficient. By exploiting concurrency and multiple instantiation of agent types (cloning capabilities) of MAS systems, and by applying data mining techniques for embedding intelligent reasoning into them, useful recommendations can be much more quickly diffused while parallelism can be applied to non-related tasks, pushing system performance even higher.

1.6.3 Research issues in data mining driven agents

Research issues on the involvement of data mining and knowledge discovery in agents can be in varying aspects. Appropriate learning models need to be studied and involved to support reasoning, adaptation and evolution in multi-agents. New communication, planning, coordination and dispatching mechanisms may be developed by discovering the interaction of proper patterns and relationships between agents. In human-agent interaction, we need to develop proper algorithms to discover user behavior patterns, so that agent-user interaction and servicing can be more effective. In distributed conditions, we need to develop distributed learning algorithms to manage the coordination of agent crossing multiple applications. In the following, we list some topics that are promising areas in data mining driven agent research.

- Collaborative learning in multi-agents
- Data mining-driven agent learning, reasoning, adaptation and evolution
- Data mining-driven multi-agent communication, planning and dispatching
- Data mining-driven user modeling
- Data mining-driven user servicing
- Data mining-driven network servicing
- Data mining-driven agent recommender
- Data mining-driven trading agents
- Data mining agent assistant
- Multi-agent reinforcement learning
- Multi-agent knowledge discovery
- Data mining enhancing agent intelligence enhancement
- Decentralized clustering in large multi-agent systems
- Distributed learning in agent coordination
- Distributed learning in multi-agent systems
- Emergent agent organization and behavior
- Information gathering agents
- Learning agents
- Web mining agents
- Self-learning agents.

1.7 Mutual Issues in Agent Mining

There are many mutual enhancement issues in both agents and data mining respectively, and during the integration. Typical issues consist of architecture and infrastructure problems, actionable capability of agent-mining symbionts, constraints in agents and mining, data intelligence in agent and mining, domain knowledge in agents and mining, evaluation issues such as technical significance and business expectation, gaps filling between technical and business expectations, human intelligence and roles in agents and mining, knowledge management in agents and mining, meta-data and meta-knowledge in agents and mining, usability, expandability, openness, organizational and social factors and issues such as business factors, processes, security, privacy, trust; services request, response, recommendation and management, and finally the meta-synthesis of relevant ubiquitous factors into an effective and integrative system for agent and data mining problem-solving. We now briefly discuss some of these issues.

1.7.1 The need to study common issues for agent mining

A typical issue is the involvement of human intelligence and human roles. Even though both communities recognize the importance of human involvement and human intelligence in problem-solving and solution development, it is challenging to effectively and dynamically include human roles in problem-solving systems. Issues arise from aspects such as the understanding and simulation of human empirical intelligence and experiences that are of critical importance to problem-solving, acquisition and representation of human qualitative intelligence in agent-mining systems, and the interaction and interfaces between humans and systems to cater for human intelligence and roles.

Organizational, environmental and social factors constitute important elements of complex problems/systems and their environments in agents and data mining fields. This consists of comprehensive factors such as business processes, workflows, business rules and human roles that are relevant to business problem-solving, organizational and social factors such as organizational rules, protocols and norms. For instance, while concepts such as organizational rules, protocols and norms have been fed into agent organizations, they are also important for data mining systems in converting patterns into operable deliverables that can be smoothly taken over by business people and integrated into business systems.

There are often gaps between technical outcomes and business expectations in developing workable agents and data mining algorithms and systems, which is due to the inconsistency and incompleteness of evaluation systems between technical and business concerns. As a result, the resulting deliverables are often not of interest to business people and are not operable for action-taking in business problem-solving. An ideal scenario is to generate algorithms and systems that care about concerns from both the technical and business aspects and from both objective and subjective perspectives.

The above case studies show that it is essential to study common issues for the benefits of the particular field. In fact, the studies can also activate the possible emergence of agent-mining symbionts. For instance, the modeling and representation of domain knowledge and knowledge management in agents and data mining may be shared. It may serve as an intrinsic working mechanism for an agent-mining symbiont that has the capability of involving domain knowledge in agent-human interaction and data mining algorithm modeling, and managing knowledge for data mining agents and agent-based systems.

To facilitate the studies of common enhancement issues, the possible methodologies and approaches needed may be highly diversified and cross-disciplinary. Bodies of knowledge that may be useful consist of subjects such as cognitive science, human-machine interaction, interaction and interface design, knowledge engineering and management, ontological engineering, evaluation systems, organizational computing, social computing, artificial intelligence, and machine learning.

1.7.2 Mutual research issues in agent mining

After years of unorganized development of the one-way effect discussed above, there is further recolonization of fundamental mutual issues in agent-mining interaction [4, 8, 45, 2], which involve common issues of both parties. The studies on these mutual issues should not only tackle problems of one-way enhancement as discussed in Sections 1.5 and 1.6, but also two-way strengthening in building a super-intelligent agent-mining symbiont. However, these issues have not attracted sufficient attention in the community.

- Architecture and infrastructure problems
- Actionable capability of agent-mining symbionts
- Constraints in agent and mining
- Data intelligence in agent and mining
- Domain knowledge in agent and mining
- Domain intelligence in agent and mining
- Evaluation issues such as technical significance and business expectation
- Gap filling between technical and business expectations
- Human intelligence and role in agent and mining
- Human-system interaction
- Intelligence meta-synthesis in agent and mining
- Knowledge management in agent and mining
- Meta-data and meta-knowledge in agent and mining
- Nonfunctional issues such as usability, expandability, openness
- Ontology issues in agent and mining
- Organizational issues such as business factors, process
- Performance issues such as effectiveness, efficiency, scalability
- Social issues such as security, privacy, trust
- Services request and response, service-oriented management
- System management.

1.8 Applications and case studies

1.8.1 Applications

As we can see from many references, the proposal of agent mining is actually driven by broad and increasing applications. Many researchers are developing agent-mining systems and applications dealing with specific business problems and for intelligent information processing. For instance, we summarize the following application domains.

- Artificial immune systems
- Artificial and electronic markets
- Auction
- Business intelligence
- Customer relationship management
- Distributed data extraction and preparation
- E-commerce

- Finance data mining
- Grid computing
- Healthcare
- Internet and network services, e.g., recommendation, personal assistant, searching, retrieval, extraction services
- Knowledge management
- Marketing
- Network intrusion detection
- Parallel computing, e.g., parallel genetic algorithm
- Peer-to-peer computing and services
- Semantic web
- Social intelligence & social network analysis
- Supply chain management
- Trading agents
- Trading optimization and support
- Text mining
- Web mining.

1.8.2 Case Studies: Developing Actionable Trading Agents

The task of *designing smart trading agents* is to endow trading agents with the capabilities of searching strategies in a constrained market environment to satisfy trader preference. We introduce two approaches to designing smart trading agents. One is to design *evolutionary trading agents* [14], which are equipped with evolutionary computing capabilities, and can search strategies from a large candidate strategy space targeting higher *benefit-cost ratio*. The other is to integrate optimal instances from multiple classes of trading strategies into one combined powerful strategy through *collaborative trading agents* [15].

Evolutionary Trading Agents for Parameter Optimization

Evolutionary trading agents have capabilities of evolutionary search computing. They can search trading strategies based on given optimization fitness and specified optimization objectives. Their roles consist of optimization requests (including base strategies and arguments), creating strategy candidates (namely chromosomes), evaluating strategy candidates, crossing over candidate strategies, mutating candidate strategies, re-evaluating candidate strategies, filtering optimal strategies, and so on.

The strategy optimization using evolutionary trading agents is as follows. A *User Agent* receives optimization requests from user-agent interaction interfaces. It forwards the request to *Coordinator Agents* and the *Coordinator Agents* check the availability and validity of the optimized *Strategy Agent* class with *strategyClassID*. If a *Strategy Agent* class is available and optimizable, *Coordinator Agents* call the *Evolutionary Agents* to perform corresponding roles, for instance, *createStrate-*

gyCandidates, *evaluateStrategyCandidates*, *crossoverCandidateStrategies*, *mutateCandidateStrategies*, *re-evaluateCandidateStrategies*, or *returnOptimalStrategies* to optimize the strategy. After the optimization process, *Evolutionary Agents* return *Coordinator Agents* the searched optimal *Strategy Agent* with *strategyID* and corresponding parameter values. *Coordinator Agents* further call the *User Agents* to present the results to traders by invoking *Presentation Agents*.

Trading Agent Collaboration for Strategy Integration

In real-life trading, trading strategies can be categorized into many classes. To financial experts, different classes of trading strategies indicate varying principles of market models and mechanisms. A trading agent often takes a series of positions generated by a specific trading strategy, which instantiates a trading strategy class. Our idea is for multiple trading agents to collaborate with each other and take concurrent positions created by multiple trading strategies to take advantage of varying strategies.

The working mechanism of *trading agent collaboration for strategy integration* is as follows. There are a few *Representative Trading Agents* in the market. Each *Representative Agent* invokes an *Evolutionary Agent* to search for an optimal *Strategy Agent* from a strategy class. *Coordinator Agents* then negotiate with these *Representative Agents* and *Evolutionary Agents* to integrate the identified optimal strategies of *Strategy Agents*. An appropriate integration method is negotiated and chosen by *Coordinator Agents*, *Representative Agents* and *Evolutionary Agents* based on a globally optimal objective.

1.8.3 F-Trade: An agent-mining symbiont

In this section, we briefly introduce an agent-mining symbiont – F-Trade[1] to illustrate the development and use of agent-mining interaction technology in tackling both research and business issues.

F-Trade [4] is the acronym of Financial Trading Rules Automated Development and Evaluation, a web-based automated enterprise infrastructure for trading strategies and data mining on stock/capital markets. The system offers data connection, management and processing services. F-Trade supports online automated plug and play, and automatic input/output interface construction for trading signals/rules and data mining algorithms, data sources, and system components. It provides powerful and flexible supports for online backtesting, training/test, optimization and evaluation of trading strategies and data mining algorithms. Users can plugin, subscribe, supervise and optimize trading strategies and data mining algorithms in a human-machine cooperated manner.

F-Trade is built in Java agent services on top of Windows/Linux/Unix. XML is used for system configuration and metadata management. A super-server functions

[1] www.f-trade.info, www.ftrade.info

as the application server, another one acts as the data warehouse. It is constructed with online connectivity to distributed data sources as well as user-specific data sources.

Major roles played by agents in F-Trade consist of agent service-based architecture, agent-driven human interaction, agent for data source management, data collection and dispatch agents roaming to remote data sources, agentized trading strategies and data mining algorithms, agent and service recommender providing optimum algorithms and rules to users, and so on. Data mining assists the system in aspects such as data mining-driven trading rule/algorithm recommender agents, data mining-driven user services, data mining-driven trading agent optimizers, mining actionable trading rules in generic trading pattern set, parameter tuning of algorithm agents through data mining, etc. Mutual issues involve ontology-based domain knowledge representation and transformation to problem-solving terminology, human involvement and agent-based human interaction with algorithms and the system for algorithm supervision, optimization and evaluation, among others.

1.8.4 Agent Academy

Agent Academy [41] is an open-source framework and integrated development environment (IDE) for creating software agents and multi-agent systems, and for augmenting agent intelligence through data mining. The core objectives of Agent Academy are to:

- - Provide an easy-to-use tool for building agents, multi-agent systems and agent communities.
- - Exploit Data Mining techniques for dynamically improving the behavior of agents and the decision-making process in multi-agent systems.
- - Serve as a benchmark for the systematic study of agent intelligence generated by training them on available information and retraining them whenever needed.
- - Empower enterprise agent solutions by improving the quality of provided services.

Agent Academy has been implemented upon the JADE and WEKA APIs (Application Programming Interfaces) and its second version is now available at Sourceforge [2]. The current release contains 237 Java source files and is spun on over 28,000 lines of code.

Agent Academy proposes a new line of actions for creating DM-enhanced agents, through a simple and well-defined workflow. In a typical scenario the developer must follow a specified methodology in order to build a new agent application. First, agent behaviors, as orthogonal as possible, are built while, in a parallel process, data mining is performed in order to generate knowledge models for the application under development. Next, everything is organized/assigned to agents which, eventually, will constitute the multi-agent system.

[2] http://sourceforge.net/projects/agentacademy

1.9 Trends and directions

As a newly emerging area, agent-mining interaction has an expansive future. Unprecedented potential is embodied through unlimited research interest points in areas of (1) data mining driven agent, (2) agent driven data mining, and (3) mutual issues in agent-mining interaction.

Research and exploitation are advancing to develop the better, tighter, and more organized integration for the systems. We raise some issues here that researchers are encouraged to pursue [12].

- Foundations, including methodologies, formal tools and processes for supporting integration of agents and data mining from multiple dimensions as outlined in Fig. 1.3
- Formulation of formal methodologies, languages, notations, for agent mining software engineering
- Methodologies and techniques supporting effective involvement and integration of domain, human, organizational and social intelligence
- The integration of agents, services, organizational and social computing with data mining for engineering agent mining symbionts [13]
- The integration of semantics, visualization, service and knowledge management for agent mining systems and applications
- Building trust, reputation, privacy and security for agent mining systems, and making sure the systems and results are sound and safe
- The development of an analytical methodology for agent retraining
- The development of a methodology for evaluating MAS efficiency, and the performance of the system as a whole as well as individual units.
- Measurements of gains achieved by the agent mining system for organization of decision
- Employing various distributed computing techniques to agent-based distributed data mining, such as the peer-to-peer model
- Ubiquitous computing, including ubiquitous knowledge discovery in ubiquitous environments for agent-mining integration and super-intelligent systems
- Privacy and security models for distributed agent systems. As they are traveling all over the system, it is necessary to guarantee privacy and security preservation, as in trust-based modeling
- The development of an agent infrastructure model that includes data mining as a key component, and vice versa
- The Web as a platform for agent mining, toward the use of future Web models, as Web 2.0 and 3.0, by considering the browser itself as a user agent which allows migration of problem domains and knowledge
- Successful systems, case studies and applications of agent mining technologies and systems for both professionals and common users.

1.10 Agent mining community development

The development of agent-mining interaction is evidenced by the following evolution process and burgeoning characteristics.

1.10.1 Fast progression in fostering the community

We draw a conclusion that agent-mining interaction and integration is emerging as a new member of the scientific family due to the following survey findings.

- An ever-increasing and respectable number of publications: an initial literature review has disclosed that there are over 250 conference and journal papers, books, proceedings, and technical reports published on topics associated with agent and mining interaction and integration. This uptrend is becoming increasingly clear from the incremental change over the past three years in the number of publications by major presses including Springer, IEEE and ISI press.
- An ever-increasing level and quality of papers: with the increase in publication numbers, publication quality has also been extremely improved. A typical trend is that there have been more and more journal papers and books/proceedings published after 2003. This is evidenced by increasing papers accepted by top-ranking conferences and journals such as AAMAS, KDD and ICDM in both communities.
- An ever-increasing number of professional activities: another typical indicator of whether a research topic is evolving into a new, separate area is the number and quality of professional activities, and the involvement of key research groups and researchers from both communities in these activities.
- An increasing transparent academic voice pursuing a separate area and a first-class citizen in the scientific family: this is evidenced by panel discussions in both ADMI workshop series and AIS-ADM series.

1.10.2 Research resources on agent-mining interaction

Agent mining communities are formed through emerging efforts in both AAMAS and KDD communities, as evidenced by continuous acceptance of papers on agent mining issues by prestigious conferences such as AAMAS, SIGKDD and ICDM. During the formation of agent mining communities, two special interest groups (SIG) have dedicated. AgentLink ALAD SIG was organized within the framework of FP5 European project KDNet Project, consolidating the data mining and agent communities. Another dedicated SIG is the Agent-Mining Interaction and Integration SIG [3], which actively links leading researchers in this emerging field, and or-

[3] AMII-SIG: www.agentmining.org

ganizes regular events for promoting and encouraging research networking in both agent and data mining communities. These task forces have spearheaded many other analogous initiatives including special sessions of the international conferences, dedicated international workshops, journal special issues, and more.

AMII-SIG now is one of the main and most active driving forces in promoting community development. It summarizes and shares information covering many issues, and has also led the annual workshop series on Agent and Data Mining Interaction (ADMI) since 2006, moving globally and attracting researchers from many different communities. It has also formed an international Steering Committee involving leading researchers from both agents and data mining communities.

The other biennial workshop series - Autonomous Intelligent Systems - Agents and Data Mining (AIS-ADM) has existed since 2005. Other events include special journal issues [16, 17], tutorials [42, 18], an edited book [9], and monographs [41, 19].

1.11 Conclusions

Agent and data mining interaction and integration has emerged as a prominent and promising area in recent years. The dialogue between agent technology and data mining can not only handle issues that are hardly coped with in each of the interacted parties, but can also result in innovative and super-intelligent techniques and symbionts much beyond the individual communities.

This chapter presents a high-level overview of the development and major directions in the area. The investigation highlights the following findings: (1) agent-mining interaction is emerging as a new area in the scientific family, (2) the interaction is increasingly promoting the progress of agent and mining communities, (3) it results in ever-increasing development of innovative and significant techniques and systems towards super-intelligent symbionts.

As a new and emerging area, it has many open issues waiting for the significant involvement of research resources, in particular practical and research projects from both communities. We believe the research and development on agent mining is very promising and worthy of substantial efforts by both established and new researchers.

Acknowledgements This work is sponsored in part by Australian Research Council Grants (DP0988016, LP0989721, DP0773412, and LP0775041).

References

1. Aciar, S., Zhang, D., Simoff, S., and Debenham, J.: Informed Recommender Agent: Utilizing Consumer Product Reviews through Text Mining. Proceedings of IADM2006. IEEE Computer Society (2006)

2. Baik,S., Cho, J., and Bala, J.: Performance Evaluation of an Agent Based Distributed Data Mining System. Advances in Artificial Intelligence, Volume 3501/2005 (2005)
3. Cory, J., Butz, Nguyen, N., Takama, Y., Cheung, W., and Cheung, Y.: Proceedings of IADM2006 (Chaired by Longbing Cao, Zili Zhang, Vladimir Samoilov) in WI-IAT2006 Workshop Proceedings. IEEE Computer Society (2006)
4. Cao, L., Wang, J., Lin, l., and Zhang, C.: Agent Services-Based Infrastructure for Online Assessment of Trading Strategies. Proceedings of IAT'04, 345-349 (2004).
5. Cao, L.: Integration of Agents and Data Mining. Technical report, 25 June 2005. http://www-staff.it.uts.edu.au/ lbcao/publication/publications.htm.
6. Cao, L., Luo, C. and Zhang, C.: Agent-Mining Interaction: An Emerging Area. AIS-ADM, 60-73 (2007).
7. Cao, L., Luo, D., Xiao, Y. and Zheng, Z. Agent Collaboration for Multiple Trading Strategy Integration. KES-AMSTA, 361-370 (2008).
8. Cao, L.: Agent-Mining Interaction and Integration – Topics of Research and Development. http://www.agentmining.org/
9. Cao, L.: Data Mining and Multiagent Integration. Springer (2009).
10. Cao, L. and Zhang, C. F-trade: An Agent-Mining Symbiont for Financial Services. AAMAS 262 (2007).
11. Cao, L., Yu, P., Zhang, C. and Zhao, Y. Domain Driven Data Mining. Springer (2009).
12. Cao, L., Gorodetsky, V. and Mitkas, P. Agent Mining: The Synergy of Agents and Data Mining. IEEE Intelligent Systems (2009).
13. Cao, L. Integrating Agent, Service and Organizational Computing. International Journal of Software Engineering and Knowledge Engineering, 18(5): 573-596 (2008)
14. Cao, L. and He, T. Developing Actionable Trading Agents. Knowledge and Information Systems: An International Journal, 18(2): 183-198 (2009).
15. Cao, L. Developing Actionable Trading Strategies, Knowledge Processing and Decision Making in Agent-Based Systems, 193-215, Springer (2008).
16. Cao, L., Zhang, Z., Gorodetsky, V. and Zhang, C.. Editor's Introduction: Interaction between Agents and Data Mining, International Journal of Intelligent Information and Database Systems, Inderscience, 2(1): 1-5 (2008).
17. Cao, L., Gorodetsky, V. and Mitkas, P. Editorial: Agents and Data Mining. IEEE Intelligent Systems (2009).
18. Cao, L. Agent & Data Mining Interaction, Tutorial for 2007 IEEE/WIC/ACM Joint Conferences on Web Intelligence and Intelligent Agent Technology (2007).
19. Cao, L., Zhang, C. and Zhang, Z. Agents and Data Mining: Interaction and Integration, Taylor & Francis (2010).
20. Brazdil, P., and Muggleton, S.: Learning to Relate Terms in a Multiple Agent Environment. EWSL91 (1991)
21. Davies, W.: ANIMALS: A Distributed, Heterogeneous Multi-Agent Learning System. MSc Thesis, University of Aberdeen (1993)
22. Davies, W.: Agent-Based Data-Mining (1994)
23. Edwards, P., and Davies, W.: A Heterogeneous Multi-Agent Learning System. In Deen, S.M. (ed) Proceedings of the Special Interest Group on Cooperating Knowledge Based Systems. University of Keele (1993) 163-184.
24. Gorodetsky, V., Liu, J., Skormin, V. A.: Autonomous Intelligent Systems: Agents and Data Mining book. Lecture Notes in Computer Science Volume 3505 (2005)
25. Gorodetsky, V.; Karsaev, O.and Samoilov, V.: Multi-Agent Technology for Distributed Data Mining and Classification. IAT 2003. (2003) 438 - 441
26. Gorodetsky, V., Karsaev, O. and Samoilov, V.: Infrastructural Issues for Agent-Based Distributed Learning. Proceedings of IADM2006, IEEE Computer Society Press
27. Han, J., and Kamber, M.: Data Mining: Concepts and Techniques (2nd version). Morgan Kaufmann (2006)
28. Kaya, M. and Alhajj, R.: A Novel Approach to Multi-Agent Reinforcement Learning: Utilizing OLAP Mining in the Learning Process. IEEE Transactions on Systems, Man and Cybernetics, Part C, Volume 35, Issue 4 (2005) 582 - 590

29. Kaya, M.and Alhajj, R.: Fuzzy OLAP Association Rules Mining-Based Modular Reinforcement Learning Approach for Multi-Agent Systems. IEEE Transactions on Systems, Man and Cybernetics, Part B, Volume 35, Issue 2, (2005) 326 - 338
30. Klusch, M., Lodi, S.and Gianluca, M.: The Role of Agents in Distributed Data Mining: Issues and Benefits. Intelligent Agent Technology (2003): 211 - 217
31. Klusch, M., Lodi, S. and Moro,G.: Agent-Based Distributed Data Mining: The KDEC Scheme. Intelligent Information Agents: The AgentLink Perspective Volume 2586 (2003) Lecture Notes in Computer Science
32. Klusch, M., Lodi, S.and Moro,G.: Issues of Agent-Based Distributed Data Mining. Proceedings of AAMAS, ACM Press (2003)
33. Letia, A., Craciun, F.; et. al.: First Experiments for Mining Sequential Patterns on Distributed Sites with Multi-agents. Intelligent Data Engineering and Automated Learning - IDEAL 2000: Data Mining, Financial Engineering, and Intelligent Agents, 19 Volume 1983 (2000)
34. Liu, J. and You, J.: Smart Shopper: An Agent Based Web Mining Approach to Internet Shopping. IEEE Transactions on Fuzzy Systems, Volume 11, Issue 2 (2003)
35. Mitkas, P.: Knowledge Discovery for Training Intelligent Agents: Methodology, Tools and Applications. Autonomous Intelligent Systems: Agents and Data Mining, Lecture Notes in Computer Science Volume 3505 (2005)
36. Lu, H.; Sterling, L. and Wyatt, A.: Knowledge Discovery in SportsFinder: An Agent to Extract Sports Results from the Web. PAKDD-99, Volume 1574 (1999)
37. Mohammadian, M.: Intelligent Agents for Data Mining and Information Retrieval, Idea Group Publishing (2004)
38. Ong, K., Zhang, Z., Ng, W. and Lim, E.: Agents and Stream Data Mining: A New Perspective. IEEE Intelligent Systems, Volume 20, Issue 3, 60 - 67
39. Rea, S.: Building Intelligent .NET Applications: Agents, Data Mining, Rule-Based Systems, and Speech Processing. Addison-Wesley Professional (2004)
40. Sian, S.: Extending Learning to Multiple Agents: Issues and a Model for Multi-Agent Machine Learning (MA-ML). In Proceedings of the European Workshop Sessions on Learning EWSL91 (Kodratroff, Y.) Springer-Verlag, (1991) 458-472.
41. Symeonidis, A., and Mitkas, P.: Agent Intelligence Through Data Mining, Springer (2006)
42. Symeonidis, A., and Mitkas, P.: Agent Intelligence Through Data Mining, *Tutorial with ECML/PKDD2006.*
43. Weiss, G.: A Multiagent Perspective of Parallel and Distributed Machine Learning. In *Proceedings of Agents'98*, 226-230, 1998.
44. Wooldridge, M.: An Introduction to Multi-Agent Systems, Wiley (2002)
45. Zhang, C.; Zhang, Z., and Cao, L.: Agents and Data Mining: Mutual Enhancement by Integration. Autonomous Intelligent Systems: Agents and Data Mining, Volume 3505 (2005)
46. Zhang, Z., and Zhang, C.: Agent-Based Hybrid Intelligent System for Data Mining. Agent-Based Hybrid Intelligent Systems, Volume 2938 (2004)
47. Zhong, N., Liu, J., and Sun, R.: Intelligent Agents and Data Mining for Cognitive Systems? Cognitive Systems Research Volume 5, Issue 3, (2004) 169-170

Chapter 2
Towards the Integration of Multiagent Applications and Data Mining

Célia Ghedini Ralha

Abstract This chapter has the objective to present research on combining two originally separated areas, agents including distributed multiagent systems and data mining, which are increasingly interrelated. Recent research has present that such interaction features are bilateral and complementary, since new approaches and techniques are developed to benefit from the synergetic enhancement of intelligence and infrastructure for information processing and systems. This chapter draws attention to illustrate agent-mining interaction with two different domain multiagent applications: BioAgents at the bioinformatics area and MADIK at the computer forensics area. The presented case studies are driving forces towards the integration of the agent-mining challenging area. As ongoing research works we discuss the prospects of both agent-mining projects.

2.1 Overview of Agents/Multiagent Systems and Data Mining Integration

In the past decade, agents/multiagent systems (MAS) and data mining (DM)/knowledge discovery have emerged as two increasingly interrelated research areas, opening space to the agent-mining interaction and integration (AMII) research field. This new field has driven efforts from both sides to find benefits and complementarity to both communities. The AMMI field has prove to be so promising in recent years that there is a special interest group, entitled the Agents and Data Mining Interaction and Integration (AMII-SIG) , which aims to foster a forum for boosting the research and development on AMII studies [4].

In general terms, an intelligent software agent (ISA) uses Artificial Intelligence (AI) in the pursuit of goals [41, 56]. Thus, an ISA is a computer system capable

Computer Science Department, University of Brasilia (UnB)
P.O.Box 4466, Brasilia/DF, Zip Code 70904-970, Brazil
e-mail: ghedini@cic.unb.br

L. Cao (ed.), Data Mining and Multi-agent Integration, DOI: 10.1007/978-1-4419-0522-2_2,

of autonomous action in some environment in order to meet its design objectives [62]. According to the given agent definition, we may cite that an agent is a computing entity with four features: autonomy, reactivity, interaction and initiative. ISA properties frequently cited in the literature include: mobility, veracity, benevolence, rationality and learning/adaption. In the literature, we find many definitions for a MAS, but mostly they agree as referring to a computational system composed by more than one agent [61, 62]. Thus, a MAS is a system where many agents interact with the environment in a cooperative or competitive way, to achieve individual or group objectives.

As we already defined, agents will be acting on behalf of users with different goals, motivations and to successfully interact, they will require the ability to cooperate, coordinate, and negotiate with each other. In summary, the agents intelligence and infrastructure for information processing and systems can enhance data and knowledge treatment. MAS technology provides a powerful method for building and managing DM in a distributed system. Agents autonomously exchange data with other agents and discovering their interesting data. All these characteristics are very important to the AMII research area.

This chapter presents research work related to the Applications and Case Study AMII topic, more specifically to the Emergent Agent-Mining Organizations and Applications topic, since it presents an overview of two SMA applications: (i) BioAgents, defined as a bioinformatics application and services topic and (ii) MADIK, an artificial system and service at Computer Forensics. Thus, this chapter draws a high-level overview of two different domain multiagent applications, with the intention to promote AMII from the practical perspective of an application-oriented approach. We also present some related publications to our research and do an evaluation of our ongoing research work, with open issues and prospects of both agent-mining projects.

2.2 Related Work

Since, the research topics in the area of AMII are quite diverse and ubiquitous, there are some works that might be related to all streams of the emerging area, which were presented at [27]: [15, 57, 19]. Giving attention to integration issues of MAS and DM we can cite [49, 43]. In [49] we find a model integration proposal *MultiAgent System for Distributed Data Mining - SMAMDD*, where agents perform mining tasks locally and merge their results into a consistent global model. In order to achieve that, agents cooperate by exchanging messages, aiming to improve the process of knowledge discover generating accurate results. A Web document database integration technique is presented in [43], where mining agents were defined to information extraction from HTML files.

Considering the MAS application domains, we now present related work to Bioinformatics. The work of Santos and Bazzan [21, 26], deals with knowledge discovery and data mining. In this work, authors propose a MAS to the Bioinformatics area, where agents are responsible for applying different machine learning algorithms and using subsets of the data to be mined, and are able to cooperate to discover knowledge from these subsets. Authors propose a case study to use cooperative negotiation to construct an integrated domain model from several sources. At the bioinformatics scenario the application of the approach was related to automated annotation of proteins' keywords. In this work, agents do not use any domain dependent information, as they just encapsulate data and machine learning algorithms used to induce models to predict the annotation, using data from biological databases. In a previous work [10, 2] wee can find efforts to generate a system for automated annotation proteins related to the *Mycoplasmataceae* family data. A similar works on automatic annotation uses symbolic machine learning techniques with a ISA [36].

Referring to the intersection of bioinformatics and DM, we may say that this is a cutting edge research topic, with important forums to discuss this interaction, e.g. [1]: [35, 39, 14, 51]. Also another new research topic is multiagent projects developed at the bioinformatics domain [33, 8, 7, 48].

Now turning our attention to Computer Forensics research domain area, there are very interesting work using DM techniques. In Gary Warner spam research, they are applying the principles of DM and Grid Computing to establish the Spam DM for Law Enforcement project, but no AMII perspective is used [60, 59]. In the spam project, they have build a large corpus of spam emails, which were analyzed and clustered to provide significant forensic and investigative data to law enforcement [44]. This project has the objective to provide tools, techniques, and training to fight CyberCrime. The projects' approach is a three-pronged one: academics, awareness and research. There are some special focus area the group is doing research: spam, phishing and malware.

A more methodological work includes the CRoss Industry Standard Process for Data Mining (CRISP-DM), which is specialized for evidence mining in [58, 22]. Authors use the term *evidence mining* to refer to the application of these techniques in the analysis phase of digital forensic investigations. Thus, this paper presents an approach to the specialization of CRISP-DM to CRISP-EM, an evidence mining methodology designed specifically for digital forensics.

Another work presents a novel DM method, called *AuthorMiner*, for determining the authorship of a malicious e-mail [32]. Since there is an alarming increase in the number of cybercrime incidents through anonymous e-mails, the problem of e-mail authorship attribution is to identify the most plausible author of an anonymous e-mail from a group of potential suspects. Most previous contributions employed a traditional classification approach, such as decision tree and Support Vector Machine (SVM) . In this paper, authors introduce an innovative DM method to capture the write-print of every suspect and model it as combinations of features that occurred frequently in the suspects e-mails. This notion is called *frequent pattern*. In summary, there are many recent research work in DM integration to Bioinformatics and to Computer Forensics domains, but to the best of our knowledge, there is a lack of AMII research in both areas. Thus, we present our application-oriented approach to AMII field with a high-level overview of two different domain multiagent applications: BioAgents and MADIK.

2.3 *BioAgents*

Enormous volume of deoxyribonucleic acid (DNA) sequences of organisms are continuously being discovered by genome sequencing projects around the world. The task of identifying biological function prediction for the DNA sequences is a key activity in genome projects. This task is done in the annotation phase, which is divided into automatic and manual. The automatic annotation has the objective of finding, for each DNA sequence identified in the project, similar sequences among millions, stored in public databases, e.g. *GenBank* [13], by using approximated pattern matching algorithms (*BLAST* [5] and *FASTA* [50]). The manual annotation is done by the biologists, that use the results produced by the automatic annotation, and their knowledge and experience, to decide the function prediction to each DNA sequence. In this way, the biologists guarantee accuracy and correctness to each sequence function prediction.

BioAgents is a MAS for supporting manual annotation [54, 9, 38]. The system simulates the biologists' knowledge and experience for annotating DNA sequences in genome sequencing projects. The MAS cooperative approach, allows to create different specialized ISA that, working together, suggest proper manual annotation.

The architecture of *BioAgents* is divided into three layers: interface, collaborative and physical [54]. The interface layer receives the requests and returns the results to users. The collaborative layer is the architecture core, it is composed by the *conflict resolutions agent* (CR), the *manager agents* (MR) and the *analyst agents* (ANL). The specialized MR are responsible for executing particular algorithms, like *BLAST* and *FASTA*, that interact with ANL for treating specific databases, like *nr* or *kog*. Note that we defined specialized agents to deal with different algorithms and specific knowledge sources (KS). At last, the collaborative layer suggests annotations to be sent to the interface layer. The physical layer consists of different local databases containing the results of the automatic annotation.

BioAgents was implemented in *Java* with development environment *Eclipse SDK*, version 3.1.2. For the agent development framework we used *Java Agent DEvelopment Framework - JADE*, version 3.4.1 [12]. The parsers used by the ANL agents were implemented by adapting some libraries of the *framework BioJava*, version 1.4. For the rule-based motor, we have used *Java Expert System Shell - JESS*, version 6.1 [23] to allow agents reasoning in *BioAgents* system. With JESS we defined the biologists knowledge through the use of production rules (declarative rules) according to the parameters defined on a specific genome project. In all experiments we used the same production rules based on *BLAST* and *FASTA* results, following the biologists recommendations.

BioAgents rules use two parameters: the *expectation-value (e-value)* and *score*. These parameters express the similarity between each sequence generated on the project with each sequence stored on the database. As lower is the *e-value* as lower is the error probability between the correspondences of both sequence of nucleotides, and as higher is the *score* more close are the sequences. The annotation of each sequence is based on the similarity between both sequences, and the hypothesis is that as much closer are the sequences as higher is the chance that both have the same biological function prediction.

In order to validate *BioAgents*, we used data from three genome sequencing projects developed at the MidWest Region of Brazil: Functional and Differential Genome from the *Paracoccidioides brasiliensis* (Pb) fungus [46], Genome Project of *Paullinia cupana* (guaraná) plant [47] and Genome Sequencing Project of the *Anaplasma marginale* rickettsia [45]. We used *BioAgents* to suggest annotations for both Genome Project Pb and Genome Project Guaraná using the results of *BLAST* and *FASTA*, and comparing the suggested annotations with the manual annotations previously done by the biologists. The results of *BioAgents* applied to the Genome Project Pb presented 44.1% of correct suggestions, while the results applied to the Genome Project Guaraná presented 45.35% of correct suggestions. Results of *BioAgents* applied on the Genome Anaplasma Project presents data with 2,759 suggested annotations for a total of 3,214 ORFs, which corresponds to 85.84% of suggestions.

2.4 MADIK

Computer Forensics consists of examination and analysis of computational systems, which demands a lot of resources due to the large amount of data involved. Thus, the success of computer forensics examinations depend on the ability to examine large amounts of digital forensic data, in search of important evidences. Forensic examination consists of several steps to preserve, collect and analyze evidences found in digital storage media, so they can be presented and used as evidence of unlawful actions. In this scenario, either distributed agent/multiagent architectures and DM techniques can be of great help. At real computer forensics cases, experts can't define at first what evidence is more relevant to the incident or crime under investigation. Thus, a pre-analysis of the suspect machines would limit the number of evidences collected for examination, reducing the time of investigation and analysis by the forensic experts. But the lack of intelligent and flexible tools to help forensic experts with the pre-analysis phase, and with a concrete cross-analysis of large number of potential correlated evidences, is a reality.

Thus, we propose the use of a MAS, to help digital forensics during forensic examination process. With our approach, it is possible to have different specialized ISA to suggest proper investigative actions, based in experts knowledge of technical domain. The process cited by [11] served as basis of analysis for the conception of new specialized agents. In real forensic examinations, a constant necessity in all investigations is the distribution and coordination of tasks amongst the team of specialists. Thus, in the proposed approach, different roles are played by different levels of agents, similarly to organizational hierarchy levels [52].

The **M**ulti-**A**gent **D**igital **I**nvestigation Tool**K**it - MADIK is a SMA to assist the experts during the forensic examination process [31, 30, 29]. The architecture of MADIK is a four-layer one, defined as a metaphor to the organizational hierarchy levels: strategic, tactical, operational and specialist. With this architecture, we define autonomous agents, each specialized in a small and distinct subset of the overall objectives and constraints: HashSetAgent, FilePathAgent, FileSignatureAgent, TimelineAgent, WindowsRegistryAgent and KeywordAgent. In this case, no centralized and rigid control is necessary, but a conflict resolution mechanism, to solve differences and keep the main objective trail. With this mechanism, agents can collaborate by observing and modifying one another's work, through the use of a common base named blackboard [42, 20, 17].

MADIK was implemented as an Open Source Software Project with a GNU General Public License (GPL) [53]. MADIK is all developed in Java [37]. MADIK uses as database environment PostgreSQL (pgsql). For the agent development framework we used *Java Agent DEvelopment Framework - JADE*, version 3.4.1. For the agents inference, we have used *Java Expert System Shell - JESS*, version 6.1. In order to validate MADIK we did two experiments using data from real investigation cases. The first case included data of 110 thousand files from two different hard drives, both belonging to the same case [30]. The second experiment is based on a retirement pension fraud investigation, involving workers in the public administration [29]. The objective was to observe the blackboard and evaluate the levels of conflict between the agents. Fourteen pieces of evidence, seized from the same location, were examined (10 hard drives and 4 removable media). The total number of files is 353,466 for a total of 75.502 GB, including recovered files but excluding free space fragments.

2.5 Evaluation and Future Work

There are many different ways to tackle the emergent AMII topic. Related to bioinformatics domain area, the MAS architecture defined to *BioAgents*, described in Section 2.3, is totally adequate to the integration of DM techniques. The existence of the physical layer, which is composed by many public biological databases, e.g. *nr*, *COG*, *GO KOG* and *Swiss-Prot*, can be automatically mined with many different algorithms defined into sets of ISA in a distributed and parallel way. Although automated annotation has not been the focus of BioAgents project so far, certainly this process can increase its performance through DM techniques integration. For this focus, a very good computational infra-structure is necessary to deal with real-world biological data scale volume, even counting on the cooperative skills of the agent society.

Another focus of our research would be related to the classification model functions using DM techniques, which can be used to classify ncRNAs (non-coding RNAs). In NONCODE [28], for example, they have concentrated classification effort on the process in which a given ncRNA takes part, along with its function in this process. In this matter, *BioAgents* can use Support Vector Machines (SVM) to learn classification and improve manual annotation recommendation, as a prediction of ncRNAs with different criteria. Some recent researches have been working in this direction [34, 40]. In the study of pathogenic organisms of [6], authors used as a target for such approach the fungus *Paracoccidioides brasiliensis (Pb)*, the ethyological agent of paracoccidioidomycosis, whose transcriptome has recently been elucidated.

Another very interesting focus is being directed in our project at the moment to improve *BioAgents*' manual suggestion, through the use of structured machine learning, as considered a emerging trend for the next ten years at the Bioinformatics area [18]. The idea is to define training annotation data for a very well annotated organisms, such as *Caenorhabditis elegans* genome for example [24]. This training data would be of the form (x, y), where x is a feature vector describing a particular site and $y = 1$ is true if the annotated organism is present at a very good public biological databases, or site, e.g. *Swiss-Prot*), and $y = 0$ otherwise. This is a standard supervised learning problem. This is an interesting manual annotation prediction problem and we'd like to know how well *BioAgents* can act in such scenario.

Related to Computer Forensics, we have proposed a distributed digital forensics toolkit - MADIK. The work of [55] has defined a distributed tool too. Our proposal benefit from the distributed nature of a MAS and we already observed the performance gains, which helped us obtain better computational resource usage and reduce the time required to perform the examination. This approach can be normally extended to improve performance using DM techniques. In [25] we find two approaches for analyzing large data sets of forensic data called Forensic Feature Extraction (FFE) and Cross-Drive Analysis (CDA). We consider CDA to be the most interesting for MADIK, since it uses statistical techniques for correlating information within a single disk image and across multiple disk images. A recent work by [16] describes a tool called Forensics Automated Correlation Engine (FACE), whose objectives are similar to ours. They also present some scenarios, where an increased level of correlation of disparate evidences was achieved.

In [3], authors develop profiles to describe user or system behavior as a useful technique employed in Computer Forensics investigations. Information found in data obtained by investigators can often be used to establish a view of regular usage patterns, which can then be examined for unusual occurrences. Events compiled from potentially numerous sources are grouped according to some criteria and frequently occurring event sequences are established. The methodology and techniques to extract and contrast these sequences are then described and discussed. In this direction, we are planing to implement in MADIK an example-based method of DM technique, as cited in Section 2.4, CBR to improve the agents' reasoning process and the quality of recommendation.

This chapter presented research work related to the Emergent Agent-Mining Organizations and Applications topic. To better illustrate MAS applications, we presented two different study cases already defined and implemented: BioAgents and MADIK. We discussed some ideas to the *BioAgents* and MADIK AMII integration case, showing that both projects can be naturally extended to include DM techniques as an ongoing research works, as demanded by the emerging AMII perspective.

References

1. International journal of data mining and bioinformatics (ijdmb). http://www.inderscience.com/browse/index.php?journalCODE=ijdmb.
2. *Proceedings of the European Conference on Computational Biology (ECCB 2002), October 6-9, 2002, Saarbrücken, Germnany*, 2002.
3. Tamas Abraham. Event sequence mining to develop profiles for computer forensic investigation purposes. In *ACSW Frontiers '06: Proceedings of the 2006 Australasian workshops on Grid computing and e-research*, pages 145–153, Darlinghurst, Australia, Australia, 2006. Australian Computer Society, Inc.
4. The Agents, Data Mining Interaction, and Integration interest group (AMII-SIG). http://www.agentmining.org/.
5. S. F. Altschul, W. Gish, W. Miller, E. W. Myers, and D. J. Lipman. Basic local alignment search tool. *J Mol Biol*, 215(3):403–410, October 1990.
6. Roberto T. Arrial, Roberto C. Togawa, and Marcelo M. Brigido. Outlining a strategy for screening non-coding rnas on a transcriptome through support vector machines. In Marie-France Sagot and Maria Emilia Telles Walter, editors, *BSB*, volume 4643 of *LNCS*, pages 149–152. Springer, 2007.
7. Ezio Bartocci, Flavio Corradini, and Emanuela Merelli. Building a multiagent system from a user workflow specification. In Paoli et al. [48].
8. Ezio Bartocci, Flavio Corradini, Emanuela Merelli, and Lorenzo Scortichini. BioWMS: a web-based Workflow Management System for bioinformatics. *BMC Bioinformatics*, 8(Suppl 1):S2, 2007.
9. Ana L. C. Bazzan, Mark Craven, and Natália F. Martins, editors. *Advances in Bioinformatics and Computational Biology, Third Brazilian Symposium on Bioinformatics, BSB 2008, Santo André, Brazil, August 28-30, 2008. Proceedings*, volume 5167 of *LNCS*. Springer, 2008.
10. Ana L. C. Bazzan, Paulo Martins Engel, Luciana F. Schroeder, and Sérgio C. da Silva. Automated annotation of keywords for proteins related to mycoplasmataceae using machine learning techniques. In *ECCB* [2], pages 35–43.
11. Nicole Beebe and Jan Guynes Clark. A hierarchical, objectives-based framework for the digital investigations process. *Digital Investigation*, 2(2):147–167, 2005.
12. Fabio L. Bellifemine, Giovanni Caire, and Dominic Greenwood. *Developing Multi-Agent Systems with JADE (Wiley Series in Agent Technology)*. Wiley, April 2007.
13. D. A. Benson, I. Karsch-Mizrachi, D. J. Lipman, J. Ostell, and D. L. Wheeler. Genbank. *Nucleic Acids Res*, 36(Database issue), January 2008.
14. Massimo Buscema and Enzo Grossi. The semantic connectivity map: an adapting self-organising knowledge discovery method in data bases. experience in gastro-oesophageal reflux disease. *Int. J. Data Min. Bioinformatics*, 2(4):362–404, 2008.
15. Longbing Cao, Chao Luo, and Chengqi Zhang. Agent-mining interaction: An emerging area. In Gorodetsky et al. [27], pages 60–73.
16. Andrew Case, Andrew Cristina, Lodovico Marziale, Golden G. Richard, and Vassil Roussev. Face: Automated digital evidence discovery and correlation. *Digital Investigation*, 5(Supplement-1):S1–S140, september 2008. The Proceedings of the Eighth Annual DFRWS Conference.
17. Daniel D Corkill. Collaborating Software: Blackboard and Multi-Agent Systems & the Future. In *Proceedings of the International Lisp Conference*, New York, October 2003.
18. Thomas Dietterich, Pedro Domingos, Lise Getoor, Stephen Muggleton, and Prasad Tadepalli. Structured machine learning: the next ten years. *Machine Learning*, 73(1):3–23, October 2008.
19. Christos Dimou, Andreas L. Symeonidis, and Pericles A. Mitkas. Evaluating knowledge intensive multi-agent systems. In Gorodetsky et al. [27], pages 74–87.
20. Rajendra Dodhiawala. *Blackboard Architectures and Applications*. Academic Press, Inc., Orlando, FL, USA, 1989.

21. Cássia Trojahn dos Santos and Ana L. C. Bazzan. Integrating knowledge through cooperative negotiation - a case study in bioinformatics. In Gorodetsky et al. [26], pages 277–288.
22. CRoss Industry Standard Process for Data Mining. Crisp-dm 2.0 special interest group. `http://www.crisp-dm.org/new.htm`.
23. Ernest Friedman-Hill. *Jess in Action : Java Rule-Based Systems (In Action series)*. Manning Publications, December 2002.
24. Ratsch G, Sonnenburg S, Srinivasan J, Witte H, Muller K-R, and et al. Improving the caenorhabditis elegans genome annotation using machine learning. *PLoS Comput Biol*, 3(2), 2008.
25. Simson L. Garfinkel. Forensic feature extraction and cross-drive analysis. *Digital Investigation*, 3(Supplement-1):71–81, 2006.
26. Vladimir Gorodetsky, Jiming Liu, and Victor A. Skormin, editors. *Autonomous Intelligent Systems: Agents and Data Mining, International Workshop, AIS-ADM 2005, St. Petersburg, Russia, June 6-8, 2005, Proceedings*, volume 3505 of *LNCS*. Springer, 2005.
27. Vladimir Gorodetsky, Chengqi Zhang, Victor A. Skormin, and Longbing Cao, editors. *Autonomous Intelligent Systems: Multi-Agents and Data Mining, Second International Workshop, AIS-ADM 2007, St. Petersburg, Russia, June 3-5, 2007, Proceedings*, volume 4476 of *LNCS*. Springer, 2007.
28. Shunmin He, Changning Liu, Geir Skogerbo, Haitao Zhao, Jie Wang, Tao Liu, Baoyan Bai, Yi Zhao, and Runsheng Chen. Noncode v2.0: decoding the non-coding. *Nucl. Acids Res.*, pages gkm1011, November 2007.
29. Bruno W. P. Hoelz, Célia G. Ralha, and Rajiv Geeverghese. Artificial intelligence applied to computer forensics. In *The 24th Annual ACM Symposium on Applied Computing, SAC'09, Computer Forensics Track*, pages 211–217, Honolulu, Hawaii, USA, 2009. ACM Press.
30. Bruno W. P. Hoelz, Célia G. Ralha, Rajiv Geeverghese, and Hugo C. Junior. A cooperative multi-agent approach to computer forensics. In *IEEE/WIC/ACM International Conference on Web Intelligence and Intelligent Agent Technology*, volume 2, pages 477–483. IEEE Computer Society, 2008.
31. Bruno W. P. Hoelz, Célia G. Ralha, Rajiv Geeverghese, and Hugo C. Junior. Madik: A collaborative multi-agent toolkit to computer forensics. In Robert Meersman, Zahir Tari, and Pilar Herrero, editors, *OTM Workshops*, volume 5333 of *LNCS*, pages 20–21. Springer, 2008.
32. F. Iqbal, R. Hadjidj, B. C. M. Fung, and M. Debbabi. A novel approach of mining write-prints for authorship attribution in e-mail forensics. *Digital Investigation*, 5(1):42–51, 2008.
33. Konstantinos A. Karasavvas, Richard Baldock, and Albert Burger. A criticality-based framework for task composition in multi-agent bioinformatics integration systems. *Bioinformatics*, 21(14):3155–3163, 2005.
34. Y. A. N. Karklin, Richard F. Meraz, and Stephen R. Holbrook. Classification of non-coding rna using graph representations of secondary structure. In *Pacific Symposium on Biocomputing 2005*. Pacific Symposium on Biocomputing 2005, World Scientific Publishing Co. Pte. Ltd., 2005.
35. I. R. Konig, J. D. Malley, S. Pajevic, C. Weimar, H-C. Diener, and A. Ziegler. Patient-centered yes&no prognosis using learning machines. *Int. J. Data Min. Bioinformatics*, 2(4):289–341, 2008.
36. Ernst Kretschmann, Wolfgang Fleischmann, and Rolf Apweiler. Automatic rule generation for protein annotation with the c4.5 data mining algorithm applied on swiss-prot. *Bioinformatics*, 17(10):920–926, 2001.
37. Java language. `http://java.sun.com`.
38. Richardson Silva Lima, Célia Ghedini Ralha, Maria Emlia Machado T.Walter, Hugo Wruck Schneider, Anderson Gray F. Pereira, and Marcelo Macedo Brgido. Bioagents: Um sistema multiagente para anotao manual em projetos de seqnciamento de genomas. In *XXVII Congresso da SBC, VI Encontro Nacional de Inteligncia Artificial (ENIA'07), Rio de Janeiro, RJ, 30 Junho a 6 Julho. Anais*, pages 1302–1310, 2007.
39. Cui Lin, Shiyong Lu, Xuwei Liang, Jing Hua, and Otto Muzik. Cocluster analysis of thalamocortical fibre tracts extracted from diffusion tensor mri. *Int. J. Data Min. Bioinformatics*, 2(4):342–361, 2008.

40. Hong-Wei Liu. Predicting non-protein-coding rna genes in escherichia coli using svm with signature descriptor. In *The Second International Symposium on Optimization and Systems Biology (OSB'08)*, pages 287–293, Lijiang, China, October 31November 3 2008.
41. George F. Luger. *Artificial Intelligence: Structures and Strategies for Complex Problem Solving*. Addison-Wesley Longman Publishing Co., Inc., Boston, MA, USA, 2001.
42. H. Penny Nii. Blackboard systems, part one: The blackboard model of problem solving and the evolution of blackboard architectures. *AI Magazine*, 7(2):38–53, 1986.
43. Ayahiko Niimi, Hitomi Noji, and Osamu Konishi. Distributed web integration with multiagent data mining. In Rajiv Khosla, Robert J. Howlett, and Lakhmi C. Jain, editors, *KES (3)*, volume 3683 of *LNCS*, pages 513–519. Springer, 2005.
44. Department of Computer and USA Information Sciences, University of Alabama at Birmingham (UAB). Cis@uab computer forensics research project. `http://www.cis.uab.edu/forensics/`.
45. Genome Sequencing Project of the *Anaplasma marginale rickettsia*. `https://www.biomol.unb.br/anaplasma/servlet/IndexServlet`.
46. Genome Project of the *Paracoccidioides brasiliensis* (Pb) fungus. `https://dna.biomol.unb.br/Pb-eng/`.
47. Genome Project of the *Paullinia cupana* (guaraná) plant. `https://dna.biomol.unb.br/GR/`.
48. Flavio De Paoli, Antonella Di Stefano, Andrea Omicini, and Corrado Santoro, editors. *Proceedings of the 7th WOA 2006 Workshop, From Objects to Agents (Dagli Oggetti Agli Agenti), Catania, Italy, September 26-27, 2006*, volume 204 of *CEUR Workshop Proceedings*. CEUR-WS.org, 2006.
49. Ana Carolina M. Pilatti de Paula, Braulio C. Avila, Edson Scalabrin, and Fabricio Enembreck. Multiagent-based model integration. In *WI-IATW '06: Proceedings of the 2006 IEEE/WIC/ACM international conference on Web Intelligence and Intelligent Agent Technology*, pages 11–14, Washington, DC, USA, 2006. IEEE Computer Society.
50. W. R. Pearson and D. J. Lipman. Improved tools for biological sequence comparison. *Proc Natl Acad Sci U S A*, 85(8):2444–2448, April 1988.
51. Tuan D. Pham. Computational prediction models for cancer classification using mass spectrometry data. *Int. J. Data Min. Bioinformatics*, 2(4):405–422, 2008.
52. S. Pinson and P. Moraïtis. An intelligent distributed system for strategic decision making. *Group Decision and Negotiation*, 6:77–108, 1996.
53. MADIK's project Web hosting page. `http://madik.sourceforge.net/`.
54. Célia Ghedini Ralha, Hugo W. Schneider, Lucas O. da Fonseca, Maria Emilia M. T. Walter, and Marcelo M. Brígido. Using bioagentsfor supporting manual annotation on genome sequencing projects. In Bazzan et al. [9], pages 127–139.
55. Vassil Roussev and Golden G. Richard III. Breaking the performance wall: The case for distributed digital forensics. In *The fourth annual Digital Forensics Research Workshop*, Baltimore, Maryland, August 2004.
56. Stuart J. Russell and Peter Norvig. *Artificial Intelligence: A Modern Approach*. Pearson Education, 2003.
57. Ichiro Satoh. Self-organizing multi-agent systems for data mining. In Gorodetsky et al. [27], pages 165–177.
58. Jacobus Venter, Alta de Waal, and Cornelius Willers. *Specializing CRISP-DM for Evidence Mining*, volume 242/2007. Springer, Boston, 1st edition edition, 2007. Book Series-IFIP International Federation for Information Processing.
59. Roger L. Wainwright and Hisham Haddad, editors. *Proceedings of the 2008 ACM Symposium on Applied Computing (SAC)*. ACM, 2008.
60. Chun Wei, Alan Sprague, Gary Warner, and Anthony Skjellum. Mining spam email to identify common origins for forensic application. In Wainwright and Haddad [59], pages 1433–1437.
61. Gerhard Weiss, editor. *Multiagent systems: a modern approach to distributed artificial intelligence*. MIT Press, Cambridge, MA, USA, 1999.
62. Michael Woolridge and Michael J. Wooldridge. *Introduction to Multiagent Systems*. John Wiley & Sons, Inc., New York, USA, 2001.

Chapter 3
Agent-Based Distributed Data Mining: A Survey

Chayapol Moemeng, Vladimir Gorodetsky, Ziye Zuo, Yong Yang and Chengqi Zhang

Abstract Distributed data mining is originated from the need of mining over decentralised data sources. Data mining techniques involving in such complex environment must encounter great dynamics due to changes in the system can affect the overall performance of the system. Agent computing whose aim is to deal with complex systems has revealed opportunities to improve distributed data mining systems in a number of ways. This paper surveys the integration of multi-agent system and distributed data mining, also known as agent-based distributed data mining, in terms of significance, system overview, existing systems, and research trends.

3.1 Introduction

Originated from knowledge discovery from databases (KDD), also known as data mining (DM), distributed data mining (DDM) mines data sources regardless of their physical locations. The need for such characteristic arises from the fact that data produced locally at each site may not often be transferred across the network due to the excessive amount of data and privacy issues. Recently, DDM has become a critical components of knowledge-based systems because its decentralised architecture reaches every networked business.

This chapter discusses a symbiont synthesising the two widely accepted research fields: data mining and agents. Readers are recommended to review surveys regarding distributed data mining in [23], [34], and [46].

3.2 Why Agents

DDM is a complex system focusing on the distribution of resources over the network as well as data mining processes. The very core of DDM systems is the scalability

L. Cao (ed.), Data Mining and Multi-agent Integration, DOI: 10.1007/978-1-4419-0522-2_3,

as the system configuration may be altered time to time, therefore designing DDM systems deals with great details of software engineer issues, such re-usability, extensibility, and robustness. For these reasons, agents' characteristics are desirable for DDM systems.

Furthermore, the decentralisation property seems to fit best with the DDM requirement. At each data site, mining strategy is deployed specifically for the certain domain of data. However, there can be other existing or new strategies that data miner would like to test. A data site should seamlessly integrate with external methods and perform testing on multiple strategies for further analysis. Autonomous agent can be treated as a computing unit that performs multiple tasks based on a dynamic configuration. The agent interprets the configuration and generates an execution plan to complete multiple tasks.

[2], [21], [14], [20], [27], and [26] discuss the benefits of deploying agents in DDM systems. Nature of MAS is decentralisation and therefore each agent has only limited view to the system. The limitation somehow allows better security as agents do not need to observe other irrelevant surroundings. Agents, in this way, can be programmed as compact as possible, in which light-weight agents can be transmitted across the network rather than the data which can be more bulky. Being able to transmit agents from one to another host allows dynamic organisation of the system. For example, mining agent a_1, located at site s_1, posses algorithm alg_1. Data mining task t_1 at site s_2 is instructed to mine the data using alg_1. In this setting, transmitting a_1 to s_2 is a probable way rather than transfer all data from s_2 to s_1 where alg_1 is available.

In addition, security, a.k.a. trust-based agents[41][19], is a critical issue in ADDM. Rigid security models intending to ensure the security may degrade the system scalability. Agents offer alternative solutions as they can travel through the system network. As in [29], the authors present a framework in which mobile agents travel in the system network allowing the system to maintain data privacy. Thanks to the self-organisation characteristic which excuses the system from transferring data across the network therefore adds up security of the data.

A trade-off for the previous discussed issue, scalability is also a critical issue of a distributed system. In order to inform every unit in the system about the configuration update of the system, such as a new data site has joined the system, demands extra human interventions or high complex mechanism in which drops in performance may occur. To this concern, collaborative learning agents[40][6] are capable of sharing information, in this case, about changes of system configuration, and propagate from one agent to another allowing adaptation of the system to occur at individual agent level. Furthermore, mobile agents as discussed earlier can help reduce network and DM application server load as in state-of-art systems[3][24].

3.3 Agent-Based Distributed Data Mining

Applications of distributed data mining include credit card fraud detection system, intrusion detection system,health insurance, security-related applications, distributed clustering, market segmentation,sensor networks, customer profiling, evaluation of retail promotions, credit risk analysis, etc. These DDM application can be further enhanced with agents. ADDM takes data mining as a basis foundation and is enhanced with agents; therefore, this novel data mining technique inherits all powerful properties of agents and, as a result, yields desirable characteristics.

In general, constructing an ADDM system concerns three key characteristics: interoperability, dynamic system configuration, and performance aspects, discussed as follows.

Interoperability concerns, not only collaboration of agents in the system, but also external interaction which allow new agents to enter the system seamlessly. The architecture of the system must be open and flexible so that it can support the interaction including communication protocol, integration policy, and service directory. Communication protocol covers message encoding, encryption, and transportation between agents, nevertheless, these are standardised by the Foundation of Intelligent Physical Agents (FIPA)[1] and are available for public access. Most agent platforms, such as JADE[2] and JACK[3], are FIPA compliant therefore interoperability among them are possible. Integration policy specifies how a system behaves when an external component, such as an agent or a data site, requests to enter or leave. The issue is further discussed in [47] and [29]

In relation with the interoperability characteristic, dynamic system configuration, that tends to handle a dynamic configuration of the system, is a challenge issue due to the complexity of the planning and mining algorithms. A mining task may involve several agents and data sources, in which agents are configured to equip with an algorithm and deal with given data sets. Change in data affects the mining task as an agent may be still executing the algorithm.

Lastly, performance can be either improved or impaired because the distribution of data is a major constraint. In distributed environment, tasks can be executed in parallel, in exchange, concurrency issues arise. Quality of service control in performance of data mining and system perspectives is desired, however it can be derived from both data mining and agents fields.

Next, we are now looking at the overview of our point of focus. An ADDM system can be generalised into a set of components and viewed as depicted in figure 3.1. We may generalise activities of the system into request and response, each of which involves a different set of components. Basic components of an ADDM system are as follows.

[1] FIPA, http://www.fipa.org/

[2] Java Agent DEvelopment Framework, http://jade.tilab.com/

[3] JACK, http://www.agent-software.com.au/products/jack/index.html

Data: Data is the foundation layer of our interest. In distributed environment, data can be hosted in various forms, such as online relational databases, data stream, web pages, etc., in which purpose of the data is varied.

Communication: The system chooses the related resources from the directory service, which maintains a list of data sources, mining algorithms, data schemas, data types, etc. The communication protocols may vary depending on implementation of the system, such as client-server, peer-to-peer, etc.

Presentation: The user interface (UI) interacts with the user as to receive and respond to the user. The interface simplifies complex distributed systems into user-friendly message such as network diagrams, visual reporting tools, etc.

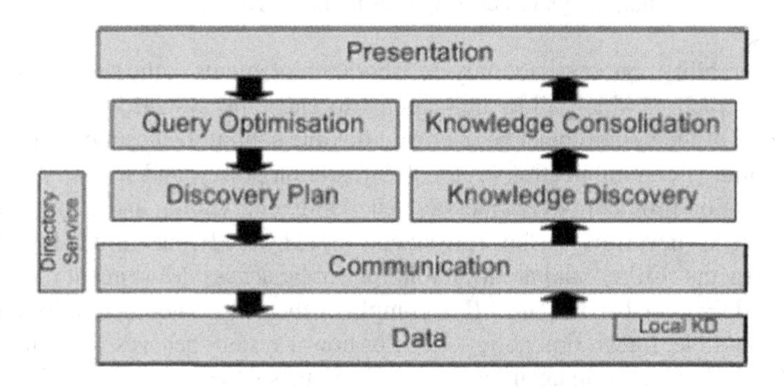

Fig. 3.1: Overview of ADDM systems

On the other hand, when a user requests for data mining through the UI, the following components are involved.

Query optimisation: A query optimiser analyses the request as to determine type of mining tasks and chooses proper resources for the request. It also determines whether it is possible to parallelise the tasks, since the data is distributed and can be mined in parallel.

Discovery Plan: A planner allocates sub-tasks with related resources. At this stage, mediating agents play important roles as to coordinate multiple computing units since mining sub-tasks performed asynchronously as well as results from those tasks.

On the other hand, when a mining task is done, the following components are taken place,

Local Knowledge Discovery (KD): In order to transform data into patterns which adequately represent the data and reasonable to be transferred over the network, at each data site, mining process may take place locally depending on the individual implementation.

Knowledge Discovery: Also known as mining, it execute the algorithm as required by the task to obtain knowledge from the specified data source.

Knowledge Consolidation: In order to present to the user with a compact and meaningful mining result, it is necessary to normalise the knowledge obtained from various sources. The component involves a complex methodologies to combine knowledge/patterns from distributed sites. Consolidating homogeneous knowledge/patterns is promising and yet difficult for heterogeneous case.

3.4 Interaction and Integration

Let us briefly outline the basic works on ADDM. A survey on agent technology for data mining can be found in [33] . [12] is one of the first works attracting attention to ADDM arguing its advantages in mining vast amount of data stored in network and using collaborative capabilities of DM agents. It studies agent-based approach to distributed knowledge discovery using Inductive Logic Programming (ILP) approach and provides for some experiment results of application of the agent-based approach to DDM.

The paper [22] proposes PADMA (Parallel Data Mining Agents) system addressing use of agent architecture to cope large scale and distributed nature of data sources as applied to hierarchical clustering. It is intended to handle numeric and textual data with the focus on the latter. Agency of DDM system consists of DM agents that are responsible for local data access and extraction of high level useful information, agent-facilitator coordinating the DM agents operation while handling their SQL queries and presenting them to user interface. Agent-facilitator gets "conceptual graphs"from DM agents, combines them and passes the results to user. The focus of the paper is to show the benefit of agent-based parallel data mining.

JAM (Java Agents for Meta-learning) system proposed in [37] is an agent based system supporting the launching of learning, classifier and meta-learning agents over distributed database sites. It uses parallelism and distributed nature of meta-learning and its possibility to share meta-information without direct access to distributed data sets. JAM [36] in constituted of distributed learning and classification programs operating in parallel on JAM sites that are linked to a network. In turn, JAM site contains one or more local data base; one or more learning agents (machine learning programs); one or more meta-learning agents, intended for combing decisions produced by local classifier agents; a repository of decisions computed locally and imported by local and meta classifier agents; a local user configuration file and graphical user interface. Once the local and meta-classifiers are generated the user manages the execution of the above modules to classify new (unlabelled) data. A peculiarity of JAM system operation is that each local site may import decisions of remote classifiers from peer JAM sites and combine these decisions with own local classifier decisions using local meta-learning agent. JAM sites are operating simultaneously and independently. Administration of JAM site local activity is performed by the user via local user configuration file. Details of the JAM system architecture and implementation can be found in [36].

The paper [42] is motivated by the desire to attack increasingly difficult problems and application domains which often require to process very large amounts of data or data collected at different geographical locations that cannot be processed by sequential and centralised systems. It is also motivated by capabilities of MAS to provide for computing processes with robustness, fault tolerance, scalability, and speed-up as well as by maturity of the computer and network technology supporting implementation of parallel and distributed information processing. Based on multi-agent learning, target system can be built as self improving their performance. In particular, the paper considers job assignment problem, the core of many scheduling tasks, aiming to find reasonable solution in a reasonable time. The nodes of partially ordered set of jobs are considered as active entities or agents and the jobs are considered as passive entities or resources to be used by the agents. The agents interact in order to find a solution meeting some predefined criteria. The solution search procedure implemented by agents is based on distributed reinforcement learning starting from an initial solution that is performed using low-level communication and coordination among agents. The paper experimentally proves some advantages of the developed multi-agent model for implementation of parallel and distributed machine learning in the problem in question.

[3] describes the developed Papyrus system that is Java-based and intended for DDM handling with clusters and meta-clusters distributed over heterogeneous data sites. Mobile DM agents of the system are capable to move data, intermediate results and models between clusters for local processing thus reducing network load. Papyrus supports several techniques for exchanging and combining locally mined decisions (predictive models) and meta-data that are necessary to describe the above models specified in terms of a special mark-up language.

[25] studies advantages and added value of using ADDM, reviews and classifies existing agent-based approaches to DDM and proposes agent-oriented implementation of a distributed clustering system. The paper explicitly formulates why agent-based approach is very perspective one for DDM (autonomy, interactivity, capability to dynamically select data sources in changeable environment, etc.). The proposed KDEC scheme addressing computing statistical density estimation and information theoretic sampling to minimise communication between sites is implemented on the basis of agent technology as distributed data clustering system. In addition, one of its distinctive features is that it preserves the local data privacy.

[10] emphasises the possible synergy between MAS and DDM technologies. It particularly focuses on distributed clustering, having every increasing application domains, e.g. in sensor networks deployed in hostile and difficult to access locations like battle fields where sensors are measuring vibration, reflectance, temperature, and audio signals; in sensor networks for monitoring a terrain, smart home, and many other domains. In these domains analysing, e.g. DDM task, are non-trivial problems due to many constraints such as limited bandwidth of wireless communication channels, peer-to-peer mode of communication and the necessity of interaction in asynchronous network, data privacy, coordination of distributed computing, non-trivial decomposability, formidable number of data network nodes, limited computing resources, e.g. due to limited power supply, etc. The authors state that the traditional framework for centralised data analysis and DM algorithms does not really scale very well in such distributed applications. In contrast, this distributed problem solving can be very well coped with the multi-agent framework supporting

semi-autonomous behaviour, collaboration and reasoning, among other perspective MAS properties. From DDM domain perspective, the paper focuses on clustering in sensor networks that offers many aforementioned challenges. This paper suggests that traditional centralised data mining techniques may not work well in the above domains and underscores, among others, and that DDM algorithms integrated with MAS architecture may offer novel very perspective synergetic information technology for these domains.

Recently some efforts were paid to development of technological issues of multi-agent data mining that can be evaluated as a sign of increasing maturity. [20] states that the core problem of distributed data mining and machine learning design does not concern particular data mining techniques. Instead of this, its core problem is development of an infrastructure and protocols supporting coherent collaborative operations of distributed software components (agents) performing distributed learning. The paper proposes a multi-agent architecture of an information fusion system possessing of DDM and machine learning capabilities. It also proposes a design technology, which core is constituted by a number of specialised agent interaction protocols supporting distributed agent operations in various use cases (scenarios) including, in particular, DDM protocol. Further development of DDM system design technology is given in [16] and [17].

[39] proposes a framework (an abstract architecture) for agent-based distributed machine learning and data mining. The proposed framework , as it is motivated by the authors, is based of the observation that "despite the autonomy and self-directedness of learning agents, many of such systems exhibit a sufficient overlap in terms of individual learning goals so that beneficial operation might be possible if a model for flexible interaction between autonomous learners was available that allowed agents to (i) exchange information about different aspects of their own learning mechanism at different levels of detail without being forced to reveal private information that should not be disclosed, (ii) decide to what extent they want to share information about their own learning processes and utilise information provided by other learners, and (iii) reason about how this information can best be used to improve their own learning performance."

The idea underlying the proposed framework is that each agent is capable to maintain meta-description of own learning processes in a form that makes it admissible, due to privacy issue, to exchange meta-information with other agents and reason about it rationally, i.e. to reason in a way providing for improving of their own learning results. The authors state that this possibility, for learning agents, is a hypothesis they intend to justify experimentally within a proposed formal framework. Actually this paper presents a preliminary research results and there are a lot of efforts to be done to reach reliable evaluation of the hypothesis used.

The last system to review in this survey is F-TRADE (Financial Trading Rules Automated Development and Evaluation) [9][32]. The system uses a dynamic approach that allows integration of external data sources and mining algorithm. The system presents, to financial traders and brokers, a testbed on actual data sets, which can help them evaluate their favourite trading strategies iteratively with confidence before investing money into the real markets. The motivation of F-TRADE is to provide financial traders and researchers, and financial data miners with a practically flexible and automatic infrastructure.

With this infrastructure, they can plug their algorithms onto the system and concentrate on improving the performance of their algorithms with iterative evaluation on a large amount of real stock data from international markets. The system services include (i) trading services support, (ii) mining services support, (iii) data services support, (iv) algorithm services support, and (v) system services support. In order to support all these services, soft plug-and-play is essential in the F-TRADE. Each system components interact through XML schema which specifies details of the components and allows agents to examine and use.

Besides those role-model systems presented earlier, there are more ADDM related works which the reader is encouraged to review: [4], [6], [7], [28], [39], [30], [31], [35], and [45].

With regards to the symbiosis, interaction and integration of the two researches have attracted the communities' attention. Many research bodies have listed agent and data mining interaction and integration (AMII) as a special interest. [8] discuss challenges of interaction and integration of agents with data mining. There is a community website dedicated to this special interest research group at *AgentMining* (http://www.agentmining.org). which provides related resources, such as list of research topics, activities, workshops and conferences, links, related publications, research groups, etc.

3.5 Open Issues and Trends

The interaction and integration between the two technologies have explore the new challenges. Considering various ingredients for the integration could be a key to rapidly enhance the development process and usability of the system, let us examine them from different perspectives.

Research Perspective: Data distribution in real-life applications are either homogeneous or heterogeneous. Data can be partitioned both vertically and horizontally, and furthermore data splitting may not be available across the sites. For examples, two related customer databases may not reflect each others in which a customer may never provide contact details but somehow appear to buy some products. The applications will require a data mining technology to pay careful attention to the distributed computing, communication, and storage of the system.

Another approach to develop ADDM is an inspiration from the nature which has proven to be promising. Swarm intelligence is closely related to intelligent agents. Recently, researchers pay attention to the possibility to implement DDM systems with swarm intelligence. Sample applications of swarm intelligence in data mining are rule-based classifiers using ants, feature selection with ant colony optimisation, data and text mining with hierarchical clustering ants, etc. Further readings can be found in [1] and [38].

Software Engineering Perspective: Expectedly, ADDM frequently requires exchange of data mining models among the data sites. Therefore, seamless and transparent realisation of DDM technology will require standardised schemes to represent and exchange models. Therefore, software engineering tools that support the design of data mining and distributed database are desired. So far, PMML, the Cross-Industry Standard Process Model for Data mining (CRISP-DM), and other related efforts are likely to be very useful.

The very basic foundation of our focus is the database. Not only full-scale database, like relational database, is taken into consideration during system integration. Desktop and light-weight database running on limited devices, such as mobile phones, can be integrated into ADDM. Mobile agents can be migrated (downloaded) and perform task on the devices and take back only a representative model for further analysis.

The second ingredient is the emergence of service oriented architecture (SOA) that enables agent-based application to integrate better than ever. SOA is a promising architecture as it is widely adapted in several applications. We cannot deny the fact that web-based applications are becoming more and more popular. Internet has become a necessary element of a computer system.

System Perspective: A novel very perspective but poorly researched application area of agents and data mining synergy is mobile, ubiquitous and peer-to-peer (P2P) computing. A specific feature of such computing systems is that the latter operate with dynamic set of information sources. E.g., the mobile devices may move and freely enter to and exit from the network thus changing the set of network nodes and communication topology, changing the set of available services as well. Examples of such application areas are, e.g., smart space and ambient intelligence. In these environments, decisions are made on the basis of fusion of information received from distributed sensors and mobile devices populating the environment. One of the objectives of such application is adaptation to multiple human habits that can be achieved through learning of multiple human profiles. On the other hand, for class of applications in question, multi-agent approach supplies for most natural framework, appropriate architecture, as well as design technology. Thus, integrating agent and data mining in ubiquitous environments like smart space, ambient intelligence, etc., could be very perspective and promising to reach high quality performance of corresponding applied systems.

In fact, ubiquitous and mobile computing form a novel and very perspective, although poorly researched, application area of agents and data mining synergy. A specific feature of such computing systems is that the latter often has to handle with dynamic set of information sources. E.g., the mobile devices may move and freely enter to and exit from the network thus changing the set of network nodes and communication topology, changing the set of available services as well. Examples of such application areas are, e.g., smart space and ambient intelligence. In these environments, decisions are made on the basis of fusion of information received from distributed sensors and mobile devices populating the environment. One of the objectives of such application is adaptation to multiple human habits that can be achieved through learning of multiple human profiles. On the other hand, for class of applications in question, multi-agent approach supplies for most natural framework, appropriate architecture, as well as sound design technology. Thus, integrating agent and data mining in ubiquitous environments like smart space, ambient intelligence, etc., could be very perspective and promising to reach high quality performance of corresponding applied systems. [15] presents a summary of challenges integrating ubicomp with MAS for data mining task.

Recently, peer-to-peer (P2P) computing has proven its excellence through its product, such as peer download software, file sharing software, which they gather users to join the service quickly. P2P is respected as one of the best scalable system, and thus it increases availability of the system as millions of peers can be attached to the network. P2P algorithm does not rely on a central server, each unit performs its own task and requests for data from others if available in order to save the redundant time. However, security is a critical issue in P2P due to exchanging information with other peers that can add a vulnerability to the network, such as denial of service or selfish behaviour. Some peers may only consume others' resources while they do not provide to others. Nevertheless, each peer must agree of terms and conditions of use before joining the network. P2P has caught researchers attention due to the compliance with multi-agent systems as appear in [18] and [11].

User Perspective: Finally human-computer interaction issues in DDM offers some unique challenges. It requires system-level support for group interaction, collaborative problem solving, development of alternate interfaces (particularly for mobile devices), and dealing with security issues.

References

1. Ajith Abraham, Crina Grosan, and Vitorino Ramos, editors. *Swarm Intelligence in Data Mining*, volume 34 of *Studies in Computational Intelligence*. Springer, 2006.
2. Sung W. Baik, Jerzy W. Bala, and Ju S. Cho. Agent based distributed data mining. *Lecture Notes in Computer Science*, 3320:42–45, 2004.
3. S. Bailey, R. Grossman, H. Sivakumar, and A. Turinsky. Papyrus: a system for data mining over local and wide area clusters and super-clusters. In *Supercomputing '99: Proceedings of the 1999 ACM/IEEE conference on Supercomputing (CDROM)*, page 63, New York, NY, USA, 1999. ACM.
4. R. J. Bayardo, W. Bohrer, R. Brice, A. Cichocki, J. Fowler, A. Helal, V. Kashyap, T. Ksiezyk, G. Martin, M. Nodine, and Others. InfoSleuth: agent-based semantic integration of information in open and dynamic environments. *ACM SIGMOD Record*, 26(2):195–206, 1997.
5. F. Bergenti, M. P. Gleizes, and F. Zambonelli. *Methodologies And Software Engineering For Agent Systems: The Agent-oriented Software Engineering Handbook*. Kluwer Academic Publishers, 2004.
6. A. Bordetsky. Agent-based Support for Collaborative Data Mining in Systems Management. In *Proceedings Of The Annual Hawaii International Conference On System Sciences*, page 68, 2001.
7. R. Bose and V. Sugumaran. IDM: an intelligent software agent based data mining environment. *1998 IEEE International Conference on Systems, Man, and Cybernetics*, 3, 1998.
8. L. Cao, C. Luo, and C. Zhang. Agent-Mining Interaction: An Emerging Area. *Lecture Notes in Computer Science*, 4476:60, 2007.

9. L. Cao, J. Ni, J. Wang, and C. Zhang. Agent Services-Driven Plug and Play in the F-TRADE. In *17th Australian Joint Conference on Artificial Intelligence*, volume 3339, pages 917–922. Springer, 2004.
10. J. Dasilva, C. Giannella, R. Bhargava, H. Kargupta, and M. Klusch. Distributed data mining and agents. *Engineering Applications of Artificial Intelligence*, 18(7):791–807, October 2005.
11. S. Datta, K. Bhaduri, C. Giannella, R. Wolff, and H. Kargupta. Distributed data mining in peer-to-peer networks. *Internet Computing, IEEE*, 10(4):18–26, 2006.
12. W. Davies and P. Edwards. Distributed Learning: An Agent-Based Approach to Data-Mining. In *Proceedings of Machine Learning 95 Workshop on Agents that Learn from Other Agents*, 1995.
13. U. Fayyad, R. Uthurusamy, and Others. Data mining and knowledge discovery in databases. *Communications of the ACM*, 39(11):24–26, 1996.
14. C. Giannella, R. Bhargava, and H. Kargupta. Multi-agent Systems and Distributed Data Mining. *Lecture Notes in Computer Science*, pages 1–15, 2004.
15. V. Gorodetskiy. Interaction of agents and data mining in ubiquitous environment. In *Proceedings of the 2008 IEEE/WIC/ACM International Joint Conference on Web Intelligence and Intelligent Agent Technology (WI-IAT'08)*, 2008.
16. V. Gorodetsky, O. Karsaev, and V. Samoilov. Multi-Agent Data and Information Fusion. *Nato Science Series Sub Series Iii Computer And Systems Sciences*, 198:308, 2005.
17. V. Gorodetsky, O. Karsaev, and V. Samoilov. Infrastructural Issues for Agent-Based Distributed Learning. In *Proceedings of the 2006 IEEE/WIC/ACM international conference on Web Intelligence and Intelligent Agent Technology*, pages 3–6. IEEE Computer Society Washington, DC, USA, 2006.
18. V. Gorodetsky, O. Karsaev, V. Samoylov, and S. Serebryakov. P2P Agent Platform: Implementation and Testing. In *Proceedings International Workshop "Agent and Peer-to-Peer Computing"(AP2PC-2007) associated with AAMAS-07. Honolulu, Hawaii*, pages 21–32, 2007.
19. V. Gorodetsky and I. Kotenko. The Multi-agent Systems for Computer Network Security Assurance: frameworks and case studies. In *IEEE International Conference on Artificial Intelligence Systems, 2002*, pages 297–302, 2002.
20. Vladimir Gorodetsky, Oleg Karsaev, and Vladimir Samoilov. Multi-agent technology for distributed data mining and classification. In *IAT*, pages 438–441. IEEE Computer Society, 2003.
21. Sven A. Brueckner H. Van Dyke Parunak. Engineering swarming systems. *Methodologies and Software Engineering for Agent Systems*, pages 341–376, 2004.
22. H. Kargupta, I. Hamzaoglu, and B. Stafford. Scalable, distributed data mining using an agent based architecture. In *Proceedings the Third International Conference on the Knowledge Discovery and Data Mining, AAAI Press, Menlo Park, California*, pages 211–214, 1997.
23. H. Kargupta and K. Sivakumar. Existential pleasures of distributed data mining. *Data Mining: Next Generation Challenges and Future Directions*, pages 1–25, 2004.
24. Hillol Kargupta, Byung-Hoon Park, Daryl Hershberger, and Erik Johnson. Collective data mining: A new perspective toward distributed data mining. In *Advances in Distributed and Parallel Knowledge Discovery*, pages 133–184, 1999.
25. M. Klusch, S. Lodi, and G. Moro. Agent-based distributed data mining: The KDEC scheme. *AgentLink, Springer Lecture Notes in Computer Science*, 2586, 2003.
26. M. Klusch, S. Lodi, and G. Moro. The role of agents in distributed data mining: Issues and benefits. In *Proceedings of the 2003 IEEE/WIC International Conference on Intelligent Agent Technology (IAT 2003)*, pages 211–217, 2003.
27. Matthias Klusch, Stefano Lodi, and Gianluca Moro. Issues of agent-based distributed data mining. In *Proceedings of the Second International Joint Conference on Autonomous Agents and Multiagent Systems (AAMAS-03)*, pages 1034–1035. ACM, 2003.
28. Raymond S. T. Lee and James N. K. Liu. ijade web-miner: An intelligent agent framework for internet shopping. *IEEE Transactions on Knowledge and Data Engineering*, 16(4):461–473, 2004.
29. Xining Li and Jingbo Ni. Deploying mobile agents in distributed data mining. *Lecture Notes in Computer Science*, 4819:322–331, 2007.

30. F. Menczer, R. K. Belew, and W. Willuhn. Artificial life applied to adaptive information agents. In *AAAI Spring Symposium on Information Gathering*, pages 128–132, 1995.
31. S. Merugu and J. Ghosh. A distributed learning framework for heterogeneous data sources. In *Conference on Knowledge Discovery in Data*, pages 208–217. ACM Press New York, NY, USA, 2005.
32. Peerapol Moemeng, Longbing Cao, and Chengqi Zhang. F-TRADE 3.0: An Agent-Based Integrated Framework for Data Mining Experiments. In *Proceedings of IEEE/WIC/ACM International Conference on Web Intelligence and Intelligent Agent Technology*, volume 3, pages 612–615, Los Alamitos, CA, USA, 2008. IEEE Computer Society.
33. L. Panait and S. Luke. Cooperative Multi-Agent Learning: The State of the Art. *Autonomous Agents and Multi-Agent Systems*, 11(3):387–434, 2005.
34. Byung H. Park and Hillol Kargupta. Distributed data mining: Algorithms, systems, and applications. In Nong Ye, editor, *Data Mining Handbook*, pages 341–358. Lawrence Earlbaum Associates, 2002.
35. M. Penang. Distributed Data Mining From Heterogeneous Healthcare Data Repositories: Towards an Intelligent Agent-Based Framework. In *Proceedings of the 15th IEEE Symposium on Computer-Based Medical Systems:(CBMS 2002): 4-7 June, 2002, Maribor, Slovenia*. IEEE Computer Society, 2002.
36. Andreas Leonidas Prodromidis. *Management of intelligent learning agents in distributed data mining systems*. PhD thesis, Columbia University, New York, NY, USA, 1999.
37. S. Stolfo, A. L. Prodromidis, S. Tselepis, W. Lee, D. W. Fan, and P. K. Chan. JAM: Java agents for meta-learning over distributed databases. In *Proceedings of the 3rd International Conference on Knowledge Discovery and Data Mining*, pages 74–81, 1997.
38. H. Van Dyke Parunak Sven A. Brueckner. Swarming distributed pattern detection and classification. In *Environments for Multi-Agent Systems*, pages 232–245. Springer, 2005.
39. J. Tozicka, M. Rovatsos, and M. Pechoucek. A framework for agent-based distributed machine learning and data mining. In *Proceedings of the 6th international joint conference on Autonomous agents and multiagent systems*. ACM New York, NY, USA, 2007.
40. Jan Tozicka, Michal Jakob, and Michal Pechoucek. Market-inspired approach to collaborative learning. *Lecture Notes in Computer Science*, 4149:213–227, 2006.
41. Frank E. Walter, Stefano Battiston, and Frank Schweitzer. A model of a trust-based recommendation system on a social network, Nov 2006.
42. Gerhard Weiss. A multiagent perspective of parallel and distributed machine learning. In *Agents*, pages 226–230, 1998.
43. M. Wooldridge and N. R. Jennings. Agent Theories, Architectures, and Languages: A Survey. *Intelligent Agents*, 22, 1995.
44. M. Wooldridge, N. R. Jennings, and D. Kinny. The Gaia Methodology for Agent-Oriented Analysis and Design. *Autonomous Agents and Multi-Agent Systems*, 3(3):285–312, 2000.
45. S. Z. H. Zaidi, S. S. R. Abidi, S. Manikam, and Cheah Yu-N. Admi: a multi-agent architecture to autonomously generate data mining services. In *Intelligent Systems, 2004. Proceedings. 2004 2nd International IEEE Conference*, volume 1, pages 273–279 Vol.1, 2004.
46. Mohammed J. Zaki. Parallel and distributed association mining: A survey. *IEEE Concurrency*, 7(4):14–25, 1999.
47. Gorodetskiy, V.; Karsaev, O.; Samoilov, V.; Serebryakov, S., Agent-based Service-Oriented Intelligent P2P Networks for Distributed Classification *Hybrid Information Technology*, 2006.

Part II
Data Mining Driven Agents

Chapter 4
Exploiting Swarm Behaviour of Simple Agents for Clustering Web Users' Session Data

Shafiq Alam, Gillian Dobbie, and Patricia Riddle

Abstract In recent years the integration and interaction of data mining and multi agent system (MAS) has become a popular approach for tackling the problem of distributed data mining. The use of intelligent optimization techniques in the form of MAS has been demonstrated to be beneficial for the performance of complex, real time, and costly data mining processes. Web session clustering, a sub domain of Web mining is one such problem, tackling the information comprehension problem of the exponentially growing World Wide Web (WWW) by grouping usage sessions on the basis of some similarity measure. In this chapter we present a novel web session clustering approach based on swarm intelligence (SI), a simple agent oriented approach based on communication and cooperation between agents. SI exploits the collective behaviour of simple agents, cooperation between the agents, and emergence on a feasible solution on the basis of their social and cognitive learning capabilities exhibited in the form of MAS. We describe the technique for web session clustering and demonstrate that our approach perform well against benchmark clustering techniques on benchmark session data.

4.1 Introduction

Data mining (DM) or Knowledge-Discovery and Data Mining (KDD), is the process of automatically searching large volumes of data for hidden, interesting, unknown and potentially useful patterns [1]. Data mining analyzes huge amounts of data for useful patterns using computational techniques from machine learning, information retrieval, computational intelligence and statistics [2]. With the rapid growth of web

Department of Computer Science, University of Auckland,
Private Bag 92019, Auckland, New Zealand
sala038@aucklanduni.ac.nz
{gill,pat}@cs.auckland.ac.nz

L. Cao (ed.), Data Mining and Multi-agent Integration, DOI: 10.1007/978-1-4419-0522-2_4,

data, web mining a sub domain of data mining has been introduced. Web mining tackles the information comprehension problem of the exponentially growing web data. Standard data mining techniques are applied to pre-process transform and extract patterns from web data. Web mining uses clustering, classification, association mining and prediction analysis to extract useful information from web documents. Web mining is further divided into web structure mining, web usage mining (WUM) and web content mining. We focus on WUM where the activities of more than 1.4 billion [1] internet users generate massive data and provide challenges for the automated discovery of interesting patterns among their usage behaviour. Organizations such as Google and Yahoo collect terabytes of data related to user activities, and analyze it for their business interests such as cross marketing, website organization, web site restructuring, recommender systems, web server performance improvement, and bandwidth management by caching and prefetching.
In recent years integration and interaction of data mining and multi agent system (MAS) has become a popular approach to tackle the problem of distributed data mining [3] [4]. The use of intelligent optimization techniques in the form of MAS has been shown to be beneficial for the performance of complex, real time, and costly data mining processes. Swarm Intelligence (SI) is one such paradigm that exploits the social and cognitive learning properties of vertebrates and insects, and models it through a multi agent system, with agents communicating with each other in a decentralized environment. The cooperative behaviour amongst the agents enables them to converge on an optimum solution. The two basic algorithms, ant colony optimization (ACO) and particle swarm optimization (PSO), have been found to be efficient in various domains of data mining. ACO is successfully implemented in classification, feature selection, rule mining and data clustering while the application of PSO can be found in data clustering, classification, pattern recognition, image processing, and recommender systems.
The main contributions of this chapter are:

- A description of web usage clustering in the context of a collective behaviour based multi agent environment
- A novel agent based technique for web session clustering based on PSO clustering, and a comparison of its performance with current techniques.

The rest of the chapter is organized as follows. Section 2 elaborates on the process of WUM and web session clustering approaches. Section 3 describes details of PSO and introduces the proposed PSO based clustering algorithm. Section 4 presents the pre-processing and clustering results. Section 5 overviews the related work in the area and section 6 introduces future work and concludes the paper.

[1] Internet usage statistics, the internet big picture http://www.internetworldstats.com/stats.htm

4.2 Web Usage Mining

Web usage mining (WUM) aims to discover interesting patterns among the fast increasing web users' activities on the WWW. It extracts hidden patterns in the visit sequence of the web users using standard data mining and KDD techniques. Web logs which record all the data related to the web users activities, needs to be passed through a sophisticated pre-processing stage. Web usage mining follows all the KDD steps; selection, pre-processing, pattern mining, post processing, and pattern analysis. Following are the main data mining techniques used to discover patterns in web usage data.

- Association rule mining
- Sequential pattern mining
- Classification rule mining
- Prediction analysis
- Clustering analysis

This section provides a detailed overview of web usage clustering practices.

4.2.1 Web Session Clustering

To understand the group behaviour of a particular class of users, an important step in web usage mining is to analyze the group behaviour of a user's sessions [5]. Clustering of web sessions is based on the data collected in the web server logs; gathered around on cache servers or in the cookies of client machines. Sometimes the process is backed by the structural and semantic information of the web pages. For web session clustering, primary attributes such as IP address, date time, page requested, page size, response and referrer are directly extracted from the web log while secondary attributes such as user visit, sessions, session length, episode, sequence of web usage and navigation, and semantic information are extracted by processing the primary attributes.

4.2.2 Session Identification

During a proper visit, web users follow a specific path related to their browsing behaviour and spend an arbitrary amount of time on each web page. The amount of time spent on a page is directly proportional to the interest of the user in that page. The sequence formed from such visits causes various hits on different pages. Such a sequence of visits is known as a web session. Identification of the session for a particular user can be by human intervention or automatic. Pseudo code of both techniques is given in Algorithm 1 and Algorithm 2 respectively. Both of these approaches have their own pros and cons. Each session must represent a single role otherwise the clustering of web sessions will be biased and have a high risk of clustering web sessions which are totally unrelated.

Algorithm 1: Time Threshold based session identification

```
1  initialize sessionStartTime, logPointer, timeOutThreshold ;
2  while there exist (moreRecordsInLog) for a particular IP do
3  |   read(nextRecord);
4  |   if recordRequestTime-lastRequestTime ¡ timeOutThreshold then
5  |   |   append(currentSession, Record);
6  |   else
7  |   |   close (currentSession);
8  |   |   createNewSession(IP);
9  |   end
10 end
```

Algorithm 2: Behaviour shift based session identification

```
1  Initialize time=sessionStartTime, logPointer=1 ;
2  while there exist (moreRecordsInLog) for a particular IP do
3  |   read(nextRecord);
4  |   if ShiftInBrowsingBehaviour == True then
5  |   |   append(currentSession, Record);
6  |   else
7  |   |   close (currentSession);
8  |   |   createNewSession(IP);
9  |   end
10 end
```

4.2.3 Web Session Clustering Techniques

Web session clustering exploits the three main dimensions of web usage data; time dimension, semantic usage behaviour dimension and browsing sequence dimension. For time dimension based clustering, the Euclidean distance is used to measure the distance between two sessions. Each session is transformed to a data vector with finite attributes representing time dimension information of a session. The Euclidean distance measure calculates the distance between two session vectors and the clustering algorithm decides in which cluster the session is to be placed.

$$d(x,y) = \left(\sum_{i}^{n} (x_i - y_i)^2 \right)^{(1/2)} \tag{4.1}$$

where x_i is the ith attribute value of data vector x and y_i is the ith attribute value of data vector y. The dimension of each data vector is from 1 to n representing the attributes of a session.
The semantic clustering of a session involves semantic information in terms of page and topic similarities. The session can be clustered on the basis of the relatedness of the pages viewed by each user during their respective session. The similarity of pages is measured using term frequency inverse document frequency $(tf - idf)$ measure shown in equation 4.2.

$$w(i,j) = tf_{i,j} * log\left(\frac{N}{df_i}\right) \tag{4.2}$$

where $tf_{i,j}$ is the frequency of term i in document j, df_i is the number of documents possessing term i, and N is the total number of documents. Some approaches perform generalization of sessions to increase the semantic coverage of the session [2]. The method in [5] first generalized the session in attribute-oriented induction according to a data structure, called page hierarchy-partial ordering of the Web pages, and then clustered using BIRCH. In [6] click stream clustering is performed using Weighted Longest Common Subsequences (WLCS).
The area which is mostly investigated by researchers is the browsing sequence based session clustering. The sequence of each session consists of a page-hit hierarchy of the web user and forms a labeled edge graph. The distance between these graphs are then calculated for clustering the related pages into identical clusters. The Levenshtein distance method gives edit distance between two sequences of navigations.

$$dG(G1,G2) = 1 - 2\left[\frac{L(S1,S2)}{(\|E(G1)\| + \|E(G2)\|)}\right] \tag{4.3}$$

where $L(S1,S2)$ is the Levenshtein distance between path $S1$ and path $S2$ and $\|E(G)\|$ shows the number of nodes in the graph. Some approaches combine the time dimension with browsing sequence to identify the relative importance of a visit. Weighted longest common subsequence (LCS) is the common subsequence among any two sessions. WLCS creates a sequence which considers the similarity of the common region weighted by time and importance of that region. The *similarity* component shows how similar the two paths are and *importance* shows how important the region of intersection is in terms of time spent on that region [6]. *Similarity* of two sequences is measured by

$$S_i = \frac{Min(Seq1_i, Seq2_i)}{Max(Seq1_i, Seq2_i)} \tag{4.4}$$

where S_i is the similarity of the ith visits of both sessions and its value is from 0 to 1. The average similarity of the two sequences is

$$S = \frac{1}{L}\sum_{i}^{L}\left(\frac{Min(Seq1_i, Seq2_i)}{Max(Seq1_i, Seq2_i)}\right) \tag{4.5}$$

where, L is the length of the longest common sub sequence. The *importance* component is calculated as

$$imp = \left[\frac{TimeOfLCS1}{TotalTimeOfSeq1}\right] \times \left[\frac{Min(Seq1_i, Seq2_i)}{Max(Seq1_i, Seq2_i)}\right] \quad (4.6)$$

Once the similarity values are calculated then the next step is to create a similarity graph of the users where similar activities are automatically grouped into identical clusters. For generalized conceptual graph clustering, the author [7] added pages similarity for clustering web sessions. In domain taxonomy based clustering, the similarities of two pages are calculated as.

$$S(c1,c2) = \frac{2 \times [depth(LCA(c1,c2))]}{(depth(c1) + depth(c2))} \quad (4.7)$$

where LCA is the lowest common ancestor of concepts c1 and c2. The equation is based on Generalized Vector-Space Model [8], where $c1$ and $c2$ are concepts in the hierarchy, and *depth* is the number of edges from the concept to the top of the hierarchy.

4.3 Swarm Intelligence

Swarm Intelligence (SI), inspired by the biological behaviour of animals is an innovative distributed intelligence paradigm for solving optimization problems [9]. It is a state of the art optimization technique based on the communication and cooperation of autonomous agents in a multi agent environment. The two main swarm intelligence algorithms; Ant Colony Optimization (ACO) and Particle Swarm Optimization (PSO) are widely used for optimization of discrete and continuous problems. Both of the techniques have been successfully used for the solution of different optimization problems such as NP hard problems, data mining, distributed systems, power systems, hybrid systems, and complex systems. In this section we discuss the details of PSO.

4.3.1 Particle Swarm Optimization

PSO is an optimization technique originally proposed by [10] and is based on the inspiration from the swarm behavior of birds, fish and bees when they search for food or communicate with each other. The particles or birds correspond to agents, the swarm a collection of particles, represents a multi agent system and the swarming behavior in the particles is like agent communication [11][12][13]. For PSO, the solution space of a problem is represented as a collection of agents where each agent represents an individual solution and the MAS represents the solution space for a particular iteration. In PSO the agents are initialized randomly to a solution set from the solution space. The velocity of the agents causes change in the agents position. The agents maintain their current velocity value and their personal best position (pBest) while moving from one position to another. The pBest maintained by every agent is the best ever position (fitness) found by that agent. The swarm also maintains a best value which is called global best position (gBest). The gBest value is the position representing best fitness value achieved for all agents of the MAS. The pBest value is calculated by equation 4.8.

$$pBest_i(t+1) = \begin{cases} pBest_i(t) & \text{if } f(X_i(t+1)) \text{ isNotBetterThan } f(pBest_i(t)) \\ X_i(t+1) & \text{if } f(X_i(t+1)) \text{ isBetterThan } f(pBest_i(t)) \end{cases} \tag{4.8}$$

where $X_i(t+1)$ is the current position of the agent, $pBest_i(t)$ is the personal best position and $pBest_i(t+1)$ is the new best position. After finding the new personal best position, the next step is to calculate the global best position, which can be extracted by $gBest(t) = argmin_{i=0}^{n} pBest_i(t)$, where i is the index of each agent ranging from 0 to the total number of agents n. The velocity of each agent is influenced by two learning components: the cognitive component $(pBest - X_i(t))$ and the social component $(gBest - X_i(t))$. The cognitive component represents learning from history and experience while the social component represents learning from the other fellow agents of the MAS. The cognitive and the social component guide the agent towards the best solution. The velocity update equation guided by the cognitive and social learning component is shown in equation 4.9.

$$V_i(t+1) = \omega * V_i(t) + q1r1(pBest - X_i(t)) + q2r2(gBest - X_i(t)) \tag{4.9}$$

where $V_i(t)$ is the current velocity, $V_i(t+1)$ is the new velocity, ω is the inertia weight, $q1$ and $q2$ are the values which weigh the cognitive and social components and $r1$ and $r2$ are two randomly generated numbers ranging from 0 to 1. The range for the velocities of the agents is from $-V_{max}$ to V_{max} . Position of the agent is updated using the position updating equation 4.10.

$$X_i(t+1) = X_i(t) + V_i(t+1) \tag{4.10}$$

After calculating the new position of each agent, the swarm looks for the global fitness which is evaluated for the final solution during a particular iteration. If the solution doesn't fulfill the specified criteria, the next generation of the swarm is iterated. The process continues until the stopping criteria i.e. maximum number of iterations or the minimum error requirement, is fulfilled. The number of agents in the system is selected according to the problem complexity. Algorithm 3 shows the pseudo code of the PSO process.

Algorithm 3: Particle Swarm Optimization

```
1  foreach particle do
2  |   initialize all parameters;
3  end
4  repeat
5  |   foreach particle do
6  |   |   calculate fitness value;
7  |   |   if fitness value is better than pBest then
8  |   |   |   set current value to pBest using (8);
9  |   |   end
10 |   |   choose the particle with the best fitness value;
11 |   end
12 |   foreach particle do
13 |   |   calculate particle velocity using (9);
14 |   |   update particle position using (10);
15 |   end
16 until stopping criteria unfulfilled ;
```

4.3.2 PSO Based Web Usage Clustering

The approach we propose in this paper cluster on time and browsing sequence dimensions of the web usage data set. We formulated sessions as particles for the particle swarm optimization algorithm using the idea of Cohen et al. [14]. The formulation of the problem for this approach is discussed in the following paragraph. We consider swarm as a multi agent system and an individual particle as an agent of the MAS. Each session vector contains attributes of a user session i.e. session length, number of pages visited during that session, and amount of data downloaded etc. All the sessions recorded for user activities represent the input data space for the clustering problem. Each agent of the MAS is initialized randomly to one of the input session vectors. Once the initialization of the entire system is completed, the next step is to iterate each agent of the system to find suitable position. After completion of the first iteration each agent is evaluated for its performance i.e. personal best position using equation 4.8. This value effects the learning of the agent from its experience. The agent uses personal best position to influence its velocity. The cognitive component is $q1r1(pBest - X_i(t))$ where $q1$ and $r1$ are the two constants which weight the cognitive component. To learn from the experience of the whole swarm, the agent takes its inspiration from the global best position called *gBest* position. To obtain the *gBest* value, the swarm evaluates each agent and selects the best single position/ fitness of all the particles and sets this value as *gBest* for the current iteration of the swarm. The social learning component $q2r2(gBest - X_i(t))$ causes the movement of the agent to be influenced from the experience of the entire swarm, $q2$ and $r2$ are the weighing constants of the social component. The self organizing component of the agents $q3r3(X_i(t) - Y_i(t))$ influence the particle movement towards the best position in its sub population. $Y_i(t)$ is the position of the session vector of a particular cluster. The social learning component, the cognitive learning component and the self organizing component decides the movement direction of the agent.

$$V_i(t+1) = \omega * V_i(t) + q1r1(pBest - X_i(t)) + q2r2(gBest - X_i(t)) + q3r3(X_i(t) - Y_i(t)) \quad (4.11)$$

For the solution of the clustering problem where the agent does not take its inspiration directly from the experience of the entire swarm and its movement is guided by the cognitive component and self organizing component only, the value of *gBest* is ignored i.e. $q2r2(gBest - X_i(t)) = 0$. In such cases a single agent represents one clustering centroid instead of a complete clustering solution. Equation 4.11 calculates the velocity of the agent which is then added to the current position value to find the new position of the agent.

$$X_i(t+1) = X_i(t) + V_i(t+1) \quad (4.12)$$

while $X_i(t+1)$ is the new position of the agent, $X_i(t)$ is the current position of the agent and $V_i(t+1)$ is the current velocity of the particle. The agents change their position with respect to their sub population while the sub population i.e. the session vector, do not change their position.

In our experiment, each agent consists of session attributes; total time of a session, number of pages visited, and amount of data downloaded in a particular session. Each agent represents a part of the clustering solution as a centroid of that cluster, while the entire swarm represents a solution to the clustering problem. Following are the main attributes of an agent, which it keeps while moving through the solution space.

- *ParticleId:* it uniquely identifies an agent or a centroid session.
- *DistanceFromEachSession:* an array which represents the distance of the agent to each session at a particular iteration. We used the Euclidean distance measure in our experiment. The closest sessions to the agent are won by that agent and added to the *WonSessionVectors*.
- *WonSessionVectors:* an array which represents the session vectors won by an agent at a given iteration. The agent organizes itself among the current won sessions. This causes the agent to learn from the neighborhood and organize itself within its sub population.
- *SessionAttributeValues:* represents the current values of the agent in each dimension in the form of a data vector. The more session attributes the easier it is to find the similarities among sessions.
- *PBest:* is the position of the nearest session to the agent achieved so far. This is obtained by keeping track of the position of the best previous session.

Once the initialization is completed, the agents are now moved from their initial position (starting session), guided by the social, cognitive and self organizing component. The cognitive component of the algorithm is encoded as $(pBest - X_i(t))$.

The social component is encoded as $(gBest - X_i(t))$ and a self organizing term as $(Y_i(t) - X_i(t))$ where $Y_i(t)$ is the current position value of the agent, *pBest* is the personal best position found by the agent so far and *gBest* is the global best position, however in this particular case the value of the social term is not as important because the agent should not follow the whole swarm but only its given sub population. So we ignore the $(gBest - X_i(t))$ for clustering the web usage session as discussed earlier. The self organizing term is more important as it causes a change in the velocity of an agent towards the current session attribute. After each iteration, the swarm changes its position by winning the nearest sessions, recalculates all its parameters, organizes itself according to the new session vector won by each agent. The process continues until there is no significant change in the position of the agents or the number of maximum iterations is approached or no movement of data vectors from one cluster to another cluster is observed.

4.4 Experimental Results

In this section, we explain the preprocessing and clustering results respectively.

4.4.1 Data, Pre-processing and Usage Statistics

For experiments and performance evaluation of our approach we chose the NASA web log file[2], which contains HTTP requests to NASA Kennedy Space Centre's web server from 1st July 1995 to 28th July 1995. There are 1891715 requests in the log and the log size is 195 MB in text format. We analyzed the logs containing one day of HTTP requests dated 1st July 1995. The log was first passed through all the pre-processing steps, data cleaning, structuring and summarization. The details of the results after the pre-processing step are given in Table 4.1. The results reveal

Table 4.1: Results of the pre-processing

Total Requests	**64578**	**Successful Requests**	**23795**
Pure Requests	25387	users having < 10 requests	4591
Distinct pages requested	1096	user having > 10 requests	408
Unsuccessful requests	1592	Average Request per user	13
Distinct user	4999	Total requests > 10 request per user	10121
Images request	61%		

Total sessions	**815**
Session > 10 request	432
Average request per session	13
Average session per user	2

[2] http://ita.ee.lbl.gov/html/contrib/NASA-HTTP.html

the fact that more than 60% of the web requests recorded in the web logs are image requests and are useless in the context of web session clustering. The importance of the pre-processing stage is verified by the ratio of successfully selected requests to the total requests i.e. 1/3 of the total request are selected for analysis.
Fig. 4.1 summarizes the usage statistics of visits, users, session and responses generated by the web server. Fig. 4.1 (a) shows the number of pages against the number of users, after the pre-processing phase. The distribution shows that most of the users are from the class where the number of pages viewed is from 5 to 15. The number of users decreases gradually with an increase in the number of pages viewed. In Fig. 4.1 (b) the number of requests is plotted against 30 minute time interval starting from July 1, 1995 12:00:01 AM to 12:00:00 PM, Fig. 4.1 (c) elaborates the number of page requested against the session length. The distribution shows that most of the sessions have an iteration number less than 5, which are known not to be representative of the real usage behavior, so they are ignored. Fig. 4.1 (d) shows the session count and percentage against time intervals. The time interval of 11 to 20 minutes gets the highest session count, which demonstrates that most of the web users have a session time between 10 to 30 minutes and longer sessions are rare in web logs.

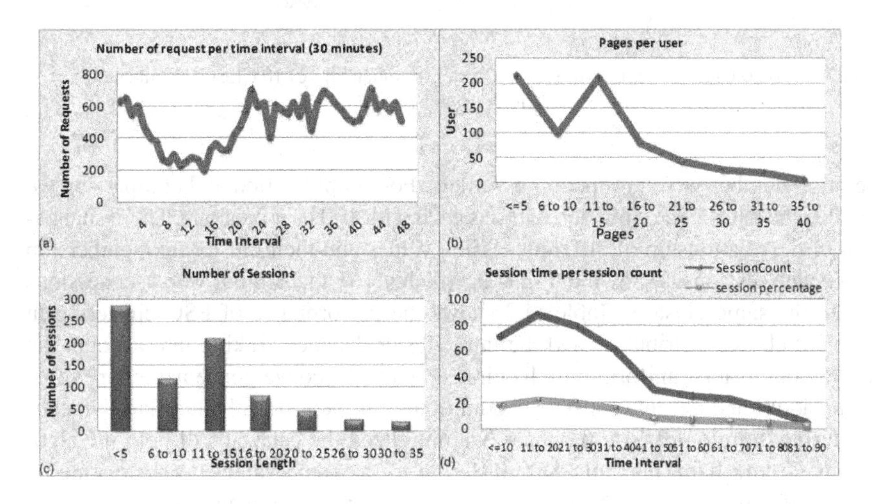

Fig. 4.1: (a) Time and request per 30 minutes distribution (b) Page viewed and number of user distribution (c) Session-number of request distribution (d) Session-time distribution

4.4.2 Clustering Results

After pre-processing and removal of sessions with < 10 requests, 432 sessions were selected for our clustering analysis. The purpose of the experiment was to group the session on the basis of the session attribute values using the agent based particle swarm optimization clustering approach for comparison of our approach with

K-means. Fig. 4.2 shows grouping of the sessions in 5, 10, 15 and 20 clusters respectively. Most of the users have similar behavior and can be grouped in 2-4 active clusters. Sessions with higher amounts of data downloaded, high visit times and larger

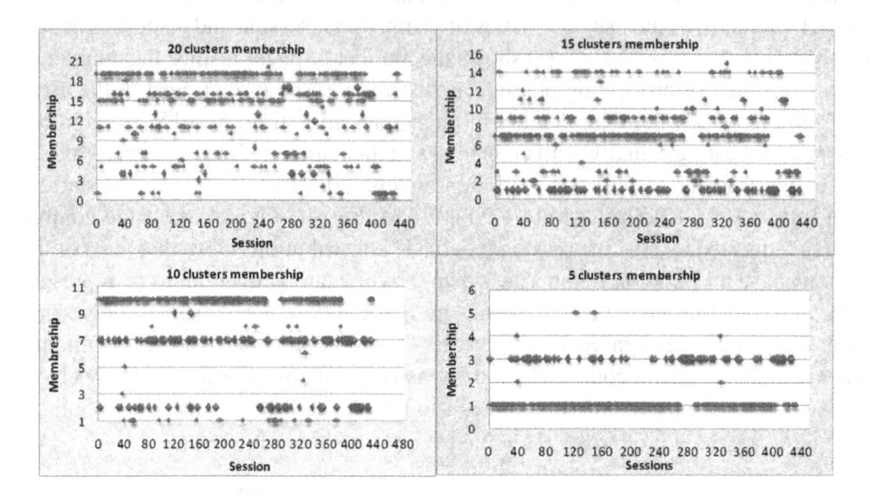

Fig. 4.2: (a) 20 Cluster Membership (b)15 Cluster Membership (c) 10 Cluster membership (c) 5 Cluster membership

iteration number were grouped into outliers. For simplification and comparison we divided the dataset into four sub datasets each with 100 user sessions. The visualization of the relationship of different session values and their clustering membership of the web sessions is shown in Fig. 4.3, which verify that similar web user sessions fall in the same clusters. Table 4.2 shows the performance of PSO clustering in terms of cluster distribution and the intra cluster distance. Taking into account the density of clusters, uniform initialization, and agent's convergence nature are some of the additional advantages. The overall fitness of the PSO is better than K-means clustering. For comparison purposes, we initialized the centroids of both the algorithms i.e. the K-means and PSO clustering to the same values. We performed a variety of experiments with initialization, parameter selection and iterations to verify the efficiency of the approach. For the PSO-clustering algorithm the parameters were set to the range $Vmax = [0.1, 0.04]$, $q1 = [.01, 0.9]$, $q2 = 0$, $q3 = [0.01, 0.9]$, $\omega = [0.01, 0.09]$, acheiving the results shown in Table 4.2. The number of iterations on which the solution is obtained in most of the cases was below 100. To access the time consistency of the approach, we scaled the number of iteration to a maximum iteration of 1000, however we have not found any inconsistency or abrupt change in execution time. The relationship between the number of iterations of PSO clustering and execution time was observed as linear as demonstrated in Fig. 4.4.

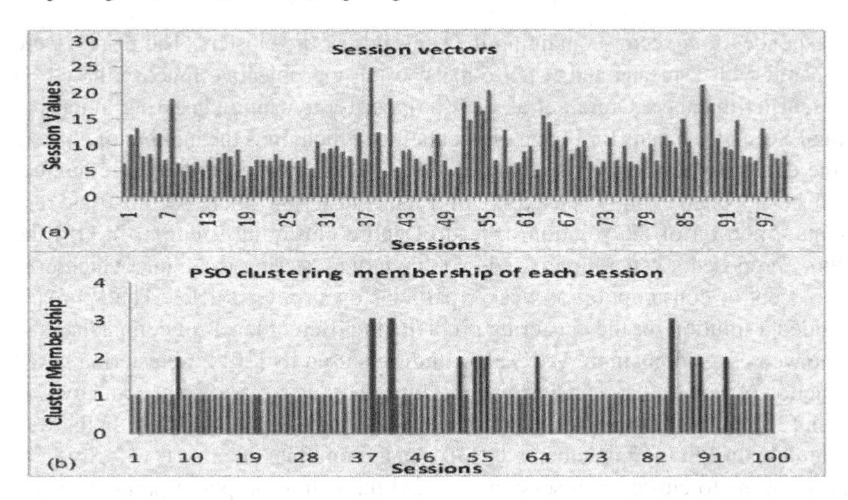

Fig. 4.3: (a) Session data (b) PSO clustering members

Table 4.2: Comparison of K-means and PSO clustering

Log	K-means		PSO	
	Mean IntraCluster Dist.	Fitness	Mean IntraCluster Dist.	Fitness
1	81.8211	245.463	81.4863	**244.459**
2	35.0334	105.1002	34.85	**104.55**
3	27.7769	55.5538	25.8365	**51.6731**
4	59.1294	118.2588	59.1367	118.2734

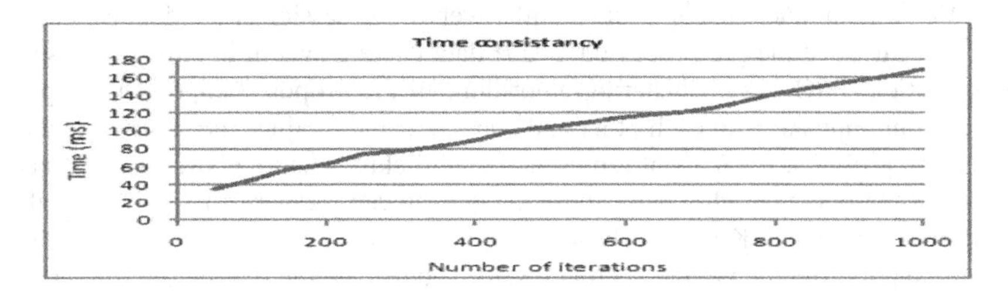

Fig. 4.4: Execution time and number of iteration

4.5 Related Work

PSO was introduced for data clustering by Van der Merwe and Engelbrecht [15] by initializing randomly created particles to a vector containing centroids of the clusters. The evaluation of the method was based on the cost function that evaluates each candidate solution based on the proposed cluster's centroids. In [16], the authors applied PSO with Self-Organizing Maps (SOM) where SOM are used for grouping and PSO optimizes the weights of the SOM. Chen and Ye [17] represented each particle

corresponded to a vector containing the centroids of the clusters. The results were compared with k-means and fuzzy c-mean using the objective function based on intra-cluster distance. Omran et al. [18] proposed a dynamic clustering algorithm using PSO and k-means for image segmentation, which finds the number of clusters in the data automatically by initially partitioning the data set into a large number of clusters. Cohen et al. [14] used PSO for data clustering where each particle represents a portion of the solution instead of entire clustering solution. In [19] the authors proposed a generation based evolutionary clustering technique which uses the concept of consumption of weaker particles by strong particles. The approach provided a solution for the clustering problem on different levels of compactness.
In the web usage domain the ACO algorithm was used by [20] for clustering based prediction of web traffic. AntClust was introduced for web session clustering by [21] and in [22], the authors proposed ant-based clustering using fuzzy logic. In the web usage domain PSO is implemented [23], which combines improved velocity PSO with k-means to cluster web sessions. In [24] the author proposed particle swarm optimization approach for the clustering of web sessions.
Recently some of the research [3] [4] and [13] have focused on data mining with multiagent integration and interaction.

4.6 Conclusion and Future Work

Integration and interaction of data mining with multi agent system is beneficial for mining the distributed nature of WWW. Web session clustering, one of the important WUM technique, aims to group similar web usage sessions into identical clusters. We clustered the pre-processed WUM data using a swarm intelligence based optimization, PSO based clustering algorithm. In the proposed approach, simple agents communicate with each other and cooperate and produce the solution to the clustering problem. Each agent represent a single cluster and the swarm of agents represent the complete clustering solution. We showed the performance of the algorithm is better than K-means clustering. The future directions in the area are the integration of different parameters for clustering, development of accurate similarity measures, PSO parameter automation and involvement of optimization algorithms in other areas of web usage mining.

References

1. Frawley, W., Piatetsky-Shapiro, G., Matheus, C. : Knowledge Discovery in Databases: An Overview. AI Magazine: pp. 213-228, (1992)
2. Edelstein, H.A.: Introduction to data mining and knowledge discovery (3rd ed). Two Crows Corp, Potomac, MD, (1999)
3. Cao, L., Gorodetski, V.: AREA OVERVIEW-Agent & data mining interaction (ADMI). In: WI-IAT 2006 IADM Workshop panel discussion, Hongkong (2006)

4. Cao, L., Luo, C., Zhang, C.: Agent-Mining Interaction: An Emerging Area, AIS-ADM07, LNAI 4476, 60-73, Springer, (2007)
5. Fu, Y., Sandhu K., Shih, M-Y.: A Generalization-Based Approach to Clustering of Web Usage Sessions, Revised Papers from the International Workshop on Web Usage Analysis and User Profiling, p.21-38, (1999)
6. Banerjee, A., Ghosh, J.: Clickstream Clustering using Weighted Longest Common Subsequence. In: Proceedings of the 1st SIAM International Conference on Data Mining: Workshop on Web Mining. (2001)
7. Nichele, C. M., Becker, K.: Clustering Web Sessions by Levels of Page Similarity. In: Pacific-Asia Conference on Knowledge Discovery and Data Mining (PAKDD '06): 346-350. (2006).
8. Ganesan, P., Garcia-Molina, H.,Widom, J.: Exploiting Hierarchical Domain Structure to Compute Similarity. ACM Transactions on Information Systems (TOIS), v.21, n.1, 64-93. (2003)
9. Abraham, A., Guo, H., Liu, H.: Swarm Intelligence: Foundations, Perspectives and Applications. Swarm Intelligent Systems, Studies in Computational Intelligence, (eds.), pp. 3-25. Springer,(2006)
10. Kennedy, J., Eberhart, R.C.: Particle swarm optimization. International Conference on Neural Networks (ICNN '95), Vol. IV, Perth, Australia (1995)
11. Engelbrecht, A.P.: Fundamentals of Computational Swarm Intelligence. John Wiley and Sons, (2005)
12. Kennedy, J., Eberhart, R.C.: Swarm intelligence. Morgan Kaufmann Publishers, (2001)
13. Altshuler, Y., Bruckstein, A.M.,Wagner, I.A.: On Swarm Optimality In Dynamic And Symmetric Environments. In: Second International Conference on Informatics in Control, Automation and Robotics (ICINCO), Barcelona, Spain, (2005)
14. Cohen, S.C. M., de Castro, L. N.: Data Clustering with Particle Swarms. In: IEEE Congress on Evolutionary Computations, Vancouver, BC, Canada, (2006)
15. Van der Merwe, D.W, Engelbrecht, A.P.: Data clustering using particle swarm optimization. In: Proceedings of IEEE Congress on Evolutionary Computation 2003 (CEC 2003), Canbella, Australia. pp. 215-220, (2003)
16. Xiao, X., Dow, E.R., Eberhart, R.C., Ben Miled, Z., Oppelt, R. J.: Gene clustering using self-organizing maps and particle swarm optimization. In: Proceedings of Second IEEE International Workshop on High Performance Computational Biology, Nice, France, (2003)
17. Chen, C.-Y., Ye, F.: Particle swarm optimization algorithm and its application to clustering analysis. In: Proceedings of IEEE International Conference on Networking, Sensing and Control. pp. 789-794, (2004)
18. Omran, M. G. H., Salman, A., Engelbrecht, A. P.,: Dynamic Clustering Using Particle Swarm Optimization with Application in Image Segmentation. Pattern Analysis and Applications, Vol. 8, pp. 2-344,(2005)
19. Alam, S., Dobbie, G., Riddle, P.: An Evolutionary Particle Swarm Optimization Algorithm For Data Clustering. In: Proceedings of IEEE International Swarm Intelligence Symposium. Missouri, USA ,(2008)
20. Abraham, A.: Natural Computation for Business Intelligence from Web Usage Mining. In: Seventh International Symposium on Symbolic and Numeric Algorithms for Scientific Computing (SYNASC'05), pp. 3-10,(2005)
21. Labroche, N., Monarch, N., Venturini, G.: AntClust: Ant clustering and web usage mining. Genetic and Evolutionary Computation. Chicago, IL. Lecture Notes in Computer Science 2723. Berlin, Heidelberg, Germany: Springer-Verlag. pp 25-36,(2003)
22. Steven Schockaert , Martine De Cock, Chris Cornelis, Etienne E. Kerre.: Clustering web search results using fuzzy ants. International Journal of Intelligent Systems, Volume 22, Issue 5, Pages 455 - 474,(2007)
23. Chen, J. Z., Huiying : Research on Application of Clustering Algorithm Based on PSO for the Web Usage Pattern. Wireless Communications, Networking and Mobile Computing, (2007)
24. Alam, S., Dobbie, G., Riddle, P.: Particle Swarm Optimization Based Clustering of Web Usage Data. In: ACM/WIC/IEEE International conference on web intelligence. Sydney, Australia,(2008)

Chapter 5
Mining Temporal Patterns to Improve Agents Behavior: Two Case Studies

Philippe Fournier-Viger[1], Roger Nkambou[1], Usef Faghihi[1], and Engelbert Mephu Nguifo[2]

Abstract We propose two mechanisms for agent learning based on the idea of mining temporal patterns from agent behavior. The first one consists of extracting temporal patterns from the perceived behavior of other agents accomplishing a task, to learn the task. The second learning mechanism consists in extracting temporal patterns from an agent's own behavior. In this case, the agent then reuses patterns that brought self-satisfaction. In both cases, no assumption is made on how the observed agents' behavior is internally generated. A case study with a real application is presented to illustrate each learning mechanism.

5.1 Introduction

Logging information about agents' behavior and analyzing it could provide useful knowledge for improving the behavior of agents. However, the amount of data to be recorded and analyzed can be huge to learn interesting information. As a solution, we propose to rely on data mining algorithms to analyze agent behaviors, and discover temporal regularities. More precisely, in line with researches indicating that data mining has the potential to improve the learning and reasoning capabilities of agents (data-mining driven agents) [25, 2], we propose two learning mechanisms that are integrated in agents deployed in real applications. The two learning mechanisms are based on the three same operation phases: (1) recording behaviors, (2) extracting temporal patterns from this data and (3) using this knowledge to improve agents' behavior. The first learning mechanism is for agents that learn a task by observing other agents performing it. In that case an agent records the behavior of

Université du Québec à Montréal, Montréal, Canada
e-mail: {fournier_viger.philippe, nkambou.roger, faghihi.usef}@courrier.uqam.ca
· Université Blaise Pascal, Clermont-Ferrand, France
e-mail: mephu@isima.fr

L. Cao (ed.), Data Mining and Multi-agent Integration, DOI: 10.1007/978-1-4419-0522-2_5,

other agents to discover patterns among this data that will then constitute its knowledge base. The second learning mechanism is for agents that learn from their own behavior. In this case, an agent records its behavior to then reuse patterns that led to self-satisfaction.

This chapter first introduces the problem of mining temporal patterns and an algorithm for mining temporal patterns, and discusses related works in agent learning. Next, the second and third sections present the two learning mechanisms based on this algorithm and describe how they are integrated in virtual agents. Finally, the last section present conclusions and preview our future work.

5.2 Mining Temporal Patterns from Sequences of Events

According to [10], there are four main kinds of patterns that can be mined from time-series data. These are trends, similar sequences, sequential patterns and periodical patterns. In this work we chose to mine sequential pattern [3], as we are interested in finding relationships between occurrences of events in agents' behavior. To mine sequential patterns, several algorithms have been proposed [10]. But to our knowledge, few works have been published on using sequential pattern mining for agent learning. For example, [14] proposed to implement sequential pattern mining in a robot playing soccer. In this case, sequential patterns are used to derive prediction rules about what actions or situations might occur if some preconditions are satisfied. This is different from the two forms of learning that we consider in this chapter.

While classical sequential pattern mining algorithms have for only goal to discover sequential patterns that occurs frequently in several transactions of a database [3], other algorithms have proposed numerous extensions to the problem of mining sequential patterns [10]. For this work, we chose a sequential pattern mining algorithm that we have developed [8], as it combines several features from other algorithms such as accepting time constraints [11], processing databases with dimensional information [17], eliminating redundancy [20, 18], and also because it offers some original features such as accepting symbols with numeric values [8]. We describe next some basic features of the algorithm. Other features will be presented through the chapter with detailed explanations about why they are important for this work. For a technical description of the algorithm the reader can refer to [8]. Moreover, the reader can download a Java implementation of the algorithm by accessing `http://www.philippe-fournier-viger.com/spmf/`.

The algorithm takes as input a database D of sequences of events. An event $X = (i_1, i_2, ...i_n)$ contains a set of items $i_1, i_2, ...i_n$ that are considered simultaneous, and where each item can be annotated with an integer value. Formally, a sequence is denoted $s = <(t_1, X_1), (t_2, X_2), ..., (t_n, X_n)>$, where each event X_k is associated to a timestamp t_k indicating the time of the event. The size, n, of a sequence is the number of events in the sequence, i.e. $|s|$. For example, the size of sequence $S1$ of Fig. 5.1 (left) contains two events. The sequence $S1$ indicates that item a appeared with a value of 2 at time 0 and was followed by items b and c with a value of 0 and 4 respectively at time 1. A sequence $sa = <(ta_1,A_1),(ta_2,A_2,...,(ta_n,A_n)>$ is said to be contained in another sequence $sb = <(tb_1,B_1),(tb_2,B_2),...,(tb_n,B_m)>$, if there exists integers $1 = k1 < k2 < ... < kn \leq m$ such that $A_1 \subseteq B_{k1}, A_2 \subseteq B_{k2},...,A_n \subseteq B_{kn}$, and that $tb_{kj} - tb_{k1}$ is equal to $ta_j - ta_1$ for each $j \in 1...m$. The relative support of a sequence sa in a sequence database D is defined as the percentage of sequences $s \subseteq D$ that contains sa, and is denoted by $sup_D(sa)$. The problem of mining frequent sequences is to find all the sequences sa such that $sup_D(sa) \geq minsup$ for a sequence database D, given a support threshold $minsup$, and optional time constraints. The optional time constraints are the minimum and maximum time intervals required between the head and tail of a sequence and the minimum and maximum time intervals required between two adjacent events of a sequence.

As an example, Fig. 5.1 illustrates a database of 6 sequences (left) and the corresponding patterns found for a minsup of 33% (right). Consider pattern $M5$. This pattern appears in sequence $S4$ and $S5$. The first event of $M5$ $(0,f)$ is contained in the first event of $S4$ $(0,f)$ and the second event of $M5$ $(2,e)$ is contained in the event of $S4$ $(2,a\{6\}e)$ that is occuring two time units after the first event of $S4$. Since the pattern $M5$ is contained in $S4$ and $S5$, it has a support of 33% (2 out of 6 sequences). Now consider patterns $M1$ and $M2$. Because the item a appears in sequence $S1$, $S2$, $S3$ and $S4$ with values 2, 2, 5 and 6 respectively the algorithm separated these values in two groups to create patterns $M1$ and $M2$ instead of creating a single pattern with a support of 66%. For each of these groups, the median (2 and 5) was kept as an indication of the values grouped. This clustering of similar values only occurs when the support is higher or equal to $2 * minsup$ (see [8] for more details).

ID	Sequences
S1	<(0,a{2}),(1,bc{4})>
S2	<(0,a{2}),(1,c{5})>
S3	<(0,a{5}),(1,c{6})>
S4	< (0,f), (1, g),(2,a{6}e)>
S5	<(0, f b{3}), (1,h),(2,ef) >
S6	<(0,b{2}),(1,d)>

->

ID	Mined Sequences	Supp.
M1	<(0,a{2})>	33 %
M2	<(0,a{5})>	33 %
M3	<(0,a{2}),(1, c{5})>	33 %
M4	<(0,c{5})>	50 %
M5	<(0,f), (2, e)>	33 %
M6	...	...

Fig. 5.1: A database of 6 sequences (left) and mined sequences (right)

5.3 Agents that Learn from Other Agents

The first form of learning that we consider for an agent is learning by observing other agents. Researchers have made various proposals for integrating learning-by-observation in agents (see [19] for a brief review). However, many of them rely on strong assumptions. For example, van Lent and Laird [19] propose a framework to learn production rules from recorded human behavior. In this approach, a human has to teach an agent by performing a task. However, this approach is tightly linked to a very specific conception of intelligence, as humans performing a demonstration are required to specify their actions as complex operators organized in a hierarchy and having goal conditions, and they must explicitly state their goals during the demonstrations. Contrarily to this view, we here address the problem of learning-by-observation for an agent by considering that it can only perceive actions of other agent, without additional information.

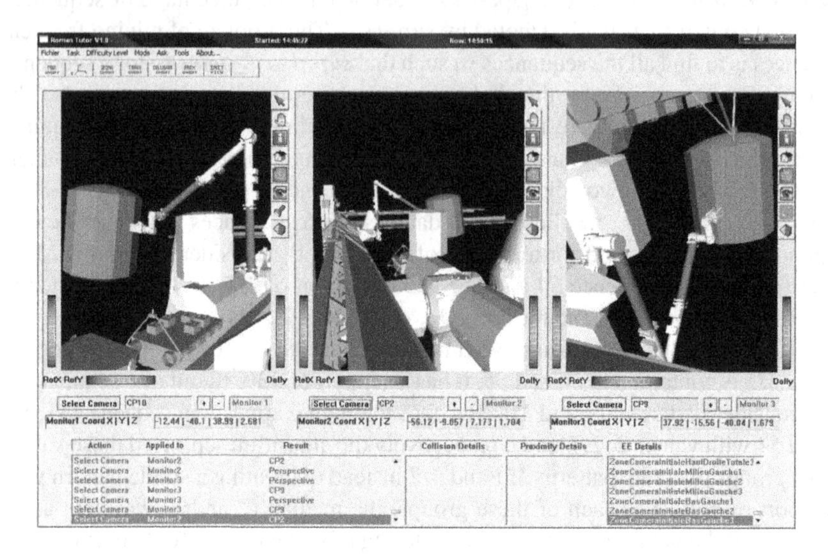

Fig. 5.2: The RomanTutor user interface

We illustrate this form of learning in the context of RomanTutor [13] (fig. 5.2), a virtual learning environment for learning how to operate the Canadarm2 robotic arm on the international space station. The main learning activity in RomanTutor is to move the arm from one configuration to another. This is a complex task, as the arm has seven joints and the user must chose at any time the three best cameras for viewing the environment from around twelve cameras on the space station, and adjust their parameters. We have integrated a tutoring agent in RomanTutor to provide assistance to learners during this task. However, there are a very large number of possibilities for moving the arm from one position to another, and because one must also consider the safety of the maneuvers, it is very difficult to define a task model for generating the moves that a human would execute [7]. For this reason, instead of providing domain knowledge to the agent, we implemented a learning mechanism which allows the agent to learn by observing the behavior of other agents performing a task (in this case, humans). The agent then uses this knowledge to provide assistance to learners. We describe next the three operation phases of the learning mechanism as they are implemented in this virtual agent, and an experiment.

5.3.1 The Observing Phase

In the *observing phase*, the virtual agent observes and records the behavior of users that attempt an arm manipulation exercise (moving the arm from an initial configuration to a goal configuration). For each attempt, the virtual agent logs a sequence of events. In this context, an event is a set of actions (items) that are considered unordered temporally. We defined 112 primitive actions that can be recorded in RomanTutor, which are (1) selecting a camera, (2) performing an increase or decrease of the pan/tilt/zoom of a camera and (3) applying a rotation value to an arm joint. Actions of types (2) and (3) are annotated with integer values that indicate respectively the number of increments/decrements applied to a camera parameter and the number of degrees that a joint is rotated. Defining actions with integer values is beneficial because, as mentioned, our algorithm can group automatically actions with similar values an treat these groups as different actions. An example of a partial sequence of actions recorded for an user is $<(0,6\{2\}),(1,63),(2,53\{4\}),(3,111\{2\}>$ which represents decreasing the rotation value of joint SP (action 6) by 2 units, selecting camera CP3 (action 63), increasing the pan of camera CP2 (action 53) by 4 units and then its zoom (action 111) by 2 units.

A problem that we faced when designing the virtual agent's observing phase is that it would also be useful to annotate sequences with contextual information such as success information and the expertise level of a user, to then mine patterns containing this information. Our solution to this issue is to take advantage of an extra feature of our algorithm (based on [17]), which is to add dimensional information to sequences. A database having a set of dimensions $D = D1, D2, ...Dn$ is called an MD-Database. Each sequence of a MD-Database (an MD-Sequence) possesses a symbolic value for each dimension. This set of values is called an MD-Pattern and is denoted $d1, d2...dn$. In the context of our virtual agent, we defined two dimensions, "success" and "expertise level", which are added manually to each sequence recorded. The left side of Fig. 5.3 shows an example of an MD-Database having these two dimensions. As an example, the MD-Sequence B1 has the MD-Pattern "true", "novice" for the dimensions "success" and "expertise level". The symbol "*", which means any values, can also be used in an MD-Pattern. This symbol subsumes all other dimension values. An MD-Pattern $Px = dx1, dx2...dxn$ is said to be contained in another MD-Pattern $Py = dy1, dy2...dym$ if $dx1 \subseteq dy1, dx2 \subseteq dy2, ..., dxn \subseteq dyn$. The problem of mining frequent sequences with dimensional information is to find all MD-Sequence appearing in an MD-Database with a support higher or equal to *minsup*. As an example, the right part of Fig. 5.3 shows some patterns that can be extracted from the MD-Database of Fig. 5.3, with a *minsup* of 2 sequences. In our virtual agent, dimensional information is very important, as it allowed to successfully identify patterns common to all expertise levels that lead to failure ("false, *"), for example.

ID	Dimensions	Sequences
B1	true, novice	<(0,a),(1,bc)>
B2	true, expert	<(0,d)>
B3	false, novice	<(0,a),(1,bc)>
B4	false, interm.	<(0,a),(1,c), (2,d)>
B5	true, novice	<(0,d), (1,c)>
B6	true, expert	<(0,c), (1,d)

->

Dimensions	Sequences
*, novice,	<(0,a)>
*, *	<(0,a)>
*, novice	<(0,a), (1,b)>
true, *	<(0,d)>
true, novice	<(0,c)>
true, expert	<(0,d)>

Fig. 5.3: An example of sequential pattern mining with contextual information

5.3.2 The Learning Phase

In the *learning phase*, the virtual agent applies the algorithm to extract frequent sequences that build its domain knowledge. For mining patterns, we setup the algorithm to mine only sequences of size two or greater, as sequence shorter would not be useful in a tutoring context. Furthermore, we chose to mine sequences with a maximum time interval between two adjacent events of two. The benefits of accepting a gap of two is that it eliminates some "noisy" (non-frequent) learners' actions, but at the same time it does not allow a larger gap size that could make it less useful for tracking a learner's actions.

A second important consideration in the learning phase is that when applying sequential pattern mining, there can be many redundant frequent sequences found. For example, in Fig. 5.1 (right), pattern M2 is redundant as it is included in pattern M3 and it has exactly the same support. To eliminate redundancy, we rely on an extra feature of our algorithm which allows mining only closed sequences. "Closed sequences" are sequences that are not contained in another sequence having the same support. Mining frequent closed sequences has the advantage of greatly reducing the size of patterns found, without information loss (the set of closed frequent sequences allows reconstituting the set of all frequent sequences and their support) [20]. To mine only frequent closed sequences, our sequential pattern mining algorithm was extended based on [20] and [18] to mine closed MD-Sequences (see [8]).

5.3.3 The Application Phase

In the third phase, the application phase, the virtual agent provides assistance to the learner by using the knowledge learned in the learning phase. Recognizing a learner's plan is the basic operation that is used to provide assistance. This is achieved by the plan recognizing algorithm (algorithm 1), described next.

The plan recognizing algorithm RecognizePlan is executed after each student action. It takes the sequence of actions performed by the student (Student_trace) for the current problem and a set of frequent actions sequences (Patterns) as inputs. When the plan recognizing algorithm is called for the first time, the variable *Pat-*

Algorithm 4: RecognizePlan

```
   Input: Student_trace, Patterns
   Output: Result
 1 Result := ∅.;
 2 foreach pattern P of Patterns do
 3 |   if Student_trace is included in P then
 4 |   |   Result := Result∪ {P}.
 5 |   end
 6 end
 7 if Result = ∅ AND size(Student_trace) ≥ 2 then
 8 |   Remove last action of Student_trace.;
 9 |   Result := RecognizePlan(Student_trace, Patterns).
10 end
11 RETURN Result.
```

terns is initialized with the whole set of patterns found during the learning phase of the virtual agent. The algorithm first iterates on the set of patterns *Patterns* to note all the patterns that include Student_trace. If no pattern is found, the algorithm removes the last action performed by the learner from Student_trace and searches again for matching patterns. This is repeated until the set of matching patterns is not empty or the size of Student_trace is smaller than 2. In our tests, removing user actions has shown to improve the effectiveness of the plan recognizing algorithm significantly, as it makes the algorithm more flexible. The next time *RecognizePlan* is called, it will be called with the set of matching patterns found by its last execution. This ensures that the algorithm will not consider patterns that have been previously rejected.

We describe next the main tutoring services that the tutoring agent provides based on the plan recognizing algorithm.

First, the virtual agent can assess the expertise level of the learner (novice, intermediate or expert) by looking at the patterns applied. If for example a learner applies 80% of the time "intermediate" patterns, then the virtual agent can assert with confidence that the learner expertise level is "intermediate". Second, the agent can guide the learner. This tutoring service consists in determining the possible actions from the current problem state and proposing one or more actions to the learner. This functionality is triggered when the student selects "What should I do next?" in the RomanTutor interface menu. The virtual agent then identifies the set of possible next actions according to the matching patterns found by *RecognizePlan* and selects the action among this set that is associated with the pattern that has the highest relative support and that is the most appropriate for the estimated expertise level of the learner. If no actions can be identified, the virtual agent can use a path planner [13] to generate approximate solutions. In this current version, the virtual agent only interacts with the learner upon request. Nonetheless, it would be possible to program the virtual agent so that it can intervene if the learner is following an unsuccessful pattern or a pattern that is not appropriate for its expertise level. Testing different tutorial strategies is part of our current work.

5.3.4 An Experiment

We conducted an experiment in RomanTutor with two exercises to qualitatively evaluate the virtual agent's capability to provide assistance. The two exercises consists each of moving a load with the robotic arm to one of the two cubes (figure 5.4.a). We asked twelve users to record plans for these exercises. The average length was 20 actions. From this data, the virtual agent extracted 558 sequential patterns with the algorithm. In a subsequent work session, we asked the users to evaluate the tutoring services provided by the virtual agent. Users agreed that the assistance provided was helpful. We also observed that the virtual agent correctly inferred the estimated expertise level of learners.

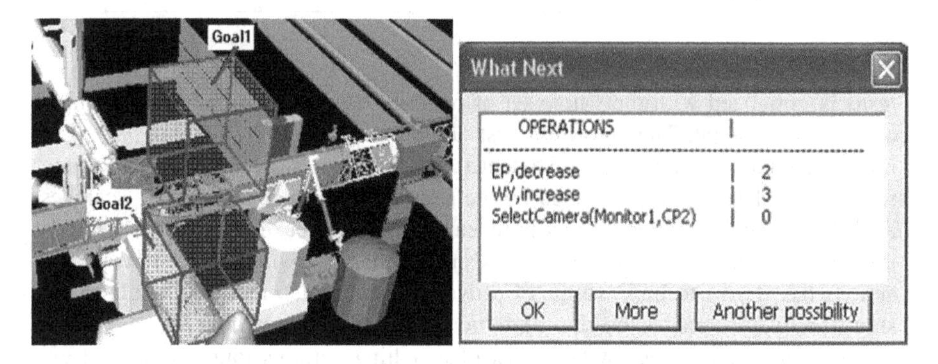

Fig. 5.4: (a) The two scenarios (b) A hint generated by the virtual agent

As an example of interaction with a learner, Fig. 5.4 illustrates a hint message given to a learner upon request during scenario 1. The guiding tutoring service selected the pattern that has the highest support value, matches the last student actions, is marked "success" and corresponds with the estimated expertise level of the learner. The given hint is to decrease the rotation value of the joint "EP" (20°), increase the rotation value of joint "WY" (30°), and finally to select camera "CP2" on "Monitor1". By default, three steps are showed to the learners in the hint window depicted in Fig. 5.4.b. However, the learner can click on the "More" button to ask for more steps or click on the "another possibility" button to ask for an alternative. The description of actions depicted in Fig. 5.4.a are an example of resources that can be used to annotate patterns.

Although the pattern mining algorithm was applied once in this experiment, it would have been possible to make the agent apply it periodically, so that the agent would continuously update its knowledge base while interacting with learners. Moreover, we have encoded only two dimensions: expertise level and success. However additional contextual information could easily be added. In future work for example, we plan to encode skills involved as dimensional information (each skill could be encoded as a dimension). This will allow computing a subset of skills that characterize a pattern. This will allow diagnosing missing and misunderstanding skill for users who demonstrated a pattern.

5.4 Agents that Learn from Their Own Behavior

The second form of learning that we consider for an agent is to learn from its own behavior. Unlike the learning mechanism implemented in the previous agent, this learning mechanism is not designed for learning new behaviors or procedural knowledge, but for making an agent reuse previously self-satisfying behaviors. We integrated this mechanism in a virtual agent named CTS [5] that we have also tested in RomanTutor to provide assistance to learners. The following subsections describe CTS, the three operation phases of the learning mechanism that was integrated in CTS, and two experiments carried in RomanTutor to validate (1) the behavior of the new CTS and (2) the behavior of the data mining algorithm with large data sets.

5.4.1 The CTS Cognitive Agent

CTS (Conscious Tutoring System) is a generic cognitive agent, whose architecture (fig. 5.5) is inspired by neurobiology and neuropsychology theories of human brain function. It relies on the functional "consciousness" [9] mechanism for much of its operations. It also bears some functional similarities with the physiology of the nervous system. Its modules communicate with one another by contributing information to its Working Memory (WM) through information codelets. Based on Hofstadter et al's idea [12], a codelet is a very simple agent, "a small piece of code that is specialized for some comparatively simple task". As in Baars theory's [4], these simple processors do much of the processing in the CTS architecture.

CTS possess two routes for processing external stimuli (cf. fig. 5.5). Whereas the "long route" is the default route, the "short route" (which will not be described here) allows quick reactions when received information is deemed important by the pseudo-amygdala, the module responsible for emotional reactions in CTS [6]. In both cases, the stimuli processing begins with percept codelets [12] that perform collective interpretations of stimuli. The active nodes of the CTS's Perception Network constitute percepts. In the long route, these percepts enter WM as a single network of codelets, annotated with an activation value. These codelets create or reinforce associations with other already present codelets and create a coalition of information codelets. In parallel, the emotional codelets situated in the CTS's pseudo-amygdala inspect the previously mentioned coalition's informational content, and if it is deemed important, infuse it with a level of activation proportional to its emotional valence. During every cognitive cycle, the coalition in the WM that has the highest activation is selected from the WM by the "Attention Mechanism" and is broadcast to all the modules in the CTS architecture. This selection process

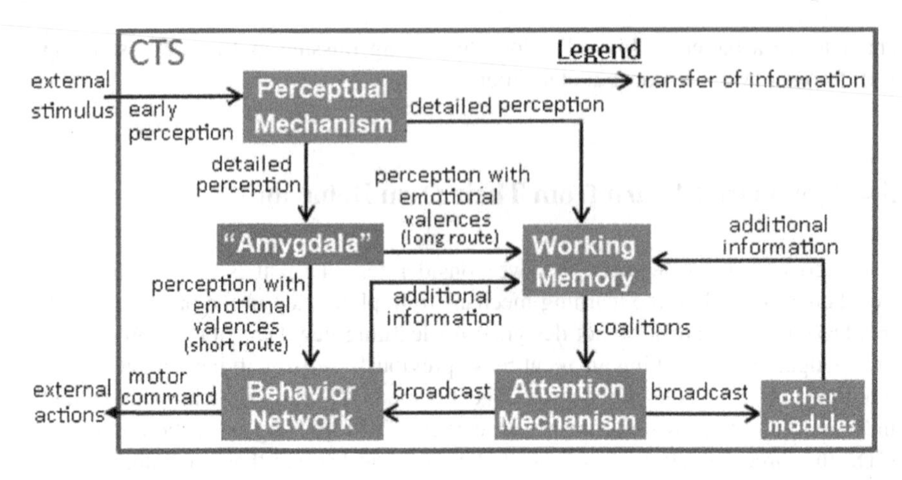

Fig. 5.5: A simplified view of the CTS architecture (see [6] for more details)

ensures that only the most important, urgent, or relevant information is broadcast in the architecture. Following a broadcast, every subsystem (module or team of codelets) that recognizes the information may react to the coalition by appending additional information to it. This process of internal publications (as suggested by Baars [4]) can continue for many cognitive cycles before an action is executed by CTS. The module responsible for action planning, selection and execution is the Behavior Network (BN) [15]. When the BN receives a broadcast coalition, it selects the appropriate action to execute. In the current CTS version, we have designed the BN using a graphical authoring tool. We have implemented in CTS the second form of learning that we consider in this article. This learning mechanism is implemented in CTS by the three operation phases, described next.

5.4.2 The Observation Phase

In the first phase, the *observation phase*, CTS records a sequence of events (as defined in section 5.2) for each of its executions. Each event $X = (t_i, A_i)$ represents one cognitive cycle. While the timestamp t_i of an event indicates the cognitive cycle number, the set of items A_i of an event contains (1) an item that represents the coalition of information-codelets that was broadcast during the cognitive cycle and (2) four optional items having numeric values indicating the four emotional valences (high threat, medium fear, low threat, compassion) associated with the broadcast coalition[1]. For example, one partial sequence recorded during our experimentation was $<(1,c2),(2,c4),(3,c8\ e2\{-0.4\})>$. This sequence shows that during cogni-

[1] CTS actually incorporates four emotions inspired by the OCC model of emotions [16]. See [6] for in-depth details about the emotional mechanism of CTS.

tive cycle 1 the coalition $c2$ was broadcast, followed by the broadcast of $c4$ during cognitive cycle 2. Furthermore, it indicates that coalition $c8$ was broadcast during the third cognitive cycle and that it generated a negative emotional valence of -0.4 for emotion $e2$ (medium fear).

5.4.3 The Learning Phase

The second operation phase consists of mining frequent patterns from the sequences of events recorded for all executions of CTS by applying our sequential pattern mining algorithm. This process is executed at the end of each CTS execution, from the moment where five sequences are available (five CTS executions). Currently, we have setup the sequential pattern mining algorithm to mine only closed sequences with more than three events and with a support higher than 25%. Applying the algorithm results in a set of frequent sequential patterns.

5.4.4 The Application Phase

The third operation phase consists in improving CTS behavior by making CTS reuse relevant patterns that carry positive emotions. This is done by intervening in the coalition selection phase of CTS. The idea is here to find, during each cognitive cycle the patterns that are similar to CTS's current execution to then select the next coalition to be broadcast that is the most probable of generating positive emotions for CTS according to these patterns. Influencing the coalitions that are broadcast will then directly influence the actions that will be taken by the CTS behavior network. This adaption of CTS could be implemented in different ways. We used the *SelectCoalition* algorithm (algorithm 2), which takes as parameters (1) the sequence of previous CTS broadcasts (*Broadcasts*), (2) the set of frequent patterns (*Patterns*) and (3) the set of coalitions that are candidates to be broadcast during a given cognitive cycle (*CandidateCoalitions*). This algorithm first sets to zero a variable *min* and a variable *max* for each coalition in *CandidateCoalitions*. Then, the algorithm repeats the following steps for each pattern p of *Patterns*. First, it computes the *strength* of p by multiplying the sum of the emotional valences associated with the broadcasts in p with the support of p. Then, it finds all the coalition $c \in$ *Candidate-Coalitions* that appear in p after the sequence of the last k broadcasts of *Broadcasts* for any $k \geq 2$. For each such coalition c, if the strength of p is higher than $c.max$, $c.max$ is set to that new value. If that strength is lower than $c.min$, $c.min$ is set to that new value. Finally, when the algorithm finishes iterating over the set of patterns, the algorithm returns to CTS's working memory the coalition c in *CandidateCoalitions* having the highest positive value for the sum $c.min + c.max$ and where $c.max > 0$. This coalition will be the one that will be broadcast next by CTS's attention mechanism. In the case of no coalition meeting these criteria, the algorithm will return a randomly selected coalition from *CandidateCoalitions* to CTS's working memory.

Algorithm 5: SelectCoalition Algorithm

```
   Input: Patterns, Broadcasts, CandidateCoalitions
   Output: SelectCoalition
 1 foreach pattern c ∈ CandidateCoalitions do
 2 |   c.min := 0;
 3 |   c.max := 0;
 4 end
 5 foreach pattern P of Patterns do
 6 |   Strength := CalculateSumOfEmotionalValences(P) * Support(P);
 7 |   for k := 2 to |P| do
 8 |   |   Sa := last k Broadcasts of Broadcasts;
 9 |   |   if Sa ⊆ P then
10 |   |   |   foreach coalition c ∈ CandidateCoalitions appearing after Sa in P do
11 |   |   |   |   c.max := maxOf(Strength, c.max);
12 |   |   |   |   c.min := minOf(Strength, c.min);
13 |   |   |   end
14 |   |   end
15 |   end
16 end
17 RETURN c ∈ CandidateCoalitions with the largest positive(c.max + c.min) and such that
   c.max > 0.
```

The $c.max > 0$ criterion is included to ensure that the selected coalition appears in at least one pattern having a positive sum of emotional valences. Moreover, we have added the $c.min + c.max$ criterion to make sure that patterns with a negative sum of emotional valences will decrease the probability of selecting the coalitions that it contains. In our experiments, this criterion has proved to be very important as it can make CTS to quickly stop selecting a coalition appearing in positive patterns, if it becomes part of negative patterns. The reader should note that algorithms relying on other criteria could have been used for other applications.

5.4.5 Testing the New CTS in RomanTutor

To test CTS's new learning mechanism, users were invited to perform arm manipulations using RomanTutor with integrated CTS. These experiments aimed at validating CTS ability to adapt its behavior to learners. During these experiments, we qualitatively observed that CTS adapted its behavior successfully to learners. Two experiments are here described. The first describes in details one situation that occurred with User 3 that illustrates well how CTS adapts its behavior thanks to the new learning mechanism. The second experiment describes how the data mining algorithm behaves when the number of recorded sequences increases.

User 3 tended to make frequent mistakes when he was asked to guess the arm distance from a specific part of the space station. Obviously, this situation caused collision risks between the arm and the space station and was thus a very dangerous situation. This situation was implemented in the CTS's Behavior Network. In this situation, CTS has to make a decision between (1) giving a direct solution such as "You should move joint SP" (Scenario 1) or giving a brief hint such as "This movement is dangerous. Do you know why?" (Scenario 2).

During the interaction with different users, the learning mechanism recorded several sequences of events for that situation, each of them carrying emotional valences. The average length of the stored sequences was of 26 events. For example, one partial trace saved when CTS gave a hint (scenario 2) to User 2 was $< (13,c11),(14,c14),(15,c15),(16,c18),\ (17,c19\ e4\{0.8\}) >$. In this trace, the positive valence 0.8 for emotion $e4$ (compassion) was recorded because the learner answered to an evaluation question correctly after receiving the hint. In another partial trace saved by CTS $< (16,c11),(17,c14),(18,c16),\ (19,c17),\ (20,c20\ e2\ \{-0.4\}) >$, User 2 received a direct solution from CTS (Scenario 1), but failed to answer correctly an evaluation question. This resulted in the valence -0.4 being associated to emotion $e2$ (medium fear). After five executions, the learning mechanism extracted ten frequent sequences from the recorded sequences, with a minimum support (minsup) higher than 0.25.

Now turning back to User 3, during the coalition selection phase of CTS, the learning mechanism evaluated all mined patterns to detect similar patterns having ended by self-satisfaction. The learning mechanism chose the pattern $< (0,c11),(1,c14),\ (3,c18),\ (4,c19\ e4\{0.8\}) >$, because it contained the most positive emotional valence, had the highest frequency, and the events $(0,c11),(1,c14)$ matched with the latest events executed by CTS. Therefore, CTS chose that it is better to give a hint (Scenario 2) than to give the answer (Scenario 1) to User 3. Concretely, this was achieved by broadcasting coalition $c18$ (Scenario 2) instead of coalition $c16$ (Scenario 1). If the emotional valence had not been as positive as was the case for previous users, CTS might have chosen Scenario 1 rather than Scenario 2. It should be noted that because the set of patterns is regenerated after each CTS execution, some new patterns can be created, while other can disappear, depending on the new sequences of events that are stored by CTS. This ensures that CTS behavior can change over time if some scenarios become less positive or more negative, and more generally that CTS can adapt its behavior to a dynamic environment. In this experiment, the learning mechanism has shown to be beneficial by allowing CTS to adapt its actions to learners by choosing between different scenarios based on its previous experience. This feature is very useful in the context of a tutoring agent, as it allows the designers to include many alternative behaviors but to let CTS learn by itself which ones are the most successful.

We performed a second experiment with the learning mechanism, but this time to observe how the data mining algorithm behaves when the number of recorded sequences increases. The experiment was done on a 3.6 GHz Pentium 4 computer running Windows XP, and consisted of performing 160 CTS executions for a situation similar to the previous one where CTS has to choose between scenario 1 and scenario 2. In this situation, CTS conducts a dialogue with the student that includes from two to nine messages or questions (an average of six) depending on what the learner answers and the choices CTS makes (similar to choosing between scenarios 1 and 2). During each trial, we randomly answered the questions asked by CTS, and took various measures during CTS's learning phase. Each recorded sequence contained approximately 26 broadcasts. Fig. 5.6 presents the results of the experiment. The first graph shows the time required for mining frequent patterns after each CTS execution. From this graph, we see that the time for mining frequent patterns was generally short (less than 6 s) and increased linearly with the number of recorded sequences. In our context, this performance is very satisfying. The second graph shows the average size of patterns found for each execution. It ranges from 9 to 16 broadcasts. The third graph depicts the number of patterns found. It remained low and stabilized at around 8.5 patterns during the last executions. The reason why the number of patterns is small is that we mined only closed patterns (c.f. section 5.3.2). If we had not mined only closed patterns, all the subsequences of each pattern would have been included in the results. Finally, the average time for executing the *SelectCoalition* algorithm at each execution. This time was always less than 5 milliseconds. Thus, the costliest operation of the learning mechanism is the learning phase.

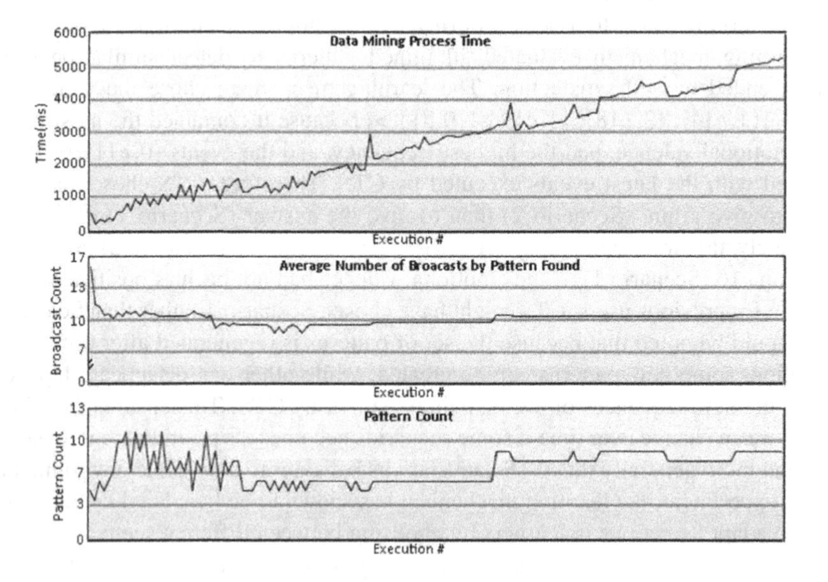

Fig. 5.6: Results from second experiment

5.5 Conclusion

In this chapter, we presented the idea of building agents that learn by extracting temporal patterns from their own behavior or the behavior of other agents. To demonstrate this idea, we proposed two learning mechanisms. While the first learning mechanism is aimed at learning new procedural knowledge by observing other agents performing a task, the second one is designed for making an agent reuse its self-satisfying behaviors. We presented each learning mechanism through the case study of a virtual agent. The two virtual agents were tested in real applications and experiments have shown positive results, as regards the capability of the agents to adapt their behavior and the performance of the learning mechanisms. The two learning mechanisms should be reusable in other agents and contexts, as they make little assumption on the architectures of the observed agents and their decision making processes, and the format for encoding behaviors is fairly generic.

In future work, we will perform further experiments to measure empirically how the virtual agents influence the learning of students. We will investigate different ways of improving the performance of our sequential pattern mining algorithm, including modifying it to perform an incremental mining of sequential patterns. We also plan to compare the two learning mechanisms with others agent learning mechanisms, and to integrate the two virtual agents in other tutoring systems.

Acknowledgements

The authors thank the Canadian Space Agency, the Fonds Québécois de la Recherche sur la Nature et les Technologies (FQRNT) and the Natural Sciences and Engineering Research Council (NSERC) for their logistic and financial support. The authors also thank current and past members of the GDAC and PLANIART teams who participated in the development of RomanTutor.

References

1. Cao, L., Luo, C., Zhang, C.: Agent-Mining Interaction: An Emerging Area. Proc. AIS-ADM 2007, LNAI 4476, pp. 60-73, 2007. Springer-Verlag, Berlin.
2. Symeonidis, A. L., Athanasiadis, I. N., Mitkas, P. A.: A retraining methodology for enhancing agent intelligence. Knowledge-Based Syst. **20**(4), 388-396 (2008)
3. Agrawal, R., Srikant, R.: Mining sequential patterns. Proc. Int. Conf. Data Eng., pp. 3-14 (1995)
4. Baars, B.J.: In the theater of consciousness. Oxford Univ. Press, Ofxord. (1997)
5. Dubois, D., Poirier, P., Nkambou, R.: What does Consciousness bring to CTS?. Proc. 2007 AAAI Fall Symp., pp. 55-60, AAAI Press (2007)
6. Faghihi, U., Poirier, P., Dubois, D., Gaha, M., Nkambou, R.: How emotional mechanism learn and helps other types of learning in a cognitive agent. Proc. WI-IAT 2008 (2008)
7. Fournier-Viger, P., Nkambou, R., Mayers, A.: Evaluating spatial representations and skills in a simulator-based tutoring system. IEEE Trans. Learn. Technol. **1**, pp. 63-74 (2008)

8. Fournier-Viger, P., Nkambou, R., Mephu Nguifo, E.: A knowledge discovery framework for learning task models from user interactions in intelligent tutoring systems. Proc. 7th Mex. Int. Conf. Artif. Intell., LNAI 5317, pp. 765-778. Springer (2008)
9. Franklin, S., Patterson, F.G.J.: the LIDA architecture: adding new modes of learning to an intelligent, autonomous, software agent. Proc. IDPT-2006 (2006)
10. Han, J., Kamber, M.: Data mining: concepts and techniques, Morgan Kaufmann Publ., San Franc. (2000)
11. Hirate, Y., Yamana, H.: Generalized sequential pattern mining with item intervals. J. **1**(3), 51–60 (2006)
12. Hofstadter, D.R., Mitchell, M.: The Copycat project: a model of mental fluidity and analogy-making. In: Barnden, J., Holyoak, K., (eds.) Advances in connectionnist and neural computation theory, pp. 31-113. Lawrence Erlbaum Associates (1992)
13. Kabanza, F., Nkambou, R., Belghith, K.: Path-planning for autonomous training on robot manipulators in space. Proc. 19th Int. Joint. Conf. Artif. Intell., pp. 1729-1731 (2005)
14. Lattner, A.D., Miene, A., Visser, U., Herzog, O.: Sequential pattern mining for situation and behavior prediction in simulated robotic soccer. Proc. Robocup 2005 Conf., pp. 118-129 (2005)
15. Maes, P.: How to do the right thing. Connect. Sci. **1**, 291–323 (1989)
16. Ortony, A., Clore, G.: cognitive structure of emotion. Cambridge Univ. Press, Camb. (1988)
17. Pinto, H., Han, J., Pei, J., Wang, K., Chen, Q.: Multi-dimensional sequential pattern mining. Proc. 10th Int. Conf. Inf. Knowl. Manag., pp. 81-88 (2001)
18. Songram, P., Boonjing, V., Intakosum, S.: Closed multidimensional sequential pattern mining. Proc. 3rd Int. Conf. Inf. Techn.: New Gener., pp. 512-517 (2006)
19. van Lent, M., Laird, J.E.: Learning procedural knowledge through observation. Proc. K-CAP 2001, pp. 179-186 (2001)
20. Wang, J., Han, J., Li, C.: Frequent closed sequence mining without candidate maintenance. IEEE Trans. Knowl. Data Eng. **19**(8), 1042–1056 (2007)

Chapter 6
A Multi-Agent System for Extracting and Analysing Users' Interaction in a Collaborative Knowledge Management System

Doina Alexandra Dumitrescu, Ruth Cobos and Jaime Moreno-Llorena

Abstract In this paper we present a Multi-Agent System (MAS) for extracting and analysing users interaction in a Collaborative Knowledge Management system called KnowCat. The proposed MAS employs Web Use Mining and Web Structure Mining techniques in order to detect the most relevant interactions of the KnowCat users and therefore should have more weight in the Knowledge Crystallization mechanism of KnowCat. More concretely, the MAS extracts the users interaction information and analyses whether they are working in the system in an organised or disorganised way. The obtained results in this study give us evidence that organised users contribute positively in a good performance of the Knowledge Crystallization mechanism of KnowCat.

6.1 Introduction

It is well known that users has a great interest and utility in most computer systems, for instance to evaluate their interface usability [2] or to adapt the system to each user. In the computer supported cooperative work systems (CSCW) the takes a special relevance, since it can help in the study of the collaborative tasks carried out [11, 17] and in the analysis of the achieved individual results by each user [10].

The problem involves diverse aspects among those that highlight the Data Mining, to process the activity registers generated by the users and to figure out structural aspects of the interface. The solution for this problem may also profit from the use of intelligent agents, which collaborate among them in order to carry out the analysis in an autonomous, flexible and less intrusive way regarding the involved systems. The low-level events that are generated by many user interfaces to analyze

Doina Alexandra Dumitrescu, Ruth Cobos and Jaime Moreno-Llorena
Department of Computer Science, Universidad Autónoma de Madrid, Spain, 28049 Madrid
Telephone: +34914972243, Fax: +34914972235
e-mail: (doina.dumitrescu, ruth.cobos, jaime.moreno)@uam.es

L. Cao (ed.), Data Mining and Multi-agent Integration, DOI: 10.1007/978-1-4419-0522-2_6,

the users interaction are also a potent and interesting alternative, mainly because it deals with part of the problem in a generic way. In great detail, many researches shows us how Multi-Agent approaches are used in the task of analysing learner's interactions in collaborative learning systems in order to improve the collaboration (i.e. [14, 16]).

In this paper the interaction of the users is studied using the Collaborative Knowledge Management system called KnowCat [1, 5]. KnowCat (acronym for "Knowledge Catalyser") is based on a mechanism called "Knowledge Crystallisation that gives us evidence about which the best contributions in the users opinion are through their interaction with the system.

In this paper, it is presented the design of a Multi-Agent system (MAS) whose purpose is to extract information on how users interact with the KnowCat system and afterwards analyse that information. The main aim of the MAS is to find what the most relevant interactions of the users are and therefore should have more weight in the Knowledge Crystallization mechanism of KnowCat.

6.2 Related Work

Agents can have a multitude of features [6, 18], but we will mention here only two of the most important: the autonomy and being part of a community of agents, characteristics that distinguish agent-based applications from the rest. , a multitude of agents that interact in order to solve problems, are a suitable approach when dealing with complex software systems that are distributed and dynamic.

For some time a great emphasis was given to the integration of agents in collaborative learning systems in order to improve the effectiveness of the learning efficiently . Furthermore represent a powerful approach that based on the main features can support the collaborative learning systems [19]. Suh and Lee [16] present a multi-agent framework where the monitoring agents collect, analyse and process the information of the user' collaborative learning activities and the facilitator agents analyse this information in order to offer learning advice. Letizia [13] is an intelligent agent that observes the users browsing behaviour in order to suggest interesting web links for the user. WebWatcher [12] based on the users behaviour and making use of learning techniques recommends the best web resource related to the users keywords. Another example of web mining agent is WebAce [9] that suggests possible interesting new pages for a user based on his browsing activity.

All these intelligent agents make use of Web Mining techniques, where Web Mining is the result of applying techniques of Data Mining to the discovery of patterns on the Web. This is in its content, usage logs or structures [8, 15, 4, 20]. Generally, Web Use Mining techniques are employed, allowing the discovery of usage patterns in the Web Logs, and Web Structure Mining techniques are also used, facilitating patterns discovering in the links structure of the Web and permitting internal structure analysing the Web sites.

The interest concerning the users activity has been fundamental in learning based on technologies context, mainly from the introduction of collaborative support in this area. Some business initiatives are also interesting, such as the techniques used by the advertising system DoubleClick (`http://www.doubleclick.com`), the distributed computing model HWME designed for macroeconomic decision-support based on intelligent information agents in [3] or the recommendation applications that monitor the user's activity and are becoming more and more common.

6.3 The proposal Context: the KnowCat System and its Client Monitor

KnowCat is a Collaborative Knowledge Management Web-based system, which is grounded on a client/server architecture and permits multiple instances of the system, called KnowCat nodes. Each KnowCat node deals with a knowledge area, and has its own knowledge repository and its own users community. The knowledge repository is fundamentally composed of these two elements: Documents, that are the atomic knowledge units in the system; and Topics, that are hierarchically organised in a Knowledge Tree. The aim of each document is to describe the topic where it is located in the knowledge tree. More concretely, when a document is added in a KnowCat node it competes against the others to become the best description on its location (topic). This competitive environment is achieved by the Knowledge Crystallisation mechanism of KnowCat (see [5] for details).

Several previous studies carried out with KnowCat have corroborated that KnowCat encourages communities to share their knowledge and, progressively, construct knowledge sites of reasonable quality [5]. The KnowCat system has been extended with a Client Monitor (CM). The aim of this extension, CM, is to store both the user data activity on the client side and on the server side in the server Web LOG file of KnowCat.

CM carries out a low-level users activity monitoring, paying attention to certain events that are considered activity indicators, such as application focus obtaining and losing, windows scrolling and keyboard pulsation and mouse movements and clicks. CM records the user activity observed on the client side in the server Web LOG file of KnowCat periodically. In the next section a more detailed view of the Web LOG will be presented. Due to this enriched Web LOG file and with the use of both Usage Web Mining and Web Interface Structure Mining techniques the users interaction with KnowCat will be deeply analysed.

6.4 A Multi-Agent System for Extracting and Analysing Users' Interaction

The main aim of the project presented in this paper is to bring out the design, development and experimental performance of a system formed of intelligent agents [18], whose purpose is to extract information on how users interact with the KnowCat system and afterwards analyse that information and report how each user behaves on certain characteristics (e.g. how an organised user is in his/her work with the system) in order to detect what interactions of the users are the most relevant ones and therefore should have more weight in the Knowledge Crystallization mechanisms of KnowCat.

A prototype of this proposed Multi-Agent system (called MAS-IA, Multi-Agent System for Interactions Analysis), which addresses whether a user is organised or not in his/her work with the KnowCat system, was designed and implemented with the idea of being extended with more agents that can be in charge of the analysis of other characteristics of the users work with the system (i.e. motivation, interest, etc.).

The technology used to implement the MAS-IA is JADE mainly because it is a set of open source libraries written in Java that offer support for the development of multi-agent systems and presents a series of important features in our case, such as: portability, openness, high scalability and a built-in messaging system. Additionally, the platforms implemented in JADE can be distributed on several machines with different operating systems [7], attribute that is useful because KnowCat allows having the KnowCat nodes distributed in different machines. Furthermore, Jade offers various FIPA-specified interaction protocols in order to support the agent communication.

The proposed MAS-IA integrates the following agent types:

- Data Extractor Agent (DEA), whose role is to obtain interaction information from the enriched Web LOG file of KnowCat for a certain user. This information is represented in an ontology that serves as means of communication between this agent and the following one.
- Organised Behaviour Interpreter Agent (OBIA), whose role is to decide whether a certain user is organised or not in his/her work with KnowCat. The organised characteristic is weighted by a numeric value that we call OrganisedValue (OV), between 0 and 1, where 0 means that the user is totally disorganised and 1 means he/she is very organised.

An overview of the process carried out is illustrated in the Fig. 6.1 left-hand side and in right-hand side we can see the ontology used to store the information extracted from users'interactions.

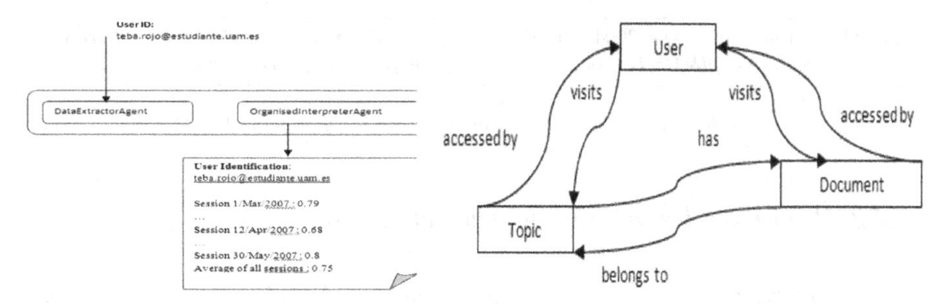

Fig. 6.1: Overview of the process conducted by the Agents (left-hand side) and the ontology used by the Agents (right-hand side).

6.4.1 Data Extractor Agent

The Web LOG file of a Web server has a line for each accessed resource in chronological order. The content of the lines of the LOG file is fixed by the server configuration, but generally they contain standard data (see Fig. 6.2). The DEA deals

```
62.151.101.192 - -[20/Apr/2004 : 20 : 02 : 47 + 0200]"GET /5.0a/controlNT2.pl?b =
KC_SOG2HTTP/1.1"...
62.36.67.34 - -[20/Apr/2004 : 20 : 02 : 35 + 0200]"GET /img/version.gifHTTP/1.1"...
```

Fig. 6.2: Typical lines LOG web server.

with the activity Web LOG file of KnowCat enriched by CM. CM registers some specific lines interspersed among the lines in the server LOG file. These LOG lines, which are registered by the MC, add to the standard LOG lines interaction information of the users work on the client side. As we can see in Fig. 6.3, this recorded activity is composed of three kinds of data: (i) identification data as field UsrID that shows the user identification; (ii) temporal data, as CntIni and NtfTmp that indicate the start registering time and the notification instant respectively; and (iii) activity data, as MseM, MseD, Scr, Fcs, Blr and KeyD, which are the numbers of the different events through out the time interval, respectively mouse movements and clicks, scroll movements, focus obtaining and losing, and Keyboard pulsations.

Furthermore, the DEA, that has as input the identification of a certain user, makes use of Web usage Mining techniques in order to analyse the LOG file syntactically and extract all the data related with the interactions of that user. In other words, the DEA identifies the users access to the knowledge elements of KnowCat (documents and topics) taking into account the chronological order of the extracted data and the structure information of the KnowCat interface, for this purpose this agent makes use of Interface Structure Mining, techniques, too.

$213.37.225.150 - -[20/Apr/2004:20:02:35+0200]"GET/5.0a/informSituacion.pl?urlrsp=$
$../5.0a/monitorKC.pl\&MseD = 0\&Scr = 0\&Fcs = 0\&Blr = 0\&KeyD = 0HTTP/1.1"...$

Fig. 6.3: An example line in web Log file of KnowCat generated by MC.

6.4.2 Organised Behaviour Interpreter Agent

Beforehand we realised an inspection of the enriched Web LOG of KnowCat for some controlled users. Knowing the behaviours and the organisation of the work of these users with KnowCat, certain patters were detected, patters that were based in the following two issues:

- Repeated accesses to an element (topic or document) by a user. A repeated access to a selected element means that this element is accessed more than once in a user session, in other words between the accesses to the selected document the user has accesses to other elements.
- Time access to documents (which is the atomic knowledge unit in KnowCat) by a user.

The characteristic of being organised or not with KnowCat is directly related with the first issue: repeated accesses to KnowCat knowledge elements. In order to calculate the OV for a certain user session, the OBIA calculates the following values whether the user has repeated accesses to topics and/or documents in the session:

$$RP_{topics} = tnrat/tnat \quad RP_{docs} = tnrad/tnad \quad RP = (RP_{topics} + RP_{docs})/2 \quad (6.1)$$

Where the repetition percentage to accessed topics, RP_{topics}, is the total number of repeated accessed topics, tnrat, divided by total number of accessed topics, tnat. Similarly it is defined the repetition percentage to accessed documents, RP_{docs}. And the repetition percentage of a user in a session is RP.

The agent distinguishes among these three cases: i) disorganised users (who repeat access to topics and documents): they receive an OV between 0-0,4 (0 means that the user is totally disorganised in his/her work with the system, and 0,4 signifies that it is getting a medium organised behaviour); ii) medium organised users (who repeat access to topics or documents): they receive a OV between 0,4-0,8 (in the realised inspection, we have noticed that these intermediate values correspond to the behaviours of the users that are not very organised and in the same time not very disorganised); iii) organised users (who dont repeat access to topics or documents): they receive an OV between 0,8-1 (1 means that the user is totally organised with his/her work with KnowCat).

Moreover, the OBIA takes into account the average access time to documents in the calculation of the OV for a certain user session. This value is, called t_{avg}, and it is used by the agent to distinguish for each previous mentioned case into these three subcases: t_{avg} is lower than 10 seconds (very short access duration to documents),

t_{avg} is higher than 60 seconds (very long access duration to documents) and t_{avg} is between 10 an 60 seconds.

The calculation of the OV for a user session is based on the algorithm shown in Fig. 6.4 (t_{avg} is in seconds):

If $(repeateddocumentsandtopics)$ then $iniValue = 0$
Else if $(repeateddocumentsortopics)$ then $iniValue = 0.4$
In other case then $iniValue = 0.8$
If $(0.76 \leq RP \leq 1)$ then
  If $(t_{avg} < 10)$ then $OrganisedValue = iniValue$
  Else if $(t_{avg} > 60)$ then $OrganisedValue = iniValue + 0.1$
  Else $OrganisedValue = iniValue + ((t_{avg} - 10)/500)$
Else If $(0.51 \leq RP \leq 0.75)$ then
  If $(t_{avg} < 10)$ then $OrganisedValue = iniValue + 0.1$
  Else if $(t_{avg} > 60)$ then $OrganisedValue = iniValue + 0.2$
  Else $OrganisedValue = iniValue + 0.1 + ((t_{avg} - 10)/500)$
Else If $(0.26 \leq RP \leq 0.50)$ then If $(t_{avg} < 10)$ then $OrganisedValue = iniVale + 0.2$
  Else if $(t_{avg} > 60)$ then $OrganisedValue = iniValue + 0.3$
  Else then $OrganisedValue = iniValue + 0.2 + ((t_{avg} - 10)/500)$
Else If $(0.1 \leq RP \leq 0.25)$ then If $(t_{avg} < 10)$ then $OrganisedValue = iniValue + 0.3$
  Else if $(t_{avg} > 60)$ then $OrganisedValue = iniValue + 0.4$
  Else then $OrganisedValue = iniValue + 0.3 + ((t_{avg} - 10)/500)$
Else If $(RP = 0)$ then If $(t_{avg} < 10)$ then $OrganisedValue = iniValue$
  Else if $(t_{avg} > 60)$ then $OrganisedValue = iniValue + 0.2$
  Else then $OrganisedValue = iniValue + ((t_{avg} - 10)/500)/2$

Fig. 6.4: The calculation of the *OrganisedValue* for a user session.

6.5 Experimentation and Results

In order to test the proposed approach, we have carried out a research study with a community of 120 students enrolled in an "Information Systems course at the Computer Science Department, at Universidad Autónoma de Madrid (UAM). This study was carried in the last three months of the course. The students tasks during this period were executed in the following two phases:

- Creation Phase: in the first month they had to contribute with 2 documents in two assigned topics. They participated like *knowledge creators*.
- Evaluation Phase: in the following 2 months they had to evaluate (with votes and annotations) all documents of others 3 assigned topics. They participated like *knowledge evaluators*.

Our study began with the following two initial research questions:

1. Are the students organised when they are interacting with the KnowCat system? How organised were the students like knowledge creators and like knowledge evaluators?

2. How does this influence the way in which students interact with the knowledge crystallisation mechanism? We wanted to know whether organised student interactions with the system had any influence in the successfulness of the knowledge crystallisation mechanism.

MAS-IA calculated for all students the following values: i) the OV like a knowledge creator: the average value of this characteristics taking into account their interactions during the creation phase; ii) the OV like a knowledge evaluator: the average value of this characteristics taking into account their interactions during the evaluation phase; and iii) the OV global: the average value of this characteristics taking into account their interactions during the whole study.

At the end of the research study, course instructors evaluated the students work in the following way: i) at first, they assigned a qualification to each student like a knowledge creator, taking into account the quality of their documents; and ii) secondly, instructors assigned a qualification to each student like a knowledge evaluator taking into account the quality of their evaluations. The qualification in both cases was a grade between 0-10, where 10 is the maximum value.

The OV global was the most significant value in order to know whether a student worked in an organised way with the system during the study. Moreover, we compared the other two calculated values and had evidence that 90% of students received a higher OV like a knowledge creator than like a knowledge evaluator. This result was expected because in the second task the students had to think about their contributions and compare them with classmates contributions, and they considered to be necessary to access to previously visited topics and documents several times.

On the one hand, most of the students with a high qualification like evaluators (greater than 7.5 in the instructors opinion), received an OV global higher than 0.5, therefore they were considered organised students. On the other hand, most of the students with a low qualification like evaluators (less than 7.5 in the instructors opinion), received an OV global lower than 0.5, therefore they were considered disorganised students.

Testing whether the organised characteristic had any influence on the successfulness of the knowledge crystallisation mechanism, the following process was made to each topic of the knowledge tree: i) a ranking of the documents was created taking into account the qualifications assigned to them by the instructors, ii) these rankings were compared with the classification offered by the KnowCat system through its Knowledge Crystallisation mechanism, which is based in the students opinions.

In 50% of topics both rankings were very similar (coincidently), all the students, who participated in these topics like evaluators, received a very high qualification like knowledge evaluators in the instructors opinion. Moreover, 90% of these students received an OV global higher than 0.5.

In 30% of topics the rankings were very different, which give us evidence that the instructors were in disagreement with the students opinions about the documents contained in these topics. 40-50% of the students, that participated in these topics like evaluators, received a qualification like a knowledge evaluator lower than 7.5 by the instructors. Moreover, 100% of these students received an OV global lower than 0.5.

The previous obtained results corroborated, at first, that the students are more organised like knowledge creators than like knowledge evaluators, and secondly, that organised students were good evaluators. Furthermore, this second result give us evidence that the perception from instructors about which student is a good or a bad evaluator is directly related to the detection by the MAS-IA about what student works in an organised way or not.

In consequence, the most important outcome of this research study was that the quality of students evaluations is directly related to whether they are organised or disorganised and this could be used to improve the Knowledge Crystallisation mechanism.

6.6 Conclusions and Future Work

In this work we have presented a Multi-agent System, called MASIA, whose purpose is to extract and analyse information about how a group of users of a Collaborative Knowledge Management system, called KnowCat, is interacting with the community knowledge. More concretely, the MAS-IA extracts the users interaction information and analyses whether they are working in the system in an organised or disorganised way.

The MAS-IA has been designed to deal with agents that can operate in both tasks: users interaction extraction and interactions analysis. The proposed approach can determine per each KnowCat user its OV for some selected sessions executed during his/her work with the system.

This MAS-IA has been tested in a research study with a community of 120 students at Universidad Autnoma de Madrid (UAM). They had to work like knowledge creators and knowledge evaluators.

The obtained results in this research study were: i) generally students are more organised like knowledge creators than like knowledge evaluators and ii) quality of students evaluations is directly related to whether they are organised or disorganised. These results corroborated the necessity of analysing the users interactions of the KnowCat system, in order to extract what the most relevant interactions are of the users and therefore should have more weight in the Knowledge Crystallization mechanism of KnowCat.

Nowadays, we are working in these areas. At first, we are designing new research studies in order to corroborate and generalise the results obtained in the research study presented in this paper (e.g. use of other educational contexts). Secondly, we are planning to modify and improve the Knowledge Crystallisation mechanisms through the use of the organised/disorganised characteristics and other possible characteristics that we will be able to obtain in the following research studies. Finally, we are designing and implementing a new set of agents that will be able to deal with more characteristics that have special relevance in the CSCW systems, such as motivation or implication level in the collaborative task.

Acknowledgements This research was partly funded by the Spanish National Plan of R+D, project number,TIN2008-02081/TIN; and by the AECID (The Spanish Agency for International Development Cooperation) project number A/017436/08.

References

1. Alamán, X., Cobos, R. KnowCat: a Web Application for Knowledge Organization. In P.P. Chen and al.(Eds.). LNCS 1727. Springer, 348-359. 1999.
2. Atterer, R., Wnuk, M., Schmidt, A., Knowing the User's Every Move - User Activity Tracking for Website Usability Evaluation and Implicit Interaction. In Procs. WWW2006, pp. 2-9, Edinburgh, 2006.
3. Cao, L., Dai, R. Human-Computer Cooperated Intelligent Information System Based on Multi-Agents, ACTA AUTOMATICA SINICA, 29(1):86-94, 2003, China
4. Chang, G., Healey, M., McHugh, J.,Wang, J., Mining the World Wide Web: An Information Search Approach, Kluwer Academic Publishers, 2001.
5. Cobos, R. Mechanisms for the Crystallisation of Knowledge, a proposal using a collaborative system. Doctoral dissertation. Universidad Autónoma de Madrid. 2003.
6. Franklin, S. and Graesser, A. Is it an agent, or just a program? A taxonomy for autonomous agents. In Procs. of the Third International Workshop on Agent Theories, Architectures, and Languages, pp 443-454 Spring-Verlag. 1996.
7. García, G., Cobos, R, ESMAP: A Multi-Agent Platform for Extending a Knowledge Management System. In Nishida, T, et.al. (eds) The 2006 IEEE/WIC/ACM WI. Hong Kong, December 2006. 59-65. 2006.
8. Garofalakis, M. N., Rastogi, R., Seshadri, S., Shim, K. Data mining and the Web: Past, present and future. In Procs. WIDM99, 43-47, KC, Mssouri, 1999.
9. E. Han, D. Boley, M. Gini, R. Gross, K. Hastings, G. Karypis, V. Kumar, B. Mobasher, and J. Moore. Webace: a web agent for document categorization and exploration. In Procs. of the 2nd AA, pp 408-415. ACM Press, 1998.
10. Heraud, J., France, L., and Mille, A. Pixed: an ITS that guides students with the help of learners' interaction log. In Procs. of the ITS, Maceio. 57-64. 2004.
11. Inaba, A. Ohkubo, R. Ikeda, M. Mizoguchi, R, An Support System for CSCL: An Ontological Approach to Support Instructional Design Process, In Procs. of ICCE'02, 358-362,2002
12. T. Joachims, D. Freitag, and T. Mitchell. Webwatcher: A learning apprentice for the world wide web. In Procs. of IJCAI97, August 1997.
13. H. Lieberman and D. Maulsby. Instructible agents: Software that just keeps getting better. IBM Systems Journal, 35(3&4), 1996.
14. Marc-André, Bélanger, S., and Frasson, C. WHITE RABBIT - Matchmaking of User Profiles Based on Discussion Analysis Using Intelligent Agents. In Procs. of the 5th ITS. Vol. 1839. Springer-Verlag, London, 113-122. 2000
15. Srivastava, J., Cooley, R., Deshpande, M., Tan, P. N. Web usage mining: Discovery and applications of usage pattern from Web data. SIGKDD Explorations, 1(2):1-12, 2000.
16. Suh, H. J. and Lee, S. W. Collaborative learning agent for promoting group interaction. ETRI Journal, 28(4), 461-474.
17. Voyiatzaki, E., Margaritis, M., Avouris, N., Collaborative : The Teachers' Perspective. In Procs. of the 6th IEEE ICALT, 345-349. 2006.
18. Wooldridge, M., Jennings, N.R. Intelligent Agents: Theory and Practice. In Knowledge Engineering Review, 10(2), 115-152, 1995.
19. Yacine, L. and Tahar, B. (2007). Learners assessment in a collaborative learning system. Asian Journal of Information Technology, 6(2), 145-153.
20. Zhang, C., Zhang, Z-. Cao, L. Agents and Data Mining: Mutual Enhancement by Integration, LNCS 3505, 2005.

Chapter 7
Towards Information Enrichment through Recommendation Sharing

Li-Tung Weng, Yue Xu, Yuefeng Li and Richi Nayak

Abstract Nowadays most existing recommender systems operate in a single organisational basis, i.e. a recommender system recommends items to customers of one organisation based on the organisation's datasets only. Very often the datasets of a single organisation do not have sufficient resources to be used to generate quality recommendations. Therefore, it would be beneficial if recommender systems of different organisations with similar nature can cooperate together to share their resources and recommendations. In this chapter, we present an Ecommerce-oriented Distributed Recommender System (EDRS) that consists of multiple recommender systems from different organisations. By sharing resources and recommendations with each other, these recommenders in the distributed recommendation system can provide better recommendation service to their users. As for most of the distributed systems, peer selection is often an important aspect. This chapter also presents a recommender selection technique for the proposed EDRS, and it selects and profiles recommenders based on their stability, average performance and selection frequency. Based on our experiments, it is shown that recommenders' recommendation quality can be effectively improved by adopting the proposed EDRS and the associated peer selection technique.

7.1 Introduction

Recommender systems are being applied in an increasing number of ecommerce sites to increase their business sales by helping consumers locate desired items to purchase. Generally, recommender systems make recommendations to users based on their implicit or explicit preferences, the preferences of other users, and item and user attributes [14, 13]. [3] suggested five different categories of recommender systems based on the information resources and the prediction algorithm employed.

Queensland University of Technology
Brisbane, QLD 4001, Australia

L. Cao (ed.), Data Mining and Multi-agent Integration, DOI: 10.1007/978-1-4419-0522-2_7,

Among these five categories, collaborative filtering and content based filtering are the two most recognized and widely applied techniques.

Collaborative filtering based methods [7, 19] take the preferences of users (e.g. user ratings and purchase histories) as the major information resources in order to aggregate opinions from users with similar preferences, and the recommendations are generated based on these aggregated opinions. On the other hand, content-based filtering methods use information retrieval related techniques to recommend items that have similar contents (or attributes) to the user preferred items.

One of the most well-known challenges for recommender systems is the cold-start problem. The cold-start problem occurs when making recommendations for new users with their preferences unknown (i.e. lack of previous rating information or transaction histories), or suggesting new items that no one has yet rated or purchased [15]. Collaborative filtering based recommenders are very vulnerable to the cold-start problem because they operate solely on the basis of the user preference information, and therefore many works propose the so called "hybrid recommenders" that combine both content-based filtering and collaborative filtering together to replenish the insufficient user preference information [3, 16].

Even though hybridization based recommenders have been widely applied against the cold-start problem, they are still not comparable to using non-hybrid recommenders with sufficient information resources [3]. For instance, assuming a new ecommerce site wants to run a recommender system with very limited user records and product catalogue in its database, it will not be able to generate quality recommendations despite the hybridization techniques are employed.

In this chapter, a novel strategy for alleviating the cold start problem is explored. The basic idea of the strategy is to increase data volume of recommenders via allowing them to share and exchange recommendations with each other over a distributed environment. As mentioned previously, most of the existing recommender systems are designed for one single organisation (i.e. business to customer (B2C) recommenders), and in general, one single organisation may not possess sufficient information or data for analysis in order to give their customers precise and high quality recommendations. Therefore, it can be beneficial if organisations can share their information resources (i.e. products and customer database) and recommendations boundlessly (i.e. build recommendation systems at an interorganisational level).

This chapter presents a framework for distributed recommendation sharing among recommenders, namely Ecommerce-oriented Distributed Recommender System (EDRS). The proposed EDRS is different from existing distributed recommender systems. While existing distributed recommender systems are mainly designed for C2C (Customer to Customer) based applications (such as file sharing applications), the proposed EDRS introduces additional B2B (Business to Business) features on top of the standard B2C recommender systems. Specifically, the goal of the EDRS is to allow the standard recommenders from existing ecommerce sites or e-shops (e.g. Amazon.com, Netflix.com) to improve their recommendation quality towards their users by sharing their information resources and recommendations with each other.

7.2 Prior and Related Work

Notwithstanding the popularity of centralised recommenders in last decades, recommender systems that operate on distributed environments or decentralised infrastructures have started to attract attention from researchers, and these systems are commonly referred to as distributed recommender systems or decentralised recommender systems [4, 10].

Generally, a distributed recommender system associates each of its users with a recommender agent (or peer recommender) on his or her personal computer (client-side machine). These recommender agents gather user profile information from their associated users, and exchange these profile information with other agents over a distributed network (e.g. internet), in the end a recommender agent makes recommendations to its associated user by utilizing the user's personal profile as well as these gathered peer profiles (i.e. profiles of other users gathered from other recommender agents) [17, 18].

There are several reasons that lead to increasing popularity of distributed recommender systems:

- The fast growing development of internet related technologies and applications (e.g. the Grid, ubiquitous computing, peer-to-peer networks for file sharing and collaborative tasks, Semantic Web, social communities, WEB 2.0, etc.) has yielded a wealth of information and data being distributed over most of nodes (i.e. web server, personal computer, mobile phone, etc.) in the internet. Hence, getting information recommended from only one single source (e.g. ecommerce site) is no longer sufficient for many users, and instead, they are thirsty for richer information from multiple sources [17]. For example, the peer-to-peer (P2P) based file sharing protocol, BitTorrent (www.bittorrent.com), has proven to be among the most competent methods to allow large numbers of users to efficiently share large volumes of data. Instead of storing files or data in a central file server (e.g. FTP server), BitTorrent stores files in multiple client machines (i.e. peers), and when a file is requested by a user (i.e. a peer), the user can download this file simultaneously from multiple peers [4]. Intuitively, as there is no central server for storing file contents and user (or peer) profiles in BitTorrent, distributed recommender systems would be more suitable to be applied to such system than centralised recommenders.
- User privacy and trust is another area that distributed recommender systems are considered superior to centralised recommender systems. In a centralised recommender system, all user information and profiles are possessed by the ecommerce site that runs the recommender system, and this can result in privacy and trust concerns. Firstly, a centralised recommender system might share users' personal information and profile in inappropriate ways (e.g. selling user information to others), and the users generally have no control over it. Secondly, a centralised recommender system owned by an ecommerce site might make recommendations for the business's own good instead of serving users' needs. For example, a

site can adjust its recommender's configuration, so it only recommends products that are overstock instead of required by the users.

- The privacy and trust issues are alleviated by distributed recommender systems. In a distributed recommender system, users' personal information and profiles are stored in their own machines, and they generally can explicitly define and set which parts of their personal data and profiles are sharable. In addition, because a recommender agent in a distributed recommender system is a piece of software that runs independently on each client's machine and it usually gathers information only from other peer agents rather than from an ecommerce site, therefore, it is less possible that the ecommerce sites can manipulate the recommendations to the users [12].
- In addition, scalability is one of the major challenges for the centralised recommender systems. It is because correlating user interests in a large dataset can be very computationally expensive (it normally require a quadratic order matching steps). Some research works, therefore, suggest implementing recommender systems in a decentralised fashion to improve the scalability and computation efficiency [12].

Most existing works on distributed recommender systems are mainly designed for peer-to-peer (P2P) or file sharing applications (which usually adhere to C2C paradigm). Awerbuch's [1] work provides a generalized view to these distributed recommenders. Awerbuch suggested a formalized model for the C2C distributed recommender systems. In Awerbuch's model, for the distribute system with m users and n items, there will be m recommender systems (i.e. agents or peers), and each of the recommender agents will associate with exactly one user. Each recommender works on behalf of the associated user either to trade recommendations with other agents or probe the items on its own. Each recommender aims to finally discover the p items preferred by the associated user, where $p \leq n$. In Awerbuch's opinion, from the perspective of the entire distributed recommender system, the goal is rather similar to the "matrix reconstruction" proposed by Drineas et al. [6]; the overall task is to reconstruct an $m \times n$ user preference matrix in a distributed fashion. It can be observed that many distributed recommender systems belong to this model.

Generally, the goal of these C2C based distributed recommenders is to avoid central server failure and protect user privacy (no central database containing information about customers) [1, 18, 16] . However, most of them are not aiming at improving their effectiveness or the recommendation quality. By contrast, the goal of the proposed EDRS is aiming at improving the recommendation quality and alleviating the cold start problem. Hence, the infrastructure of the proposed distributed recommender system is different from Awerbuch's model as well as many other existing systems. EDRS contains a set of classical recommenders, and each of them serves their own set of users. Our goal is to improve the recommendation quality of these recommenders by allowing them making recommendations for others in a decentralised fashion. Thus, for the profiling and selection problem, we proposed a more sophisticated strategy rather than random sampling for recommender peers to explore others.

Moreover, recommender systems and information retrieval (IR) systems are generally considered similar research fields [13], since both of them try to satisfy users' information needs by either retrieving the most relevant documents or recommending the most preferred items to users. Information retrieval retrieves documents based on users' explicit queries, while recommender systems recommend items or products based on users' previous behaviour. In distributed IR [2, 5] , the entire document collection is partitioned into subcollections that are allocated to various provider sites, and the retrieval task then involves:

- Querying minimal number of subcollections (to improve the efficiency), and ensure the selected subcollections are significant to uphold the retrieval effectiveness.
- Merging the queried results (fusion problem) that incorporates the differences among the subcollections in such a way that no decrease in retrieval effectiveness is effectuated with respected to a comparable non-distributed setting.

For distributed recommender systems, the recommender peer selection and recommendation merging are also two important tasks. In fact, one of the major focuses of the works presented in this chapter is to design an effective recommender peer profiling and selection strategy. The selection criteria for distributed IR including the: efficiency (selecting minimal number of subcollections) and effectiveness (retrieving the most relevant documents) is similar to the criteria for the proposed distributed recommender system. However, in distributed IR, the collection selection is content based [5] and it requires the subcollections provide or use sampling techniques to get subcollection index information (eg. the most common terms or vocabularies in the collection) and statistical information (eg. document frequencies). By contrast, the proposed selection technique requires no content related information about recommender peers (assuming recommender peers share minimal knowledge to each other), the proposed selection algorithm is based on the observed previous performance (i.e. how well a recommender peer's recommendations satisfy the users) about each of the recommender peers.

7.3 Ecommerce-Oriented Distributed Recommender System

As mentioned earlier, the goal of the proposed distributed recommender system is to allow standard recommenders to overcome cold-start problem and improve recommendation quality by cooperating, interacting and communicating with recommenders of other parties (e.g. other ecommerce sites). Hence, the proposed system is designed to contain of a set of recommenders from different sites and each of these recommenders is associated with their own users. Note, it is possible that a user might visit multiple sites, and therefore two or more recommenders may share common users. Similar to the centralised paradigm, each recommender peer in the proposed system still serve its own users in a centralised fashion (i.e. the recommender stores all its user and product data in a central place within the recommender). However, in the proposed system, the recommender peers can enrich their information sources by communicating and cooperating with each other. A general overview of the proposed system is depicted in Fig.7.3.

Because the proposed distributed recommender system is designed to benefit ecommerce sites (rather than focusing on helping users to gain more controls on recommenders), we therefore name our system as "Ecommerce-oriented Distributed Recommender System", and abbreviate it to EDRS. We also abbreviate the standard Distributed Recommender System to DRS and Centralised Recommender System to CRS in order to clarify and differentiate the three different system paradigms.

Before explaining the proposed distributed recommender framework in more detail, some general differences among the EDRS, DRS and CRS are investigated. In particular, these systems are compared according to the following aspects:

- **Ecommerce Model:** based on the general ecommerce activities and transactions involved in the recommenders' host application domains, we can roughly categorize them into three different models, namely, Business-to-Business (B2B), Business to Customer (B2C) and Customer to Customer (C2C). In B2B model, activities (e.g. transactions, communications, interactions, etc.) mainly occur among businesses. In the B2C model, activities are mainly between businesses and customers, and the most typical example is activities of E-businesses serving end customers with products and/or services. Finally, the C2C model involves the electronically-facilitated transactions between consumers. A typical example is the online auction (e.g. eBay), in which a consumer posts an item for sale and other consumers bid to purchase it.
- **Architectural Style:** an architectural style describes a system's layout, structure, and the communication of the major comprising system modules (or software components). Over past decades, many architectural styles have been proposed, such as, Client-Server, Peer-to-Peer (P2P), Pipe and Filter, Plugin, Service-oriented, etc. Client-Server and Peer-to-Peer are the two major architectural styles related to our work, and therefore will be explained in more details. The Client-Server architecture usually consists of a set of client systems and one central server system, client systems make service requests over a computer network (e.g. internet) to the server system, and the server system fulfils these requests. Peer-to-Peer architec-ture consists of a set of peer systems interacting with each other over a computer network, and it does not have the notion of clients and servers, instead, all peer systems operate simultaneously as both servers and clients to each other.
- **Communication Paradigm:** based on how two types of entities communicate with each other within a system, three major communication paradigms have been proposed, and they are: One-to-One, One-to-Many and Many-to-Many communication paradigms (or relationships). In One-to-One communication paradigm, communication occurs only between two individual entities, example applications include: e-mail, FTP, Telnet, etc. By contrast, a website that displays information accessible by many users is considered having a One-to-Many relationship. In Many-to-Many paradigm, entities communicate freely

with many others, example applications include: file sharing (multiple users to multiple users), Wiki (multiple authors to multiple readers), Blogs, Tagging, etc.

Figure 7.1 shows a general overview of a standard centralised recommender system (i.e. CRS). The host application of CRS is usually an ecommerce site (e.g. Amazon.com, Netflix.com, etc.) which possesses all user/product relevant information, and the recommender then utilizes all the information from the site to make personalized recommendations to the site's users and further create business values to the ecommerce site. As the nature of the CRS is to serve the users (i.e. customers) and to satisfy the users' information needs to the ecommerce site (i.e. business), it can be considered as adhering to the B2C paradigm. It is usually implemented based on the Client Server architecture because the entire recommendation generation process occurs only within the central server, and users interact with the recommender though thin clients (e.g. web browsers) whose major functions are presenting users the recommendations generated from the server and sending users' information requests to the server. In the most common case, all users of a site are served by a single recommender, therefore, the communication paradigm between recommenders and users in CRS is considered as One-to-Many.

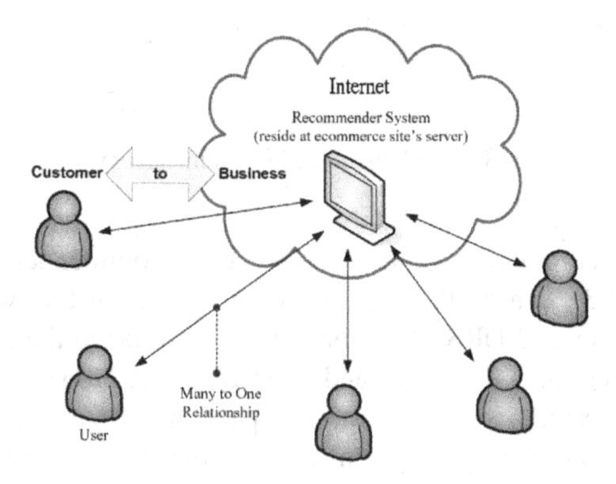

Fig. 7.1: Classical centralised recommender system

The standard distributed recommender system (DRS), as depicted in Fig. 7.2, differs from CRS in all of the three mentioned aspects. First of all, it emphasizes users' privacy protection by preventing personal user data being gathered and used (or misused) by ecommerce site owners (or businesses), hence adheres to the Customer-to-Customer model (as Business entities are evicted from the system for privacy protection). It is shown in Fig. 7.2 that, a standard distributed recommender system associates every user in the system with a recommender peer serving the user's personal information needs, hence the relationship between the user and recommender peer is considered as One-to-One. On the other hand, in order to make better recommendations to its user, a recommender peer might need to communicate with other peers to exchange its user's data (in a privacy protected way) with other peers or to get recommendations from other peers because there is no central place for storing all users' data. The relationship among recommender peers in the DRS is considered as Many-to-Many, as a peer can both communicate to and be communicated by many other peers. Finally, because all recommender peers are equipped with similar set of functionalities (i.e. gather information from others and making recommendation to its user) and operate independently and autonomously from others, therefore they are commonly modelled and implemented using the Peer-to-Peer architectural style.

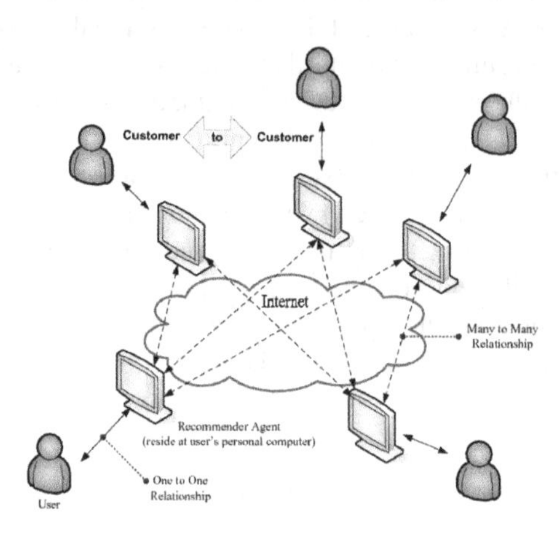

Fig. 7.2: Standard distributed recommender system

The proposed Ecommerce-oriented Distribute Recommender System (EDRS) (depicted in Fig. 7.3), can be thought as a combination of the two systems (centralised recommender and DRS) described above. Similar to the DRS, EDRS consists of a set of recommender peers and a set of users. However, while one user is associated with exactly one recommender peer in the standard distributed recommender system, the proposed system can be considered as a set of centralised recommender systems cooperate together to serve their own set of users, and therefore each recommender peer needs to interact (i.e. make recommendations to) with multiple users. Moreover, it is also possible that in our system a user is associated with more than one recommenders (i.e. he or she can visit multiple sites); for instance, a book reader might try to find a book in both Amazon.com and Book.com. Because a recommender peer in our system can serve multiple users and a user can make recommendation requests to multiple recommender peers, the relationship between users and recommender peers is considered as Many-to-Many. As mentioned previously, the recommender peers in EDRS might interact and cooperate with each other to improve their recommendation quality, and hence, apart from the Many-to-Many relationship between users and recommender peers, another Many-to-Many communication relationship exists among the peers.

Because EDRS is still designed for normal ecommerce sites, such as e-book stores like Amazon.com, its major ecommerce model is therefore same as CRS, that is, Business-to-Customer. Besides, since EDRS introduces additional commu-

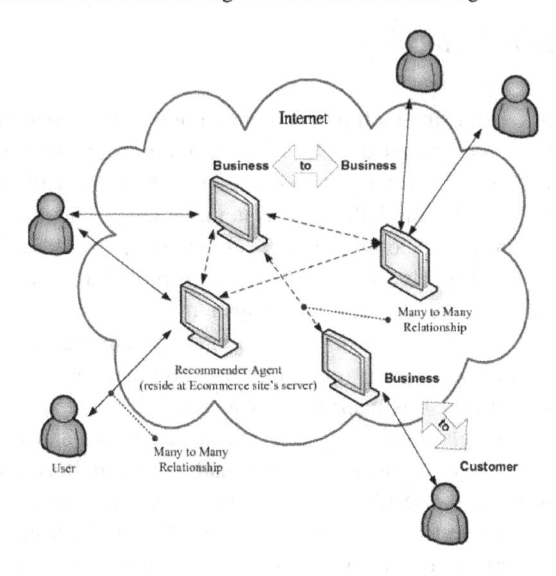

Fig. 7.3: Proposed distributed recommender system

nication and cooperation for recommenders of different sites, it is expected that the cooperation of these recommenders (also their sites) will confirm to the Business-to-Business based model.

The implementation of the proposed EDRS involves both Peer-to-Peer and Client-Server architectural styles. Client-Server architecture is employed to model a recommender peer (i.e. the server) and its users (i.e. the clients). Similar to the centralised recommender, the entire recommendation generation process is done by the recommender situated at the server side, and the users make requests to the recommender through thin clients such as web browsers. The architectural style for the network among the recommender peers is modelled with Peer-to-Peer architecture. As mentioned before, Peer-to-Peer based architecture assumes that the peers are independent and autonomous from each other, and especially they should be loosely coupled. Such definition is suitable for modelling the relationship between the recommender peers' host sites, as they are both logically and physically independent and autonomous from each other (as they are different e-commerce sites and organisations). While both DRS and the proposed EDRS can be modelled with the Peer-to-Peer architecture, the recommender peers in EDRS are more strongly coupled together than in standard DRS, because the recommender peers in EDRS need to gather/distributed information and suggestions from/to each other in a timely and effective fashion to achieve their common goal (i.e. satisfy their users' information and recommendation needs).

To the best of our knowledge, the concept of the proposed EDRS has not yet been mentioned and investigated by other works. Also, it is different from existing recommender systems (both centralised and distributed ones) at several high level aspects.

7.3.1 Interaction Protocol

As mentioned earlier, the interaction, communication and cooperation of the recommender peers in the proposed EDRS can be modelled with the Peer-to-Peer based architectural style. In particular, the "Contract Net Protocol" (CNP) is employed as the foundation for modelling the system, which provides the basis for coordinating the interaction and communication among the recommender peers. Contract Net Protocol is a high level communication protocol and system modelling strategy for Peer-to-Peer architectural based systems (or other distributed systems) [9, 20] Weiss, 1999. In CNP, peers in the distributed system are modelled as nodes and the collection of these nodes is referred to as a contract net. In CNP based systems, the execution of a task is dealt with as a contract between two nodes, each node plays a different role, one of them is the manager role and the other is the contractor role. The role of a manager is responsible for monitoring the execution of a task and processing the results of its execution. On the other hand, the role of a contractor is responsible for the actual execution of the task. Note, the nodes are not designated a priori as contractors or managers, rather, any nodes may take on either roles dynamically based on the context of their interaction and task execution [20]. A contract is established by a process of mutual selection based on a two-way transfer of information. In general, available contractors evaluate task announcements made by managers and submit bids on those for which they are suited. The managers evaluate the bids and award contracts to the nodes (i.e. contractors) that they determine to be most qualified [20].

In the case of the proposed EDRS, the recommender peers are modelled as the nodes in the contract net. Depending on difference circumstances, each recommender peer plays manager role and contractor role interchangeably. When a recommender peer makes requests for recommendations to other peers, it is considered as a manager peer. On the other hand, the recommender peer that receives a request for recommendations and provides recommendations to other manager peers is considered as a contractor peer. The roles of the manager peer and the contractor peer and their interactions are depicted in Fig. 7.4.

The communication steps involved in the interaction are indicated by the numbers in Fig. 7.4 and explained as follows:

1. User sends a request for recommendations. The recommender peer who received the request and is responsible for making the recommendation to the user is considered as in manager role.
2. Based on the user's request and profile, the manager peer selects suitable peer recommenders to help it on making better recommendations to the user.
3. The manager peer makes requests to the peers for recommendation suggestions. The request message may only contain the user's item preferences (i.e. the user's rating data); however the identity of the user is remain anonymous for privacy protection.
4. Each contractor peer generates recommendations based on the received request.

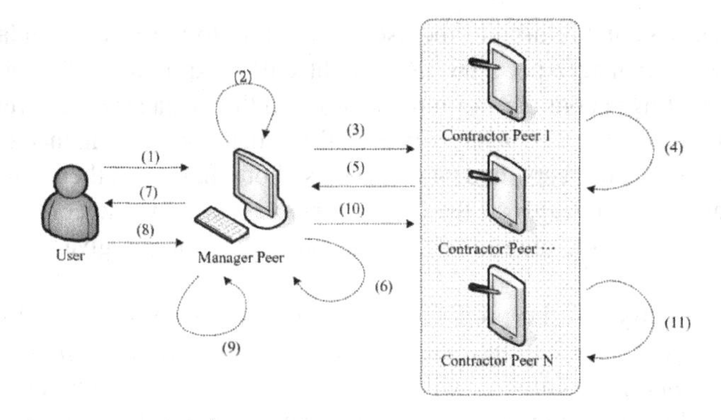

Fig. 7.4: High level interaction overview for EDRS (based on contract net protocol)

5. The contractor peers send back their recommendation suggestions to the manager peer.
6. After the manager peer received the suggestions from the contractors, it then synthesizes and merges these recommendation suggestions.
7. Based on the synthesized recommendation suggestions from the contractor peers (might also include the manager peer's own recommendations) the manager peer generates the item recommendations to the user.
8. When the user received the recommendations, he or she might supply implicit or explicit ratings to the recommendations. That is, the user might provide indications about whether he or she likes or dislikes one or more items in the recommendation list.
9. Based on the user rating feedbacks, the manager peer can objectively evaluates each of the peers' (i.e. contractors') performances to the recommendation suggestions they supplied and update its profiles about these peers.
10. The manager peer sends feedbacks and rewards to the contractor peers based on their performances to the task.
11. When the contractor peers received feedbacks about the performances of their recommendation suggestions, they then update their profiles about the manage peer in order to improve their future suggestions.

From Fig. 7.4, it can be seen that when a recommender peer is requested to make recommendations for a user, it acts as a manager peer. In the role of a manager peer, the recommender first generates a strategy about how and what to recommend to the user based on the user's profile and request, then chooses a set of recommender peers (in this context, they act as contractor peer) based on the profiles of peer recommenders, and finally makes requests for recommendations to these selected contractor peers. When these selected contractor peers received the requests, they then construct and return their recommendation suggestions based on the requests received and the manager peer's profile (e.g. preferences, domain of interests, trustworthiness and etc.). After the manager peer received the recommendations returned from the contractor peers, it then merges the recommendations (also include its own recommendations) and returns to the user. According to the recommendations received from the manager peer, the user might either explicitly or implicitly give feedbacks or ratings about the recommendations to the manager peer. After receiving the user's feedback, the manager peer will evaluate the performance of each of the selected contractor peers, update its profiles about them, and then construct the feedbacks and make rewards to the contractor peers. Finally, the contractor peers will update theirs profile about the manager peer based on the given rewards and feedbacks.

In order to carry out the proposed interaction described above, tasks such as *recommender peer selection, recommendation generation, recommendation merge, peer feedback and profile update* will need to be considered. Among all these formentioned tasks, recommender peer profiling and selection is the major focus of this chapter, and a novel contractor peer profiling and selection strategy is proposed, discussed and investigated in Sect. 7.4.

7.4 Peer Profiling and Selection

Part of the major contributions in this chapter includes a recommender profiling scheme (for manager peers to profile contractor peers) and a recommender selection algorithm designed for the proposed EDRS. In particular, the recommender peer selection problem is modeled as the classical exploitation vs. exploration (or k-armed bandit) problem [11], in which the recommender selection for the manager peer has to be balanced between choosing the best known contractor peers to keep users satisfied and selecting other unfamiliar contractor peers to obtain knowledge about them. The proposed recommender selection algorithm is based on evaluating the Gittins Indices [11] for every recommender peer, and the indices reflect the average performance, stability and selection frequency of the recommenders (i.e. contractor peers).

7.4.1 System Formalization for EDRS

We envision a world with a set of users and items, and they are denoted by $U = \{u_1, u_2, ..., u_n\}$ and $T = \{t_1, t_2, ..., t_m\}$ respectively. The proposed distributed recommender system (EDRS) denoted as Φ contains a set of l recommender peers $\phi_1, \phi_2, ..., \phi_l$, i.e. $\Phi = \{\phi_1, \phi_2, ..., \phi_l\}$. The number of recommender peers is much smaller than the number of users in our system, i.e. $l \ll n$. Each recommender peer $\phi_i \in \Phi$ has a set of users denoted as $U_i \in U$, and a set of items denoted as $T_i \in T$, where $U = \bigcup_{\phi_i \in \Phi} U_i$ and $T = \bigcup_{\phi_i \in \Phi} T_i$. It should also be noted that some users and items can be owned by more than one recommender peers.

7.4.2 User Clustering

Intuitively, a large set of users can be separated into a number of clusters based on the user preferences. Since users within the same cluster usually share similar tastes [3] and a cluster with a large number of users and a high degree of intra-similarity can better reflect the potential preferences of the users belonging to the cluster, a collaborative filtering based recommender can improve its recommendation quality by searching similar users within clusters rather than the whole user set [13]. However, different user clusters often vary in quality. The performance of such clustering based collaborative filtering system is strongly influenced by the quality of the clusters [13]. For a given recommender, some users might be able to receive better recommendations if they belong to a cluster with better quality (the cluster has a large number of users and a high intra-similarity), whereas some other users may not be able to get constructive recommendations because the cluster to which they belong is small and has a low intra-similarity. This situation is closely related to the cold-start problem [15] which happens when a recommender makes recommendations based on insufficient data resources. Therefore, even for the same recommender, the recommendation performance might be different for different clusters of users if different user clusters have different quality. In order to provide good recommendations to various users, the proposed EDRS allows its recommender peers (i.e. manager peers) to choose peers (i.e. contractor peers) for recommendations to the current user based on their performances to a particular user cluster to which the current user belongs. We expect this design to solve the cold start problem because a recommender which is making recommendations to a user who belongs to a weak cluster can get recommendations from recommender peers who have performed well to that group of users.

In the proposed EDRS, every recommender peer has its own set of user clusters, and we denote the set of user clusters owned by $\phi_i \in \Phi$ as $UC_i = \{uc_{i,1}, uc_{i,2}, ..., uc_{i,m_i}\}$, such that $uc_{i,j} \subseteq U_i$. In addition, for the simplicity of the system, all user clusters are assumed to be crisp sets. Because different recommender peers have different user sets and different clustering techniques, the size of their cluster set might vary as well.

7.4.3 Recommender Peer Profiling

In this section, we present our approach to profile the recommender peers within the proposed EDRS. To begin with, the performance evaluation of the recommender peers is explained. The performance of a recommender peer is measured by the degree of user satisfactory to the recommendations made by the recommender [8]. In our system, a recommender peer ϕ_i makes recommendations to a user with a set of k items $P_i = \{p_{i,1}, p_{i,2}, ..., p_{i,k}\}$ where $P_i \subseteq T_i$. Once having received the recommendations, the user then input his or her evaluations to each of the k items. We use r_a to denote the user's rating to item $p_{i,a} \in P_i$. The value of r_a is between 1 and 0 which indicates how much the user likes item $p_{i,a}$. When r_a closes to 1, it indicates the user highly prefers the item, by contrast when r_a closes to 0, the user dislikes the item. Hence, each time a recommender peer generates a recommendation list (eg. P_i) to a user, it will get feedback $R = \{r_1, r_2, ..., r_k\}$ from the user, where $r_a \in (0,1)$. With R, we can compute the recommender peer's current performance χ to the user by:

$$\chi = \frac{\sum_{r \in R} r}{|R|} \tag{7.1}$$

Equation (7.1) measures the current performance of a recommender peer to a particular user in the current recommendation round. We can use the average performance of the recommender to the users in the same cluster to measure its performance to this group of users. The average performance measures how well the recommender averagely performed in the past. However, the average performance doesn't reflect whether the recommender is generally reliable or not. Hence, we employed the standard deviation to measure the stability of the recommender. Another factor that should be taken into account for profiling a recommender is the selection frequency which indicates how often the recommender has been selected before. In our system, we profile each recommender peer from the three aspects: recommendation performance, stability, and selection frequency. As mentioned previously, a recommender will seek for recommendations from other peers when it receives a request from a user. Broadcasting the user request to all peers is one solution, but obviously it is not a good solution since not all of the peers are able to provide high quality recommendations. In EDRS, the recommender peers (i.e. manager peers) will select the most suitable peers (i.e. contractor peers) for recommendations based on their profiles. Therefore, each recommender peers in EDRS keeps profiles to each of the other recommender peers.

A recommender peer may perform differently to different user clusters. Therefore its performances to different user clusters are different. For recommender $\phi_i \in \Phi$ which has m_i user clusters, that is, $UC_i = \{uc_{i,1}, uc_{i,2}, ..., uc_{i,m_i}\}$, we use $Q^i_{j,h}$ to denote the average performance of peer $\phi_j \in \Phi$ to ϕ_i's user cluster $uc_{i,h}$. Hence, we can use a $z \times m_i$ matrix $Q^i = \{Q^i_{j,h}\}_{zm_i}$ to represent the average performance of each of the other peers to each of ϕ_i's user clusters, where $z = |\Phi| - 1$ and $m_i = |UC_i|$. Q^i is called as the peer average performance matrix of ϕ_i. Similarly, we use S^i and F^i to represent the stability and selection frequency of other peers to ϕ_i. $S^i = \{S^i_{j,h}\}_{zm_i}$ and $F^i = \{F^i_{j,h}\}_{zm_i}$ are called as the peer stability matrix and peer selection frequency matrix respectively. In summary, a recommender ϕ_i's peer profile is defined as $\mathbb{P}^i = \{Q^i, S^i, F^i\}$ which consists of the three matrixes representing peer recommender's average performance, stability, and selection frequency, respectively.

Initially, the Q^i, S^i and F^i of ϕ_i are all zero matrixes, because ϕ_i has no knowledge about other peers. These matrixes will be updated when a recommender peer ϕ_j helped ϕ_i (i.e. ϕ_j is in contractor role and ϕ_i is in manager role) to make a recommendation P_i for a user belonging to (or being classified to) a ϕ_i's user cluster $uc_{i,h}$. Suppose that R_j is the recommendation list returned by ϕ_j. Ideally, R_j is expected to be a subset of T_i. But usually $R_j \nsubseteq T_i$ since ϕ_i and ϕ_j may have different item sets. In the proposed EDRS, only the items which are in T_i are considered by ϕ_i. Let P_i be the final recommendation list made by ϕ_i to the user and $P_j = \{t | t \in R_j \cap T_i \text{ and selected by } \phi_i\}$ be the recommendation list made by ϕ_j and selected by ϕ_i during the merging process (the major focus of this selection is on peer profiling, other aspects of the proposed EDRS such as merging recommendations from different peers will be explained in latter sections). P_j should be a subset of P_i. After the recommendation P_i is provided to the user, ϕ_i will get a feedback list (i.e. the actual user ratings to the recommended items) R about P_i from the target user. With the user feedback R, Equation 7.2 will be used to compute ϕ_j's performance χ for the recommendation of this round (only the items in P_i are taken into consideration when compute the χ for ϕ_j) which is ϕ_i's observation about ϕ_j's performance to user cluster $uc_{i,h}$. The methods for updating the average quality, stability and selection frequency in ϕ_i's peer profile $\mathbb{P}^i = \{Q^i, S^i, F^i\}$ are given below, where $\tilde{Q}^i_{j,h}$, $\tilde{S}^i_{j,h}$, $\tilde{F}^i_{j,h}$ are the updated value for peer ϕ_j and cluster $uc_{i,h}$ in the three matrixes, respectively:

$$\tilde{Q}^i_{j,h} = \frac{Q^i_{j,h} \times F^i_{j,h} + \chi}{F^i_{j,h} + 1} \tag{7.2}$$

$$\tilde{F}^i_{j,h} = F^i_{j,h} + 1 \tag{7.3}$$

$$\tilde{S}^i_{j,h} = \begin{cases} 0 & \text{, if } F^i_{j,h} < 2 \\ \sqrt{\dfrac{[(F^i_{j,h}-1) \times S^i_{j,h}]^2 + \frac{(\chi - Q^i_{j,h})^2}{F^i_{j,h}+1}}{F^i_{j,h}}} & \text{, otherwise} \end{cases} \tag{7.4}$$

7.4.4 Recommender Peer Selection

In this section, a novel technique is proposed that allows manager peers to effectively and efficiently select contractor peers based on the proposed recommender peer profiles described in Section 7.3 for assistances in making quality recommendations. The proposed peer selection strategy is based on the famous Gittins Indices technique [11] developed for solving the exploitation vs. exploration problem, as such, it enables the manager peers to efficiently learn their contractor peers as well as maintain their recommendation quality to the users.

7.4.4.1 Gittins Indices

The Gittins indices [11] is developed for the k-armed bandit problem (which is a subset of the exploitation vs. exploration problem) that deals with a slot machine with k arms. An amount of reward will be given when an arm is pulled. However, in each time period, only a limited number of arms can be pulled (normally one arm). Different arms have different reward distributions, and the reward distributions for the arms are initially unknown. The objective is to choose which arms to pull that will maximize the total rewards over time based on previous experience and obtained rewards as well. Formally, the k-armed bandit problem is to schedule a sequence of pulls maximizing the expected present values of

$$\sum_{t=1}^{\infty} \alpha^t R(t) \tag{7.5}$$

where t indicates the time points, $R(t)$ denotes the sum of the rewards obtained by pulling a set of arms at t, and α is a fixed discount factor where $0 < \alpha < 1$.

Traditionally, dynamic programming was the preferred framework for solving the bandit problem. It requires analysis of all possible combinations of the pulling sequences. However, Gittins has developed a solution in 1972 that requires computation only on the current states of the individual arms. Gittins suggests comparing each potential action (i.e. a pull) against a reference arm with a known and constant reward instead of to compare all possible actions against each other [11]. Gittins proved it is optimal to select actions with expected rewards equal to the reference actions with the highest equivalent rewards (i.e. Gittins index values) for each pull [11].

Specifically, a Gittins index value of an arm is computed based on the average and standard deviation of the rewards generated from the arm as well as the number of times the arm has been pulled. The application of the Gittins indices for solving the multi-armed bandit problem is therefore straight forward: we simply compute the Gittins index values for every arms (based on their current average and standard deviation of the rewards generated and the number of times each of them are pulled), and pull the arm with the highest index value. As the arm selection task involves only the current states of the arms (i.e. current average and standard deviation of the rewards and number of the times being pulled), it is therefore both memory and computationally efficient (when comparing to dynamic programming based solutions).

Note, the theorem background and the relevant index value generation techniques of the Gittins Indices technique are detailed in [11], this work mainly focuses on the application of the Gittins indices in the context of the recommender peer selection task.

Given an arm which has been pulled for n times, and generated an average reward $\bar{x}$ with a standard deviation $\hat{s}$, Gittins denotes the index value for the arm as $v(\bar{x}, \hat{s}, n)$, and he also proved that:

$$v(\bar{x}, \hat{s}, n) = \bar{x} + \hat{s} \times v(0, 1, n) \tag{7.6}$$

where $v(0, 1, n)$ is the index value for an arm being pulled for n times with a zero average reward and a standard deviation of 1. Gittins has calculated the value of $v(0, 1, n)$ for different combination of α and n in [11]. Gittins suggested that by selecting the arms with the highest index value (i.e. (7.6)) in every selection round, the overall accumulated total reward can be optimized.

7.4.4.2 Selection Strategy for EDRS

Based on Sect. 7.3 when a manager peer ϕ_i wants to find a best contractor peer ϕ_i to make a recommendation to a user $u \in uc_{i,h}$ the following equation is used to select the most suitable peer:

$$\phi_j = \text{argmax}_{\phi_j \in \Phi \ \{\phi_i\}} Q^i_{j,h} + S^i_{j,h} \times v(F^i_{j,h}) \tag{7.7}$$

where$v(F^i_{j,h})$ is the Gittins index function that maps $F^i_{j,h}$ (i.e. selection frequency) to the corresponding $v(0,1,F^i_{j,h})$. In (7.7), ϕ_i firstly calculates the average performance, stability and selection frequency of the available peers to the user cluster that u belongs to (i.e. $uc_{i,h}$). Then ϕ_i computes the index values for every peers based on (7.7). Finally, the most preferred peerϕ_j will be the one which has the highest index value. By setting up a cutoff for the index value, multiple recommender peers with index values higher than the cutoff can be selected.

7.5 Experiments and Evaluation

In this experimentation, multiple recommenders with different capability in making recommendations are constructed, and we allow them to interact with each other based on the proposed EDRS framework. Essentially, these recommenders employ the proposed peer profiling and selection strategy presented in Sect. 7.4 to learn from and select each other in order to improve their recommendation making. Our main focuses are to examine whether incorporating helps from other recommenders can indeed improve recommenders' recommendation quality and also evaluate the effectiveness of the proposed profiling and selection strategy.

7.5.1 Data Acquisition

In this work, the "Book-Crossing" dataset (http://www.informatik.unifreiburg.d e/ cziegler/BX/) is chosen to conduct the experiments. The "Book-Crossing" dataset is collected by Cai-Nicolas Ziegler in a 4-week crawl (August / September 2004) from the Book-Crossing community (http://www.bookcrossing.com/) with kind permission from Ron Hornbaker, CTO of Humankind Systems. It contains 278,858 users (anonymized but with demographic information) providing 1,149,780 ratings (explicit / implicit) about 271,379 books. In the user ratings, 433,671 of them are the explicit user ratings, and the rest of 716,109 ratings are implicit ratings. The book taxonomy and book descriptors for the experiments are obtained from Amazon.com. Amazon.com's book classification taxonomy is tree-structured (i.e. limited to "single inheritance") and therefore is perfectly suitable to the proposed technique. The average number of descriptors per book is around 3.15, and the taxonomy tree formed by these descriptors contains 10,746 unique topics.

7.5.2 *Experiment Setup*

As the main purpose of this experiment is to evaluate the proposed interaction protocol and the peer profiling and selection technique (rather than evaluating a new recommendation technique or algorithm) in a distributed recommender system, therefore the overall setup of this experiment is different from the setup for non-distributed recommender systems.

In this experiment, it is required to simulate the interactions (i.e. profiling and selection) among the recommenders from different organisations, and therefore the first step in the experiment setup process is to construct multiple recommenders with different capabilities and underlying knowledgebase (i.e. datasets). Next, the testing dataset is constructed for evaluating the recommenders' recommendation quality. Importantly, the recommendation quality comparison between recommenders utilizing the proposed EDRS framework (i.e. getting helps from other recommenders) and standalone recommenders (i.e. making recommendations based on their own efforts) are carried out. Moreover, the effectiveness of the proposed peer profiling and selection technique is also examined by comparing it with other peer selection strategies. Note, the proposed peer profiling strategy requires the manager peers to get user feedbacks for all of their recommendations so they can determine their contractor peers performances based on the feedbacks and then update their peer profiles. Hence, it is necessary to provide a way to allow the user feedbacks in the experiment. The tasks involved in this experiment setup are detailed in the following subsections.

7.5.2.1 Constructing Recommender Peers

In this experiment, four recommenders of different organisations are constructed to simulate the proposed recommender peer interactions. These four recommenders are named as ORG1, ORG2, ORG3 and ORG4, and they are equipped with different datasets but use the same underlying recommendation technique.

By evaluating the performances of the recommenders with the same recommendation technique and different underlying datasets, we can evaluate the performance of the recommenders based on their available information sources (i.e. their underlying datasets and also collaboration from other recommender peers) without the impact from using different recommendation techniques. Moreover, the results from the experiments can also be used to verify the proposed solution to the cold-start problem (i.e. enriching the information resources from other parties).

The recommendation technique employed by the four recommenders is the standard item-based collaborative filtering technique, for detailed implementation please refers to [3]. The use of the state-of-the-art recommendation technique ensures that our experiment can be compared and verified with other works. Moreover, it also suggested that the proposed EDRS framework and peer profiling and selection strategy can be easily adopted by existing recommenders. The main differences among the four recommenders are in their underlying datasets, specifically, they all have different customer sets (or user sets). We firstly select 6500 users from the Book-Crossing Dataset and then cluster them into 20 user clusters based on their item preferences (i.e. explicit item ratings). We denote the overall user set as U and the 20 user clusters as $uc_1, uc_2, ..., uc_20$.

From these 6500 users in U, 5000 users are selected as the training user set $\hat{U}$ (i.e. for forming the underlying datasets of the recommender peers) and the rest of 1500 users then forms the testing user set $\check{U}$, where $U = \hat{U} \cup \check{U}$ and $\hat{U} \cap \check{U} = \emptyset$. Furthermore, we denote the set of training users within cluster uc_i as $\hat{uc}_i$ and the set of testing users within cluster uc_i as $\check{uc}_i$. Importantly, the users in U are divided into the clusters first, and the 1500 users in the testing set U are then selected from each of the clusters. This process allows us to keep track of the percentages of the different user types (i.e. users in different clusters) in the testing user set.

7.5.2.2 Evaluation Metrics

The classification accuracy metrics such as Precision, Recall and F1 metrics are chosen for the performance evaluation of the recommenders against the users in the testing user set. The classification accuracy metrics are mainly based on comparing the recommended item list and the set of user preferred items. In this experiment, for each testing user $u_i \in \check{U}$, we divide the set of items explicitly rated by u_i (denoted as $\check{R}_i$) into two halves denoted by Y_i and T_i. For the two item sets Y_i and T_i, Y_i and the associated item ratings are used to represent u_i's user profile (i.e. the recommenders make recommendations to u_i based on u_i's ratings to the items in Y_i), and the items in T_i, on the other hand, are used to form the user preferred item list for evaluating the recommendations made to u_i. However, not all the items in T_i are preferred by the user u_i. The items with low rating values should not be considered as the user's preferred items because u_i has specifically indicated that they are disliked. Hence, the final testing item set $\check{T}_i$ is constructed by removing all items with ratings below u_i's average rating from T_i. For evaluating the recommenders' recommendation quality to a given testing user $u_i \in \check{U}$, the recommenders are firstly provided with u_i's profile (i.e. Y_i and the associated ratings), then the recommenders generate their recommendations to u_i, finally, the recommendations generated from the recommenders (i.e. P_i) are evaluated against the testing item set $\check{T}_i$ by utilizing the classification accuracy metrics (i.e. Precision, Recall and F1).

7.5.2.3 Benchmarks for the Peer Profiling and Selection Strategy

As mentioned earlier, one of the objectives of this experiment is to evaluate the effectiveness of the proposed peer profiling and selection technique described in Sect. 7.4. Hence, it is important to include other profiling and selection techniques as baselines in order to conclude the significance of the proposed technique. However, to the best of our knowledge, there are no other existing works available for the recommender peer profiling and selection tasks required for the proposed EDRS. As there are no existing standard baseline techniques available in distributed recommender systems, we therefore have adapt techniques from other research domains that are reasonably applicable to the required peer profiling and selection task. In this experiment, the following five peer profiling and selection strategies are compared:

- *Gittins:* the proposed recommender peer profiling and selection technique as described in Sect.7.4.
- *BPP:* Best Past Performances. It is the most fundamental and intuitive strategy being used for the profiling and selection related tasks in many research domains (e.g. the collection selection task in distributed information retrieval) . The basic idea behinds BPP is to select recommender peers with the best average past performances to the target users' belonging clusters.
- *Rand:* the manager peers based on this strategy keep no knowledge about other peers and select contractor peers at random. This strategy is included in this experiment to show the significance of having a reasonable peer profiling and selection strategy in the proposed EDRS.
- *Gittins_NC* this selection strategy is a simplified version of the proposed strategy Gittins. Essentially, Gittins_NC assumes all users belong to one cluster. Even Gittins_NC still profiles recommender peers based on their average performance, stability and selection frequency, and the selection is also based on the combined Gittins scores as described in Sect. 7.4, it does not profile the recommender peers by considering the performance differences for users in different clusters.
- *BPP_NC:* similar to Gittins_NC, this profiling and selection strategy does not differentiate peers' performance differences for users in different clusters, and it employs only the average past performances of the recommender peers to make selections (i.e. as similar to BPP). The main purpose of having Gittins_NC and BPP_NC included in this experiment is to empirically demonstrate that different recommenders have different performances towards users in different clusters.

7.5.3 Experimental Results

Each of the four standalone recommenders (i.e. ORG1, ORG2, ORG4 and ORG4) can run by itself using its own dataset. However, the performance of the individual recommenders may not be satisfactory due to the insufficiency of the dataset. The EDRS framework proposed in this chapter can improve the performances of all involved participant recommenders by allowing them to share datasets and recommendations. Therefore, it is expected that the distributed recommendation system with a reasonable peer selection strategy outperforms the individual recommenders. Fig. 7.5, Fig. 7.6 and Fig. 7.7 present the precision, recall and F1 results obtained from running the four standalone recommenders (i.e. ORG1, ORG2, ORG4 and ORG4) and the distributed recommendation system with five peer selection strategies described in Sect. 7.5.2.3 (i.e. Rand, BPP_NC, Gittins_NC, BPP and Gittins), respectively.

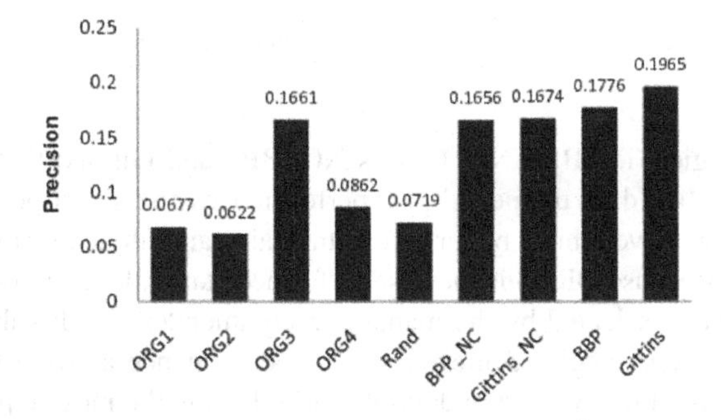

Fig. 7.5: Precision results for different recommendation settings

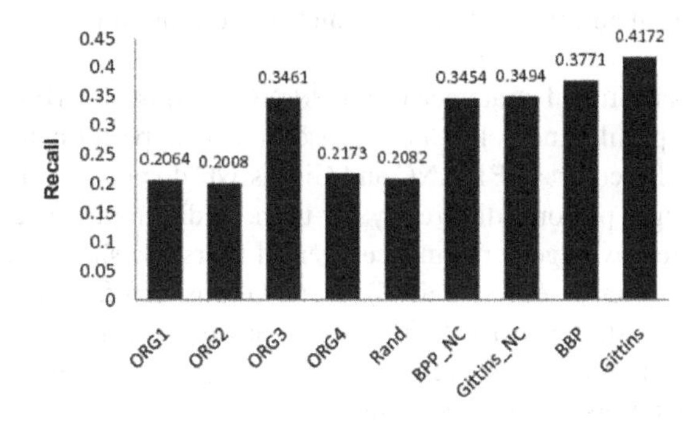

Fig. 7.6: Recall results for different recommendation settings

Let's firstly take a look at the performance of the distributed recommender system with the five different profiling and selection strategies (i.e. Rand, BPP, Gittins, BPP_NC, and Gittins_NC). Among these five strategies, Rand is the only strategy that does not have profiles for the recommender peers, and it randomly selects peers for making recommendations. Based on the experiment results shown in above figures, Rand performed the worst among all of the five strategies, and it even performed worse than two of the stand-alone recommenders ORG3 and ORG4 which make recommendations only based on their own datasets. By contrast, the

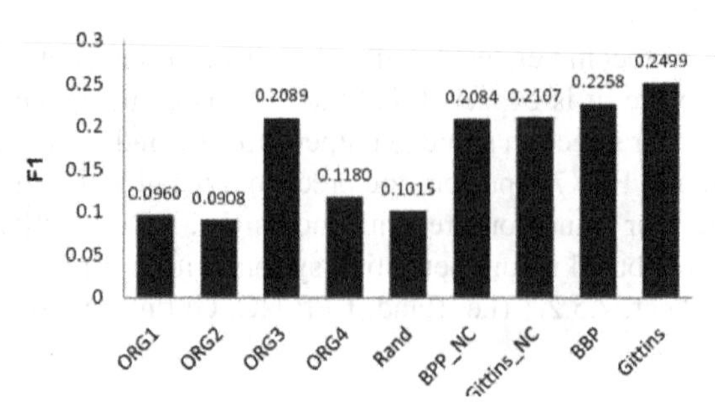

Fig. 7.7: F1 results for different recommendation settings

other four strategies (i.e. BPP_NC, Gittins_NC, BPP and Gittins) that profile recommender peers based on the peers' past performances and select peers' based on their profiles all achieved much better results than all stand-alone recommenders except for ORG3. Because ORG3 is the best performed stand-alone recommender and therefore very often selected by the manager recommender, the distributed system with some of these strategies achieved similar performance as what ORG3 does. This result suggests that by sharing datasets and selecting the most appropriate recommender to make recommendations, the distributed recommendation system can greatly improve recommendation quality. Particularly, for those peers which suffer from the cold-start problem (such as ORG1 and ORG2), the amount of improvement is significant, for instance, the performance of both ORG1 and ORG2 can be improved by more than 50% if they adapt any of the four strategies to profile and select peers.

Among the four rational strategies (i.e. BPP_NC, Gittins_NC, BPP and Gittins), BPP and Gittins profile and select peers based on their performance to users in different clusters. By contrast, BPP_NC and Gittins_NC do not consider the fact that different peers might perform differently for users in different clusters and profile peers based on their average performance over all users. As shown in 7.5, Fig. 7.6 and Fig. 7.7, the cluster based strategies BPP and Gittins significantly outperformed the noncluster based strategies BPP_NC and Gittins_NC. This is because the cluster based strategies can find the best recommender peers for making recommendations based on the target users' belonging clusters. By contrast, BPP_NC and Gittins_NC select recommender peers based on their average past performances to all users. Therefore, they will select peers performed averagely best in the past despite that these peers might be unable to produce good recommendations for some target users in certain clusters.

Finally, the experiment results show that the Gittins indices based strategies (i.e. Gittins and Gittins_NC) performed better than that of the standard performance based strategies (i.e. BPP and BPP_NC). Specifically, Gittins outperformed BPP and Gittins_NC outperformed BPP_NC. This result suggests that by combining the selection frequency and recommendation stability into peer profiling and selection process, the best performed peers can be more accurately identified than only based on the peers' average past performances.

7.6 Conclusions

In this chapter, we suggested a new distributed system paradigm for recommenders, namely, Ecommerce-oriented Distributed Recommender System (EDRS). EDRS is designed to allow the recommenders from different organisations or parties to share datasets and recommendations with each other, so that all of them can achieve better recommendation quality and provide better services to their users. Also, as the recommenders within the proposed EDRS no longer make recommendations solely on their own efforts, they are therefore more resistant to the cold-start problems. In order to facilitate the interaction among the recommenders in the EDRS, a novel peer profiling and selection strategy is proposed in this chapter. The proposed strategy profiles and selects recommender peers based on their past recommendation performance, stability and selection frequency in cluster level, and our experiment results show that the proposed strategy allows recommender peers to effectively learn from each other and select the most appropriate peers to provide satisfactory recommendations to their users.

References

1. Baruch Awerbuch, Boaz Patt-Shamir, David Peleg, and Mark Tuttle. Improved recommendation systems. In *SODA '05: Proceedings of the sixteenth annual ACM-SIAM symposium on Discrete algorithms*, pages 1174–1183, Philadelphia, PA, USA, 2005. Society for Industrial and Applied Mathematics.
2. Christoph Baumgarten. A probabilistic model for distributed information retrieval. In *SIGIR '97: Proceedings of the 20th annual international ACM SIGIR conference on Research and development in information retrieval*, pages 258–266, New York, NY, USA, 1997. ACM.
3. Robin Burke. Hybrid recommender systems: Survey and experiments. *User Modeling and User-Adapted Interaction*, 12(4):331–370, November 2002.
4. Maarten Clements, Arjen P. de Vries, Johan A. Pouwelse, Jun Wang, and Marcel J. T. Reinders. Evaluation of neighbourhood selection methods in decentralized recommendation systems. In *Workshop on Large Scale Distributed Systems for Information Retrieval (LSDS-IR)*, 2007.
5. Owen de Kretser, Alistair Moffat, Tim Shimmin, and Justin Zobel. Methodologies for distributed information retrieval. In *International Conference on Distributed Computing Systems*, pages 66–73, 1998.
6. Petros Drineas, Iordanis Kerenidis, and Prabhakar Raghavan. Competitive recommendation systems. In *STOC '02: Proceedings of the thiry-fourth annual ACM symposium on Theory of computing*, pages 82–90, New York, NY, USA, 2002. ACM Press.
7. David Goldberg, David Nichols, Brian M. Oki, and Douglas Terry. Using collaborative filtering to weave an information tapestry. *Commun. ACM*, 35(12):61–70, December 1992.
8. Jonathan L. Herlocker, Joseph A. Konstan, Loren G. Terveen, and John T. Riedl. Evaluating collaborative filtering recommender systems. *ACM Trans. Inf. Syst.*, 22(1):5–53, January 2004.

9. Jin Li. On peer-to-peer (p2p) content delivery. *Peer-to-Peer Networking and Applications*, 1(1):45–63, March 2008.
10. Pingfeng Liu, Guihua Nie, Donglin Chen, and Zhichao Fu. The knowledge grid based intelligent electronic commerce recommender systems. In *SOCA '07: Proceedings of the IEEE International Conference on Service-Oriented Computing and Applications*, pages 223–232, Washington, DC, USA, 2007. IEEE Computer Society.
11. Brian P. McCall. Multi-armed bandit allocation indices (J. C. Gittins). 33(1), 1991.
12. Bradley N. Miller, Joseph A. Konstan, and John Riedl. Pocketlens: Toward a personal recommender system. *ACM Trans. Inf. Syst.*, 22(3):437–476, July 2004.
13. B. M. Sarwar, G. Karypis, J. Konstan, and J. Riedl. Recommender systems for large-scale e-commerce: Scalable neighborhood formation using clustering. In *Fifth International Conference on Computer and Information Technology (ICCIT 2002)*, 2002.
14. J. B. Schafer, J. A. Konstan, and J. Riedl. E-commerce recommendation applications. *Data Mining and Knowledge Discovery*, 5(1-2):115–153, January-April 2001.
15. A. Schein, A. Popescul, L. Ungar, and D. Pennock. Methods and metrics for cold-start recommendations. In *Proceedings of the 25th Annual International ACM SIGIR Conference on Research and Development in Information Retrieval (SIGIR 2002)*, pages 253–260, 2002.
16. Christoph Sorge. A chord-based recommender system. In *LCN '07: Proceedings of the 32nd IEEE Conference on Local Computer Networks*, pages 157–164, Washington, DC, USA, 2007. IEEE Computer Society.
17. Amund Tveit. Peer-to-peer based recommendations for mobile commerce. In *WMC '01: Proceedings of the 1st international workshop on Mobile commerce*, pages 26–29, New York, NY, USA, 2001. ACM.
18. Josã M. Vidal. *A Protocol for a Distributed Recommender System.* ACM Press, 2005.
19. Jun Wang, Arjen P. de Vries, and Marcel J. T. Reinders. Unifying user-based and item-based collaborative filtering approaches by similarity fusion. In *SIGIR '06: Proceedings of the 29th annual international ACM SIGIR conference on Research and development in information retrieval*, pages 501–508, New York, NY, USA, 2006. ACM Press.
20. Gerhard Weiss, editor. *Multiagent systems: a modern approach to distributed artificial intelligence.* MIT Press, Cambridge, MA, USA, 1999.

Chapter 8
A Multiagent-based Intrusion Detection System with the Support of Multi-Class Supervised Classification

Mei-Ling Shyu and Varsha Sainani

Abstract The increasing number of network security related incidents have made it necessary for the organizations to actively protect their sensitive data with network intrusion detection systems (IDSs). IDSs are expected to analyze a large volume of data while not placing a significantly added load on the monitoring systems and networks. This requires good data mining strategies which take less time and give accurate results. In this study, a novel data mining assisted multiagent-based intrusion detection system (DMAS-IDS) is proposed, particularly with the support of multi-class supervised classification. These agents can detect and take predefined actions against malicious activities, and data mining techniques can help detect them. Our proposed DMAS-IDS shows superior performance compared to central sniffing IDS techniques, and saves network resources compared to other distributed IDS with mobile agents that activate too many sniffers causing bottlenecks in the network. This is one of the major motivations to use a distributed model based on *multiagent* platform along with a *supervised classification* technique.

8.1 Introduction

The growing importance of network security is shifting security concerns towards the network itself rather than being just host-based. Security services are evolving into network-based and distributed approaches to deal with heterogeneous open platforms and support scalable solutions. Intrusion detection is the process of identifying network activities that can lead to a compromise of security policy . Intrusion detection systems (IDSs) must analyze and correlate a large volume of data collected from different critical network access points [12]. This task requires an IDS to be able to characterize distributed patterns and to detect situations where a sequence of intrusion events occur in multiple hosts. In addition, intrusion prevention techniques

Mei-Ling Shyu and Varsha Sainani
Department of Electrical and Computer Engineering, University of Miami, Coral Gables, FL 33124, USA, e-mail: shyu@miami.edu, v.sainani@umiami.edu

L. Cao (ed.), Data Mining and Multi-agent Integration, DOI: 10.1007/978-1-4419-0522-2_8,

such as user authentication, authorization, encryption, defensive programming and IDSs are often used as another wall to protect computer systems.

The two main intrusion detection techniques are *misuse detection* and *anomaly detection* . Misuse detection systems [2][10] use patterns of well known attacks or weak spots of the system to match and identify known intrusions. Misuse detection techniques in general are not effective against novel attacks that have no matched rules or patterns yet. On the other hand, *anomaly detection* systems observe flag activities that deviate significantly from the established normal usage profiles as anomalies or in other words as intrusions. Anomaly detection techniques can be effective against unknown or novel attacks since no prior knowledge about specific intrusions is required. However, anomaly detection systems tend to generate more false alarms than misuse detection systems because an anomaly can just be a new normal behavior [1][9].

As accuracy is the essential requirement for an IDS, its extensibility and adaptability are also critical in today's network computing environment. There can be multiple weak points for intrusions to take place in a network system. For example, at the network level, malicious IP packets can crash a host, and at the host level, vulnerabilities can occur in system software which can be exploited to execute an illegal root shell. Since malicious activities at different intrusion points are normally recorded in different data sources, an IDS needs to be extended to incorporate additional modules that specialize in certain components of the network systems. Hence, IDSs need to be adaptive in such a way that frequent and timely updates are possible.

Our research aims to develop a more systematic and automated approach for building IDSs. We have developed a set of tools that can be applied to a number of tasks such as capturing data, extracting features, classifying them into known and unknown attack categories, and ultimately stopping the ongoing malicious activity. We take a data-centric point of view and consider intrusion detection as a data analysis process. The central theme of our approach is to apply data mining techniques to the extensively gathered data to compute a model that accurately captures the actual behavior and patterns of the intrusions and normal activities. This approach significantly reduces the need to manually analyze and encode intrusion patterns, as well as the guesswork in selecting statistical measures for normal usage profiles. The resultant model is more effective because it is computed and validated using a large amount of network data. This necessity of analyzing large amounts of data and finding its nature requires data mining strategies. The instant at which the sign of an attack is concluded, necessary actions are required which can be accomplished by intelligent agents. Hence, the integration of agent technology and data mining techniques makes an IDS more autonomous and efficient.

The remaining part of this chapter is organized as follows. Existing work is discussed in Section 2. Section 3 presents the design of our proposed DMAS-IDS. Section 4 describes our experimental setup, and the results of the performance analysis are given in Section 5. Finally, Section 6 concludes the paper.

8.2 Existing Work

Various distributed intrusion detection architectures using the multiagent design methodology and the data mining techniques have been developed. These approaches widely range from being comprised entirely of mobile agents like the MANET system [5][13], being merely a collection of static agents as in [17], or a combination of both as in DIDMA system [7].

IDSs have undergone rapid development in both power and scope in the last few years. Recently, the agent concept has been widely used in distributed environments because it provides many favorable characteristics including scalability, adaptability, graceful degradation of service, etc. as compared to the non-agent based systems. Most of the distributed agent-based IDSs introduced more traffic into their residing network, and therefore the communication protocol between various entities is also an important aspect that has to be considered. At the same time, most of the designed agent-based IDSs require comparatively high processing power in local machines to run the agents and other supportive software. Hence, a lightweight agent system with low network traffic generation requirements is needed. This can be accomplished with the use of appropriate data mining strategies.

Data mining techniques such as classification can be useful for both misuse detection and anomaly detection. In network intrusion detection, classification can be applied to classify network data consisting of malicious behaviors, and several existing approaches such as RIPPER, Naive Bayes, and multi-Bayes classifiers have been successfully used to detect malicious virus code. There are alternative classification approaches which can be effectively utilized for intrusion detection purposes. With intrusions, it is observed that over the time, the user establishes profile based on the numbers and types of commands they execute. Data mining classifier approaches like SOM (Self Organizing Maps) and LQM (Learning Vector Quantization) can be utilized for reducing dimensionality of these numbers. Furthermore, nearest neighbor classifier approaches based on SOM and LQM can be used to refine the collected network data in intrusion detection. A number of such systems have been developed which utilized the fast and efficient computation and pattern matching strategies of data mining to compliment the low cost and lightweight agent system architectures.

One of the well known examples of applying distributed agent design methodology in the intrusion detection domain is the Distributed Intrusion Detection System (DIDS). DIDS attempts to build a distributed system based on monitoring agents that reside at every host in the network. Distributed systems present both advantages and disadvantages. On one hand, the system utilizes the real-time traffic information from various sources, in the form of data from various host monitors or to assess the security status of its residing network. However, on other hand, the systems' scalability is poor for large networks as an increasing number of hosts monitoring the network also significantly increases the work load of the DIDS director agent. Additionally, the data flow between host monitors and the director agent may generate significantly high network traffic overheads. In [7], a system called DIDMA (Distributed Intrusion Detection using Mobile Agents) attempted to overcome the scalability issues inherent in the original DIDS architecture by employing mobile agents in the data analysis task. Thus, by decentralizing data analysis, DIDMA hoped to significantly neutralize the effects of the scalability issues.

Another well known system called *BODHI* [8] was designed for heterogeneous data sources based on the techniques such as supervised inductive distributed function learning and regression. It focuses on the guarantee for correct local and global data models having least network communication. It was implemented in Java and offers message exchanges and runtime environments of agent systems for the execution of mobile agents at each local site. A central facilitator agent takes care of initializing and coordinating the data mining tasks.

A Java-based multi-agent system (*JAM*) [18] is designed to be used for meta-learning distributed data mining. In this system, each site agent builds a classification model, where different agents built their classifiers using different techniques. JAM also provides a set of meta-learning. Once combined together, the classifiers are computed with the central JAM system coordinating the execution of these modules to classify data sets at all data sites simultaneously.

In [19], a distributed agent-based IDS analyzes anomalies to detect and identify the denial-of-services (DoS) and data theft attacks. It also attempts to respond to intrusions in real time by sending out alerts to the designated network administrator when network intrusions are detected. One of its main drawbacks is the design complexity of its comprising agents, since each agent must take on almost all work load of network traffic sniffing, data parsing, and intrusion detection. In addition, its data mining techniques are less powerful since they are capable of detecting only a limited number of attacks.

8.3 The Proposed DMAS-IDS Architecture

In this study, we present a novel data mining assisted multiagent-based intrusion detection system (DMAS-IDS) architecture. DMAS-IDS integrates a multi-class supervised classification algorithm and the agent technology in network intrusion detection. It utilizes high accuracy and speed response of the *Principal Component Classifier (PCC)* [16][20] at the first layer of our proposed architecture. Once the results from PCC classification are obtained, agents communicate them to the second layer. The second layer of the DMAS-IDS architecture is integrated with the *Collateral Representative Subspace Projection Modeling (C-RSPM)* classifier [14] which includes collateral class modeling, class ambiguity solving, and classification. Results from this stage of classification are further analyzed by the agents and policies are derived which are communicated to the next layer using our own designed low cost and low response time agent communication protocol. The architecture is further elaborated in the following sections.

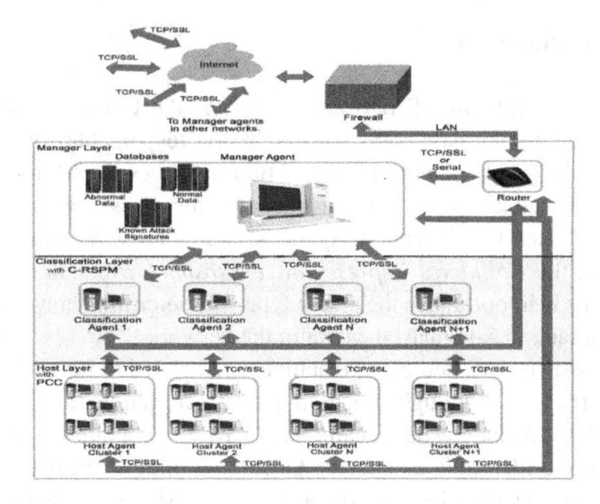

Fig. 8.1: The system architecture of the proposed DMAS-IDS

8.3.1 Agent Architecture

Fig. 8.1 presents our hybrid layered multiagent-based IDS architecture [15] which constitutes three layers called Host, Classification, and Manager layers. Each of these layers comprises of deliberative agents which are well aware of each other's presence and are capable of communicating with each other using our developed communication scheme.

8.3.1.1 Host Layer

This layer marks the entry point of our proposed architecture. These end-user machines are workstations constituting the network, and they also act as the host agents inspecting each incoming network connection. Virtually, every machine in a network can be considered as a Host Agent. Host Agents collect network connection information and classify these connections as 'normal' or 'abnormal'. Here, the *Principal Component Classifier (PCC)* is used [16][20].

Each Host Agent belongs to one Classification Agent (in the second layer), to which it reports the connections that PCC classifies them as 'abnormal'. The responsibilities of these agents are (i) capturing network traffic, (ii) detecting abnormal activities in these connections, (iii) passing the classification results to its Classification Agent, (iv) properly responding to these abnormal activities for intrusion detection, and (v) passing a subset of the normal connection instances and the abnormal connection instances to the Manager layer to be saved in a database for the purpose of re-training the classifier at a later time.

8.3.1.2 Classification Layer

The second layer of the DMAS-IDS architecture is the Classification Layer. The responsibilities of the Classification Agents are (i) responsible for a set of Host Agents, (ii) classifying the abnormal connection instances found in their host machines into known attack types, and (iii) passing the classification results to the Manager Agent. Each Classification Agent is facilitated with a misuse detection algorithm called the *Collateral Representative Subspace Projection Modeling (C-RSPM)* [14]. This is important as the attack type will determine how the IDS should respond to the attack to safeguard the data in the network.

Unlike the Host Agents, dedicated machines are needed to run the Classification Agents so that they have enough processing power to handle all classification requests of their Host Agents. Additionally, they have to generate 'policies' upon the instances which are identified as attacks. Once the policy is created, it is communicated to the next layer called the Manager Layer. All the Classification Agents present in the network add their policies to the policy repository present in the 'Database' in the Manager Layer.

8.3.1.3 Manager Layer

This layer, in terms of contemporary agent models, is the same as the planning layer. The Manager Agent is in charge of the entire system, performing several support tasks for the system. Its responsibilities include (i) assigning a Host Agent to the specific Classification Agent, (ii) assisting the Classification Agents in managing their host machines and the tasks related to them, and (iii) managing the routers and firewalls in the network.

The main task performed by the Manager Agent is to take the policies from the Classification Agents throughout the network. After receiving the policies, the Manager Agent broadcasts them to every agent present in the network. Once the policy is received by all Host Agents, the corresponding Host Agent who initiated the request implements the policy. This functionality is important as the Classification Agents can prevent or lessen the effects of a possible attack by managing resources in those nodes that they expect to be affected by the incoming attack, such as bandwidth, communication ports, and connection authorization.

8.3.1.4 Agent Communication

A simple and manageable communication scheme that utilizes a discrete number of *KQML* performatives is designed to accommodate the goals of all the aforementioned agents. Agents use these messages to register with the upper level of agents and communicate the results of the tasks they are responsible of performing.

As soon as the Classification Agent comes online, it registers itself with the Manager Agent using the "REGISTER" performative. The same registration process applies to the Host Agent when it has to register with the Classification Agent. However, since the Host Agent needs to know to which the Classification Agent it belongs as soon as it comes online, it uses the "RECOMMEND-ONE" performative to ask the Manager Agent. The Manager Agent then uses the "TELL" performative to reply to the Host Agent with the available Classification Agent's IP. As mentioned earlier, the Host Agent diagnoses each incoming connection using PCC and uses the "EVALUATE" performative to communicate this result to the Classification Agent. The Classification Agent further classifies any abnormal connection into a known attack type and derives a policy based on it. It uses the "REPORT" performative to communicate the corresponding policy to the Manager Agent. Finally, the Manager Agent broadcasts this policy to all other agents present in the network using the "BROADCAST" performative. In our proposed DMAS-IDS architecture, the TCP/IP Secure Socket Layer (SSL) was adopted for the implementation of the secure communication channel among the agents to provide total privacy and authentication capabilities. The adoption of cryptographic communication services is important for making the distributed multiagent IDS architecture immune against attacks.

8.3.2 Principal Component Classifier (PCC)

In the classification module of C-RSPM, each classifier is called the Principal Component Classifier (PCC) [20] (as shown in Fig. 2(a)). PCC will classify a connection as either normal (non-intrusion) or abnormal (possible intrusion). PCC basically goes through four basic steps of classification: (i) preprocessing, (ii) Principal Component Subspace Projection, (iii) Automatic Representative Component Selection, and (iv) Establishment of the Decision Rules.

In the preprocessing step, the average and standard deviation of the instances in the normal class are calculated, and these statistical characteristics are used to normalize the data. Let L be the set of normalized training data instances, $i = 1,2,\ldots,p$, $j = 1,2,\ldots,L$, $\bar{\mu}_i$ and s_{ii} be the sample mean and the variance of ith row of the trimmed matrix $\mathbf{X}$ respectively, and $\mathbf{x}_{ij}$ (i=1,2,...,p, j=1,2,...,L) be the elements in matrix $\mathbf{X}$. Define the normalized un-trimmed data set consists of p features as shown in Equation (8.1) and its corresponding column vectors as presented in Equation (8.2), where Equation (8.3) is used for normalization. In order to decide the percentage of the data instances that can be regarded as outliers, a Parzen window is used to decide which data instances are retained and which are removed as outliers according to a defined 'rtd' factor. The 'rtd' factor is chosen corresponding to the center of a Parzen window where the maximum accuracy is reached. If there is a tie, then the first one is chosen as the default one.

$$\mathbf{Z} = \{\mathbf{z}_{ij}\},\ i = 1,2,\ldots,p,\ j = 1,2,\ldots,L. \quad (8.1)$$

$$\mathbf{Z}_j = (\mathbf{z}_{1j},\mathbf{z}_{2j},\ldots,\mathbf{z}_{pj})',\ j = 1,2,\ldots,L. \quad (8.2)$$

$$\mathbf{z}_{ij} = \frac{\mathbf{x}_{ij} - \bar{\mu}_i}{\sqrt{s_{ii}}}. \quad (8.3)$$

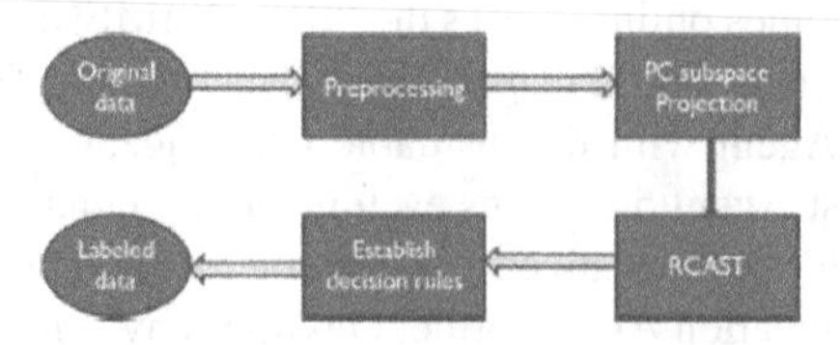

(a) Principal Component Classifier Architecture

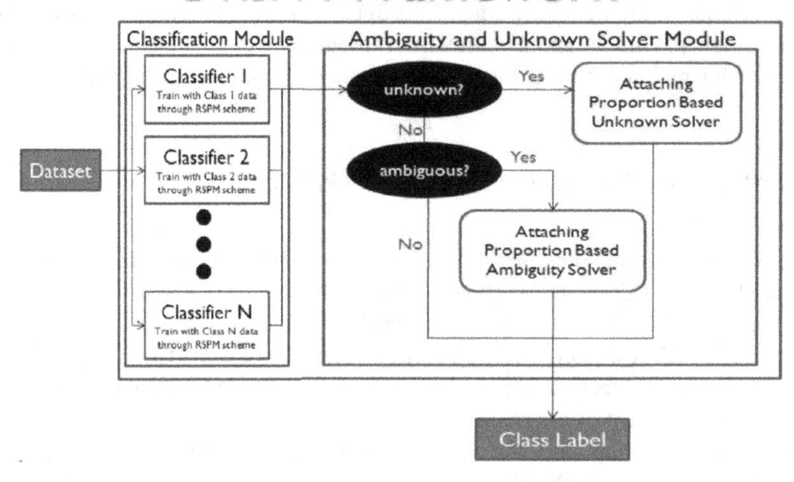

(b) Collateral representative subspace projection modeling for supervised classification architecture

Fig. 8.2: PCC and C-RSPM

Next step is to automatically select the representative principal components (PCs). In the PC space, only those dimensions that have positive eigenvalues are selected. Before this, we need to perform the projection for the data instances from the original space to the PC space. That is, each retained training data instance is projected to a subspace by using those PCs derived from the normal class. Let $\mathbf{E}_i = (e_{i1}, e_{i2}, \ldots, e_{ip})'$ be the ith eigenvector, and $(\lambda_1, \boldsymbol{E}_1), (\lambda_2, \boldsymbol{E}_2), \ldots, (\lambda_p, \boldsymbol{E}_p)$ be the p eigenvalue-eigenvector pairs of the robust correlation matrix $\mathbf{S}$. Also, let $\mathbf{Y}$ be the projection of $\mathbf{Z}$ onto the p-dimensional eigenspace (shown in Equation (8.4)) consisting of $\boldsymbol{Y}_j$ column vectors (given in Equation (8.5)). Then, a sample score value of the normalized training data instance vector can be computed using Equation (8.6). We define the p score row vectors of $\mathbf{Y}$, representing the distribution of the p eigenspace features of all the un-trimmed, normalized, and projected training instances, as $\boldsymbol{R}_i$=($\mathbf{y}_{i1}, \mathbf{y}_{i2}, \ldots, \mathbf{y}_{iL}$), i=1,2,...,p. Following this, a lambda score value is computed for each PC.

$$\mathbf{Y} = \{\mathbf{y}_{ij}\},\ i = 1,2,\ldots,p\ j = 1,2,\ldots,L. \quad (8.4)$$

$$\boldsymbol{Y}_j = (\mathbf{y}_{1j},\mathbf{y}_{2j},\ldots,\mathbf{y}_{pj})',\ j = 1,2,\ldots,L. \quad (8.5)$$

$$\mathbf{y}_{ij} = \mathbf{E}_i'\boldsymbol{Z}_j = e_{i1}\mathbf{z}_{1j} + e_{i2}\mathbf{z}_{2j} + \ldots + e_{ip}\mathbf{z}_{pj}. \quad (8.6)$$

In the attempt to generate a better predictive model, the set of available score row vectors is refined by eliminating those possessing extremely insignificant or null variability, i.e., extremely little standard deviation. Next, a distance measure is defined as shown in Equation (8.7).

$$\mathbf{c}_j = \sum_{m \in M} \frac{(\mathbf{y}_{mj})^2}{\lambda_m}. \quad (8.7)$$

Finally, the classification decision rules can be established, where the decision rules to classify each of the data instances $\boldsymbol{X}'_j$, j=1,2,...,N', are based on the selected threshold value C_{thresh}. The details of the steps of establishing the decision rules and determination of C_{thresh} can be found in [14]. In summary, we classify the jth testing data instance as abnormal if $\mathbf{c}'_j > C_{\text{thresh}}$.

8.3.3 Collateral Representative Subspace Projection Modeling (C-RSPM)

Among various data mining techniques, *supervised classification* has become an essential tool that has been applied successfully in diverse research areas including *network intrusion detection systems*. Since it shows promising results with a number of data sets as compared to other available algorithms, it is utilized in our proposed framework. The system architecture of our previously developed C-RSPM classifier [14] is illustrated in Fig. 2(b). As can be seen from this figure, it includes the *Classification* module and *Ambiguity Solver* module.

The Classification module is composed of an array of deviation classifiers (i.e., one PCC for each class) which are executed collaterally. That is, each of the classifiers receives and classifies the same testing instance simultaneously [14]. The basic idea of the C-RSPM classifier is that each classifier is trained with the data instances of a known class in the training data set. Thus, training the C-RSPM classifier consists basically of training each individual classifier to recognize the instances of each specific class. In the ideal case, a testing data instance will be classified as 'normal' to only one classifier's training data instances. However, in realistic situations, a testing data instance may be classified as 'normal' by multiple classifiers or none of the classifiers considers it as 'normal'. To address these two scenarios, the *Ambiguity Solver module* is introduced. Ambiguity Solver captures and coordinates classification conflicts and also provides an extra opportunity to improve the classification accuracy by estimating the true class of an ambiguous testing data instance.

There can be one more scenario for class ambiguity, which arises with a simple fact that no classifier can ensure the 100% classification accuracy, as in most cases, a data set would contain classes with very similar properties. Thus it makes it difficult for a classifier to identify the correlative differences between these classes with small feature differences.

Unlike many other algorithms which upon encountering such issues simply resort to solutions such as the random selection of a class label from among the ambiguous class labels, the C-RSPM classifier attempts to properly address the issues by defining a class-attaching measure called *Attaching Proportion* for each of the ambiguous classes. The goal of solving such an ambiguity issue via attaching proportion measure is to label an ambiguous testing data instance with the label of the classifier exhibiting the lowest *Attaching Proportion* value. Additionally, the Attaching Proportion can be viewed as a measure of the degree of normality of a data instance with respect to a class, indicating the percentile of normality the data instance under analysis is associated to the corresponding class.

8.3.4 Policy Derivation at Classification Agents

Ideally, an IDS should aim at not only detecting the intrusions but also stopping them as soon as they are detected. Our proposed architecture takes this requirement of the IDSs into consideration as well. As mentioned above, the Classification Agent derives a policy after it concludes the type of the abnormal connection, and then reports this policy, rather than the attack type, to the Manager Agent.

Policy derivation plays a very important role in the scenarios when the same attacker generates different attacks to different hosts in the same cluster at the same time instant or at different time instants. If the Classification Agent just reports the attack types, it does not serve the purpose, as the Manager agent may recommend a common measure for all different attacks at different hosts. In such situations, the policies prove their importance since policies provide the solutions to the ongoing attacks. This could be better explained with the example as follows. If an attacker A makes one attack called 'attack1' at host HA1 and the other attack called 'attack2' at host HA2, and both hosts belong to Classification Agent CA. Understanding the properties of 'attack1', it is enough to block the access of the attacker to a particular port; whereas for 'attack2', it may be required to block the access of the attacker to a particular server in the network. It is very much expected that in real case scenarios, we cannot just go blocking each doubtful machine. Hence, it is required to have case specific solutions, and policies can well provide acceptable solutions for this. Policies add more dynamicity and adaptability to an IDS by providing attack specific solutions.

To demonstrate how such types of policies can be established, in our experiments, three types of policy options are defined. Of course, more policies can be defined to accommodate different network intrusions. Following gives the three pieces of policies that are defined in our experiments.

1. Block;
2. Block on a port; and
3. Block on a server.

As soon as the Classification agent concludes the attack type, it looks at the features of the attacks, chooses one of the three policy options, and then reports this to the Manager Agent. Fig. 8.3 shows a screen dump where incoming instances are being blocked to the port which has been diagnosed as a penetration point in the network.

```
Problems  Javadoc  Declaration  Console  Debug
ClassificationAgent [Java Application] C:\Program Files\Java\jre1.5.(
blocking 10.0.0.53 on 1037
 ca.POLICY-REPORT Message Listing!!!
(REPORT
 Sender IP:129.171.217.5
Sender Port: 1235
Ontology: Agent Communication
Content: Block on port:1037
reply_with: 78306
reply_to: 32995
)
Getting proxy
Message Listing!!!
(BROADCAST
 Sender IP:129.171.217.5
Sender Port: 1234
Ontology: Agent Communication
Content: Block on port:1037
reply_with: 47189
reply_to: 78306
)
blocking 10.0.0.6 on 1037
blocking 10.0.0.67 on 1037
blocking 10.0.0.67 on 1037
Creating Client
blocking 10.0.0.53 on 1037
```

Fig. 8.3: Policy derivation and implementation

8.4 Experimental Setup

In order to assess the overall performance of our DMAS-IDS together with the our previously developed communication protocol in a realistic scenario, a prototype of the proposed architecture was implemented using Java RMI package in the particular lipe RMI [11], which allows a number of machines to communicate with each other at the TCP level. We have conducted evaluations in terms of scalability-related criteria such as network bandwidth and system response time for the analysis of our implemented protocol and proved its performance in our previous study [15]. Fig. 8.1 shows our DMAS-IDS network testbed, where the different types of agents were placed in mutually exclusive machines within the testbed in a manner such that any communication among the agents could only be realized through the generation of network traffic rather than local traffic within the same machine.

In order to test our framework, various experiments were organized to assess the performance of (i) the response time characteristics of the proposed architecture in terms of agent communication with supervised classification scheme, and (ii) the accuracy of classification and policy derivation of the agents. Both the training and testing data sets were acquired from network traffic data generated in our own testbed. To evaluate the performance of the proposed framework, the classifiers were trained offline with the data generated from the testbed. Here, ten types of traffic were generated, which include 5,000 normal connections and 100 connections for each type of abnormal traffic classes. The generated 'normal' connections provide a proper quantity of data and transfer them during a 5-second interval; whereas typical abnormal connections generate a large amount of data in a short span of times (varying for all types of attacks) continuously. For example, connections sending extremely huge packets are used to simulate the 'ping of death' attacks; connections with a lot of packets in a short time duration simulate the 'mail bombing' attacks; connections which try to access the reserved ports are simulated as the 'Trojan infections'; connections transmitting a number of large packets in a short time are used to simulate the 'buffer-overflow' cases, etc.

For processing these data sets, a number of tools were employed including our own previously developed tools. As mentioned above, in JAVA's JPCAP [3][4][6], packages were used in our program to capture the packets from the network interface card and TCP trace was used to transform these packets into data instances. This traffic was generated using our own developed traffic generator, which is capable of generating a number of myriad attacks by simply varying its input parameters. The TCP trace extracts 88 attributes from each TCP connection like elapsed time, number of bytes, and segments transferred on both ways, etc. We have also utilized our own developed feature extraction technique to extract 46 features from the output of TCP trace, which are useful for the proposed DMAS-IDS. These features include some basic, time based, connection-based, and ratio-based network features. Out of these 46 features, 14 are real time rates which are employed for our experiments.

The focus of our testbed based experiments is on network attacks based on the TCP network protocol, since a great majority of the attacks are either executed via or rely on a certain degree on the TCP protocol. This is due mostly to TCP's frangibility and instability. Please note, however, that our proposed architecture is not limited to the detection of simply TCP based network attacks. The network protocol is simply one of many categorical features employed to describe a network connection.

Having our tools installed in all of the machines in our testbed, we start the traffic generator on one of the machines serving as the sender and have another machine as the receiver where we have our Host Agent running. This Host Agent captures the incoming packets by reading it from the network card in the machine and caches it in the memory at the interval of every 5 seconds. Next, it calls the TCP trace software which reads the dump files and converts them to the data instances. Having the data instance from the TCP trace, the Host Agent calls our feature extraction tool to extract the required features from the data instance. After retrieving the resulting features from each data instance (i.e., a network connection), the Host Agent classifies it using PCC.

We had saved all the classification parameters for each of the classes into text files in the training phase. This helps us classify data instances in real time. PCC [20] classifies the data instances only in 'normal' and 'abnormal' classes and returns this label to the Host Agents. If the data instance is classified as 'normal', then it is saved with the Host Agent, but if it is classified as 'abnormal', then the Host Agent further sends it to its Classification Agent on another machine using the 'EVALUATE' message. Upon receiving the 'EVALUATE' message with the abnormal instance information, the Classification Agent calls C-RSPM [14] to further analyze this data instance to classify it to a predefined known attack class. This label is used by the Classification Agent to derive its policy by looking at the features of that attack. The Classification Agent then places this policy into the 'REPORT' message and connects it to the Manager Agent which is located on another different machine. Upon receiving the message from the Classification Agent, the Manager Agent broadcasts it to all the Classification Agents in the network which further send it out to its Host Agents using the 'BROADCAST' message.

8.5 Results and Analysis

Our implemented agent architecture (excluding C-RSPM) has made its mark by proving its scalability, low cost, and low response time in our previous study [15]. It was evaluated with a number of experiments and a total of 506 agents were simulated altogether. We simulated 500 host agents and 5 classification agents along with one manager agent. The experiments over this setup were conducted in order to evaluate its performance in terms of its bandwidth requirements and response time. From [15], it can be observed that the system scaled linearly in terms of the attack response time. The highest response time was of 95 seconds for 100 host agents, 5 classification agents, and 1 manager agent, which is absolutely negligible. This also proved how fast our agents respond and hence how fast they can deliver the task assigned. Also, the bandwidth consumption results of this system were less than 1 MBit/sec, which is almost negligible as well. This demonstrates our communication protocol enables information exchange with low overhead and system scalability. This makes our proposed system low cost which is definitely a desirable feature for any distributed system. It is also clear from the results that in our proposed architecture, the performance of the system will not deteriorate too much with the increase in the number of attacks, which is justified by its low bandwidth consumption and quick response time behavior.

Next, the performance of C-RSPM was evaluated in [14]. As mentioned above, it is capable of performing high accuracy supervised classification and outperforms many other classification algorithms. C-RSPM has shown excellent performance with our testbed data and it has shown that it achieved 100% accuracy for some of the classes for our attack data generated from the testbed, and 99.97% on average for other data with the standard deviation varying only up to 0.09. This was motivating enough for us to integrate it with our agent system and build the proposed DMAS-IDS architecture.

Having proven the performance of our prior developed sub-systems, we next move to demonstrate the performance of the entire DMAS-IDS architecture together. As described above, we generated the attack data for 5900 instances and each of these instances was classified in real time, being captured from the network data card of the machine and then going through each agent layer and respective classifiers. Fig. 4(a) shows the times consumed by each of the 100 instances continually going through the whole cycle of classification and communication, differentiated by the attack types. As can be seen from this figure, the data instances from all types of attacks, on average, result in the same response time irrespective of the attack types. This is definitely a desirable feature. If an IDS performs good for some of the attack types and not in the same way for the others, it can not be considered as a good system. Also, the longest time taken is 40 seconds in our experiments, which shows that our proposed DMAS-IDS architecture shows its promise by addressing an ongoing attack within 40 seconds after it has been detected as an intrusion. The time is measured from when an attack is detected and until the last 'BROADCAST' message has reached the Host Agent.

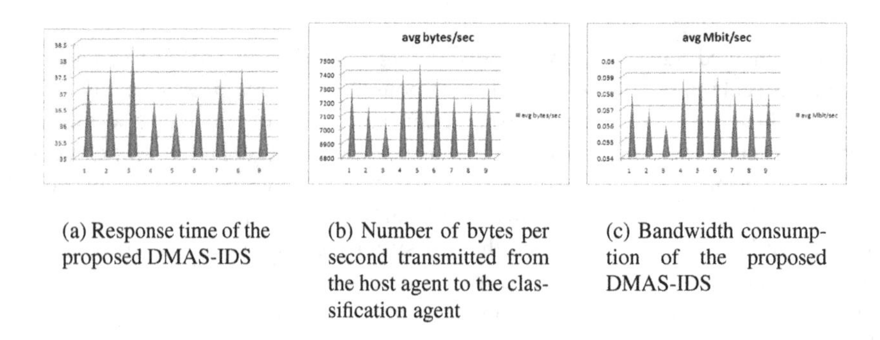

(a) Response time of the proposed DMAS-IDS

(b) Number of bytes per second transmitted from the host agent to the classification agent

(c) Bandwidth consumption of the proposed DMAS-IDS

Fig. 8.4: Experimental results

Let's now look at the bandwidth consumed by this architecture. As depicted by Fig. 4(b) and Fig. 4(c), the maximum bandwidth consumed by the system is 0.06 Mbits/sec, which is very low as well. This proves that even if the number of attacks increases exponentially, our system will still result in considerably fair performance. Also, if more machines are connected to the network, our proposed DMAS-IDS architecture will still withstand the load and deliver results. Therefore, it can be clearly seen that our proposed DMAS-IDS architecture is scalable, fast, and efficient. It provides an option over contemporary firewalls which are only good at blocking traffic for pre-known malicious connections.

8.6 Conclusion

In this paper, a novel distributed multiagent IDS architecture called DMAS-IDS is presented, which incorporates the desirable features of the multiagent design methodology with highly accurate, fast, and lightweight *PCC* Classifier and *C-RSPM* schemes. Even as a larger number of agents are introduced into the network, our proposed DMAS-IDS architecture provides effective communication between its two comprising layers, through an efficient agent communication scheme that requires only a small and manageable generation of network traffic overhead as shown in the experimental results. A key concept in the design of the proposed DMAS-IDS architecture was to show the integration of the agent technology with the data mining strategies, and to prove how both compliment each other. To test the feasibility of a realistic employment and all the salient features of the proposed DMAS-IDS architecture, a private LAN testbed is built to facilitate both the generation of realistic normal and anomalous network traffic data and some common network attack generation tools, to appropriately validate the performance of proposed DMAS-IDS when compared to other well-known anomaly detection and supervised classification methods, and to assess the important scalability aspects of the proposed architecture such as system response time and agent communication generated network traffic overhead. From our experimental results, it can be concluded that the proposed DMAS-IDS architecture yields promising results, indicating a linear scalability and low response time including both agent communication and classification response times. These results are indicative that our proposed DMAS-IDS architecture provides many favorable characteristics such as being lightweight, good scalability, and least degradation of service.

References

1. Garuba M., Liu C., Fraites D.: Intrusion techniques: Comparative study of network intrusion detection systems. Fifth International Conference on Information Technology, New Generations, 2008.
2. Ilgun K., Kemmerer R. A., Porras P. A.: State transition analysis: A rule-based intrusion detection approach. IEEE Trans. Softw. Eng. 21, 3, pages 181-199, 1995.
3. JAMA (2008) Available at: `http://math.nist.gov/javanumerics/jama/`
4. Java Agent Development Framework (2008). Available at: `http://jade.tilab.com/`
5. Jin X., Zhang Y., Zhou Y., Wei Y.: A novel IDS agent distributing protocol for MANETs, V.S. Sunderan et al. (Eds.), ICCS 2005, LNCS 3515, pages 502-509, 2005.
6. JPCAP (2008) Available at: `jpcap.sourceforge.net/javadoc/index.html`
7. Kannadiga P., Zulkernine M.: DIDMA: A distributed intrusion detection system using mobile agents, Proceedings of Sixth International Conference on Software Engineering, Artificial Intelligence, Networking and Parallel/Distributed Computing and First ACIS International Workshop on Self-Assembling Wireless Networks, pp. 238-245, 2005.
8. Kargupta H., Park B., Hershberger D., Johnson E.: Advances in distributed and parallel knowledge discovery, chapter 5, Collective Data Mining: A New Perspective Toward Distributed Data Mining. AAAI/MIT Press, 2000.
9. Klusch M., Lodi S., Moro G.: The role of agents in distributed data mining: Issues and benefits. Proceedings of the IEEE/WIC International Conference on Intelligent Agent Technology (IAT'03),2003.
10. Kumar S., Spafford E. H.: A software architecture to support misuse intrusion detection. In Proceedings of the 18th National Conference on Information Security. 194-204, 1995.
11. lipeRMI (2006). Available at http: `//lipermi.sourceforge.net/`
12. Marhusin M., Cornforth D., Larkin H.: An overview of recent advances in intrusion detection. CIT, 2008.
13. Pahlevanzadeh, B., Samsudin, A.: Distributed hierarchical IDS for MANET over AODV+, IEEE International Conference on Telecommunications and Malaysia International Conference on Communications, pages 220-225, May 14-17, 2007.

14. Quirino T., Xie Z., Shyu M.-L., Chen S.-C., Chang L.: Collateral representative subspace projection modeling for supervised classification. The Proceedings of 18th IEEE International Conference on Tools with Artificial Intelligence (ICTAI'06), pages 98-105, 2006.
15. Sainani V., Shyu M.-L.: A hybrid layered multiagent architecture with low cost and low response time communication protocol for network intrusion detection systems. The IEEE 23rd International Conference on Advanced Information Networking and Applications, Accepted for publication, 2009.
16. Shyu M.-L., Chen S.-C., Sarinnapakorn K., Chang L.: Principal component-based anomaly detection scheme. Foundations and Novel Approaches in Data Mining, pages 311-329, Springer-Verlag, Vol. 9, 2006.
17. Spafford E., Zamboni D.: Intrusion detection using autonomous agents. Computer Networks 34, 4, 547-570,2000.
18. Stolfo S., Prodromidis A., Tselepis S., Lee W., Fan D., Chan P.: JAM: Java agents for meta-learning over distributed databases. Proceedings of KDD-97, pages 74-81, Newport Beach, California, USA, 1997.
19. Vaidehi K., Ramamurthy B.: Distributed hybrid agent based intrusion detection and real time response system. Proceedings of the First International Conference on Broadband Networks, pages 739-741, 2004.
20. Xie Z., Quirino T., Shyu M.-L.: A distributed agent-based approach to intrusion detection using the lightweight PCC anomaly detection classifier. Proceedings of the IEEE International Conference on Sensor Networks, Ubiqquitous, and Trustworthy Computing (SUTC'06), pages 446-453, 2006.

Chapter 9
Automatic Web Data Extraction Based on Genetic Algorithms and Regular Expressions

David F. Barrero[1], David Camacho[2], and María D. R-Moreno[1]

Abstract Data Extraction from the World Wide Web is a well known, unsolved, and critical problem when complex information systems are designed. These problems are related to the extraction, management and reuse of the huge amount of Web data available. These data usually has a high heterogeneity, volatility and low quality (i.e. format and content mistakes), so it is quite hard to build reliable systems. This chapter proposes an Evolutionary Computation approach to the problem of automatically learn software entities based on Genetic Algorithms and regular expressions. These entities, also called *wrappers*, will be able to extract some kind of Web data structures from examples.

9.1 Introduction

Flexible and scalable mechanisms are needed for the integration of information in order to obtain the necessary data from available sources. However, if these sources are not structured, for instance being relational-based, or no design has been previously made by an expert (i.e. a database designer) it is usally difficult to build, and maintain those mechanisms. The previous situation becomes a critical issue when talking about the World Wide Web, considered as a highly heterogeneous data source.

Web Data Extraction (WDE) is a well known and unsolved problem. Also it is related to the extraction, management and reuse of a huge amount of Web data available. These data usually has a high heterogeneity, volatility and low quality.

[1]Computer Science Department
Universidad de Alcalá, Madrid, Spain
e-mail: {dfbarrero,mdolores}@aut.uah.es
·[2]Computer Science Department
Universidad Autónoma de Madrid, Madrid, Spain
e-mail: david.camacho@uam.es

L. Cao (ed.), Data Mining and Multi-agent Integration, DOI: 10.1007/978-1-4419-0522-2_9,

One popular approach to address this problem is related to the concept of *wrappers*. The wrappers [5] are specialized programs that automatically extract data from documents and convert the stored information into a structured format.

The main contribution of this work is a novel approach to the WDE based on Genetic Algorithms (GA) [3] which are used to automatically evolve wrappers. The main difference with other closer approaches [1, 4, 6] is the utilization of regular expressions using a multiagent system to generate them and extract information [2]. A regular expression, or simply *regex*, is a powerful way to identify a pattern in a particular text. Any regex is written in a formal language, that is translated into a particular syntax like POSIX or Perl, and later processed by a regex engine such as Perl, Ruby or Tcl. Regular expressions are used by many text editors, utilities, and programming languages to search and manipulate text based on patterns.

This approach considers the basic (evolved) regex as the atomic extraction element. The representation, genetic operators and fitness function are designed in order to obtain simple extraction elements that are later used, shared, and integrated by a set of information extraction agents. A multi-agent semantic integration plataform named Searchy [1] is used to deploy and test the evolved regex. This approach has to find answers for two important questions. First, how the regex can be represented taking into account its particular features, i.e. vocabulary, syntaxis, grammar and semantic relationships between the grammatical syntaxis and the patterns that it can extract. Second, how once a particular individual is found, it can be combined, or integrated, with others to build a new data extraction (regex).

The following corresponds to the structure of this chapter. Section 2 describes the basic concepts in GA and its application to wrappers and regular expressions. Section 3 explains how a variable length population of agents can support the evolution of regular expressions. Section 4 shows how a specific information agent uses simple gramatical rules to combine, and integrate, the evolved atomic regular expressions. Section 5 shows the experimental results obtained for a set of web documents. Finally, some conclusions and future lines of work are outlined.

9.2 Genetic Algorithms and Its Application in Wrappers and Regular Expressions

This section briefly explains basic concepts related to GA and regex that later will be used to automatically obtain the wrappers.

9.2.1 Genetic Algorithms

From the AI point of view, GA can be seen as a stochastic search algorithm inspired in the biological evolution. GA code the solution of a problem using a string called chromosome or individual, each chromosome represents a point in the search space [3]. If the GA is successful, the individuals will evolve exploring the search space until a global solution is found and the individuals will converge in that solution. The success or failure of a GA depends on the four principal parameters: Genome codification, genetic operators, selection strategy and fitness function.

Genome codification is a key subject in any GA. Each chromosome contains genetic information that codes a solution, therefore it will need a mechanism to mappings between the solution (phenotype) and the gentic code (genotype). Fig. 9.1 represents an example of binary fixed-length coding. The individual represented is the string *[rc]at*. Each attribute in the individual (i.e. each character in the string) is coded by four bits in the chromosome. The piece of chromosome that codes one attribute is called *gen*. Thus, in the example one *gen* codes one character using four bits.

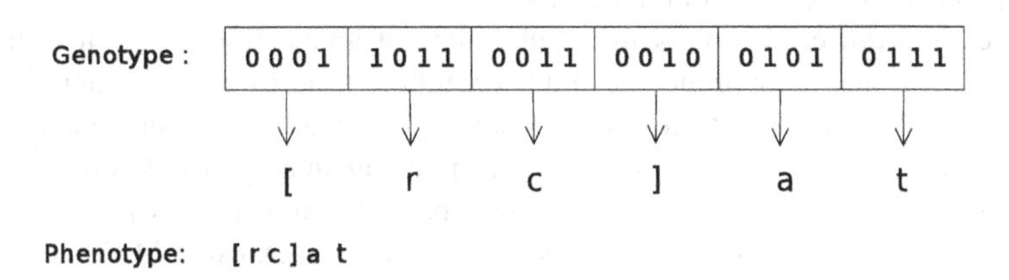

Fig. 9.1: Chromosome coding example

Once the coding has been defined, it is necessary to modify the genetic code of the population in order to explore the solution space. Genome modifications are done by the *genetic operators*. There are two main types of genetic operations: recombination and mutation. Recombination, also known as *crossover*, aims to imitate biological sexual reproduction. It consists of interchanging the genetic code of two individuals. A simple recombination algorithm is the one-point crossover, that is, it interchanges two chunks of chromosomes cutting them in a random point. Another key genetic operator is *mutation*, it introduces random changes in the genome that can generate new attributes in the phenotype not presented previously.

It is necessary to introduce a *selection strategy* in order to improve the population in the successive iterations of the GA (generations). It is analogous to the biological natural selection. The goal of the selection strategy is to generate a selective pressure, this means to force good chromosomes to have more probabilities to reproduce than bad ones. However, there is not a clear non-ambiguous meaning for "good" and "bad" yet.

Goodness and badness are two fuzzy concepts that cannot be used in a scientific context without a precise definition. GA defines good and bad using a *fitness function*. It is a basic piece of any GA, and it is usually one of the most challenging problems that must be faced in order to successfully implement a GA. In some cases defining a fitness function is a trivial issue; however in other problems the definition of the fitness function is more complex. This is the case of the regex evolution.

9.2.2 Regular Expressions

Regular expressions [7] conform a powerful tool to define string patterns. Then, using regex makes possible to manipulate strings according to a potentially quite complex pattern. An extended and well known use of regex is to define sets of files in many user interfaces. The string *rm *.jpg* means in a UNIX shell delete all the files whose name ends in *.jpg*. Actually **.jpg* is a regex representing the set of all strings that ends in *.jpg*.

Many practical applications have been found for regex, especially in the UNIX community, that has achieved a long experience using this tool. Indeed, regex is a basic feature of shell commands like *ls grep* and some programming languages largely used by the UNIX community like Perl or AWK.

Regex is a powerful tool, with a wide range of applications but generation of regex is a tedious, error prone and time consuming task, especially when dealing with complex patterns that require complex regex. Reading and understanding a regex, even if it is not very complex, is far from being an easy task. In order to ease regex generation, several assistant tools have been developed, but writing regex is still a problematic task. An automatic way to generate regex using Machine Learning techniques is a desirable goal that could likely exploit the potential that regex provides.

Our approach proposses two stages for the generation of regex. In a first stage a multiagent system is used to evolve a variable length regex able to extract data from documents that follow a known pattern. Then, a second stage that uses two or more evolved specialized regex to compose a complex regex able to extract and integrate several types of data.

9.3 How Agents Support Data Mining: Variable Length Population

The first stage of the data extraction mechanism proposed in this paper deals with the automatic generation of a basic regex. The aim is to use supervised learning to automatically generate a regex in an evolutionary fashion. A multiagent system (MAS) is used in order to generate basic regex able to extract information to conform a pattern, such as phone numbers or URLs. The agents share a training set composed by positive and negative examples that are used to guide the evolutionary process until the regex is generated.

Extraction capabilities from a regex are closely related to its length, and the length of the regex is determined by the length of the chromosome. Traditional fixed-length GA introduce an arbitrary constrain to the size of the evolved regex that should be avoided for many reasons. The GA should be able to self adapt its genome length without human intervention. One solution might be the use of variable-length genomes, though our interest is an intrinsic parallel solution like a MAS.

Regex are generated by a MAS that unfolds a variable length genome, where subsets of agents use a fixed length genome, as it is shown in fig. 9.2. Each agent runs a GA containing a population whose individuals own a chromosome of fixed length, and can evolve by its own with a high degree of independence that conform a microevolution. Agents are not isolated thus their populations are influenced by other populations by means of emigration: a part of the population can emigrate from one agent to another agent every generation, so the evolution of one subpopulation is affected by the evolution of other agents. The result is that the total population of the MAS presents a macroevolution. Microevolution and macroevolution are different problems that must be addressed individually.

9.3.1 Macroevolution

The MAS is actually a way to implement variable-length GA, in which the sets of agents containing populations of different length evolve as a whole. The mechanism that makes this macroevolution possible is the population interchange among agents. An agent containing a population with a chromosome of length n always clones a number of individuals to a population with a chromosome size $n+k$, where k is the gen size (see fig. 9.2). Thus, the genetic operation that modifies the length of the chromosome is performed when the individual is emigrating, adding a new chunk in a random position of the chromosome. The chunk that is inserted into the genome is a non-coding region, i.e., a chunk that codes an empty character and therefore, it does not affect the phenotype. Otherwise the potentially good genetic properties of the individual might be lost.

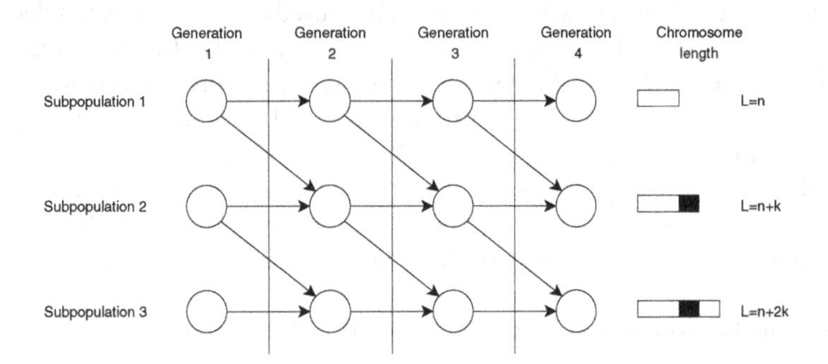

Fig. 9.2: Population interchange among agents

An important issue is the selection process of the individuals that emigrate and the selection of the individuals that are replaced in the target population. Both selections are done using a tournament, however, there is a difference. The tournament of the individuals that emigrate is won by the individual with the best fitness, meanwhile the tournament between the individuals to be replaced is won by the individual with the worst fitness. In this way, individuals with high fitness in a population have more chances to emigrate than individuals with a bad fitness. On the other hand, chromosomes with low fitness will likely be replaced by better immigrant chromosomes.

Each agent in the MAS contains chromosomes of different sizes and they send or receive a bunch of individuals each generation, which unfolds a population using the same parameters, strategy and fitness function.

9.3.2 Microevolution

A classical fixed-length GA is run within each agent, evolving a population of chromosomes in a microevolution.It can be described in terms of genome codification, genetic operators, selection strategy and fitness function.

The GA implemented in the MAS uses a binary genome divided in several gens of fixed length. Each gen represents a symbol from an alphabet composed by a set of valid constructed regular expressions. It is important to point out that the alphabet is not composed by single characters, but by any valid regex. These simple regular expressions are the building blocks of all the evolved regex and cannot be divided, so, we will call them atomic regex.

Genetic operators used in the evolution of regular expressions are the mutation and crossover. Since the codifications rely in a binary representation the mutation operator is the common inverse operation meanwhile the recombination is performed with a one-point crossover. These genetic operators do not modify the genome length; chromosomes modify their length only when an individual is migrating to another agent. The selection mechanism used is the tournament selection.

From a formal point of view, the fitness function is defined as follows. Given a set of positive examples, P, with M elements, and a set of negative examples, Q, with N negative examples, let $p \in P$ be an element of P, and $q \in Q$ an element of Q, we define $\Omega = \{\omega_0, \omega_1, ..., \omega_n \mid \omega_i \in P \cup Q, n = N + M\}$ as the set of elements contained by P and Q; therefore any element of P or Q belongs also to Ω. Let the set of all regular expressions be R, and r an element of R. Then, we can define a function φ_r, $\varphi_r(\omega): \Omega \times R \longrightarrow \mathbb{N}$ as the number of characters of ω that are matched by the regex r.

Finally the fitness function $\mathfrak{F}: Q \longrightarrow \mathbb{R} \in [-1, 1]$ is defined as:

$$\mathfrak{F}(r) = \frac{1}{M} \sum_{p_i \in P} \frac{\varphi_r(p_i)}{\mid p_i \mid} - \frac{1}{N} \sum_{q_i \in Q} M_r(q_i) \tag{9.1}$$

where $\mid \omega_i \mid$ is the number of characters of ω_i and $M_r(q_i)$ is:

$$M_r(q_i) = \begin{cases} 1 \; if \; \varphi_r(q_i) > 0 \\ 0 \; if \; \varphi_r(q_i) = 0 \end{cases} \tag{9.2}$$

Since true positives are calculated based on the characters, and false positives have no intermediate values, the fitness function presents an intrinsic bias: it is more sensible to false positives than to true positives. It is important to point out that the maximum fitness that an individual can achieve is 1 and it is given when all the positive examples have been completely retrieved and no negative example has been matched. If a chromosome obtains a fitness of 1, it will be named an ideal chromosome.

From the evolution of each specialized MAS we obtain a basic regex able to extract strings matching a pattern. Each MAS requires a training set and, eventually, an appropriate alphabet of atomic regex. Once the basic regex has evolved, it is possible to build more complex regex in the second stage of the extraction process.

9.4 Composition of Basic Regex

We use the grammatical rules provided by the regex notation in a composition agent to integrate the basic evolved regex from the first stage. The composition agent uses a manually created rule database to integrate two or more basic regex. The agent applies the grammatical rules to the input regex obtaining a set of composed, potentially complex, regex. They might be not suitable to extract information properly, so, regex created by the grammatical rules have to be filtered in order to select the valid ones. We use the traditional data mining F-measure to automatically evaluate its extraction capabilities and select the composed regex.

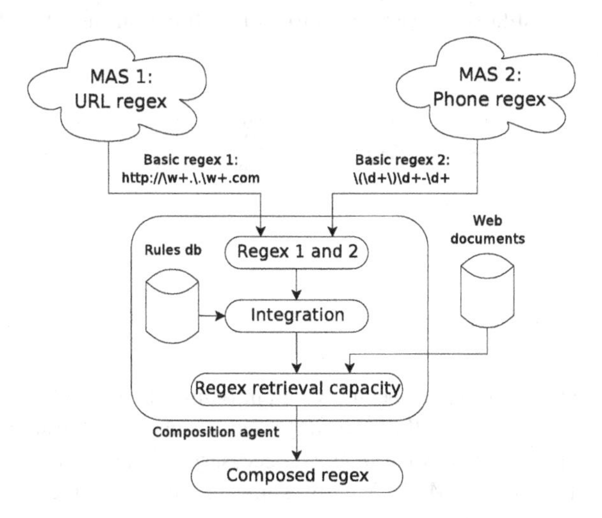

Fig. 9.3: Composition agent architecture

A graphical representation of the composition process can be seen in fig. 9.3, where an example of regex composition is depicted. The composition agent takes two basic regex from the output of two evolutive MAS and applies a set of grammatical rules stored in a database to generate a set of composed regex.

Suppose that the composition agent takes *http://\w+.\.\w+.com* and *\(\d+\)\d+-\d+* as basic regex, and the aim is to compose them using a subset of regex operators, for example, |, *(*, *)*, + and *?*, then it is possible to define a database of grammatical rules in the composition agent such as:

```
Rule 1: X|Y Rule 2: XY Rule 3: X+Y? Rule 4: (Y)+|((X)+|foo)
```

Where X and Y are *http://\w+.\.\w+.com* and *\(\d+\)\d+-\d+*. The composition agent applies the grammatical rules to the input regex generating the following set of composed regex:

```
Composed 1:http://\w+\.\w+\.com|\(\d+\)\d+-\d+ Composed 2:
http://\w+\.\w+\.com\(\d+\)\d+-\d+ Composed 3:
http://\w+\.\w+\.com+\(\d+\)\d+-\d+? Composed 4:
(\(\d+\)\d+-\d+)+\((http://\w+\.\w+\.com)+|foo)
```

Since the regex composition has used a brute force approach, not all the composed rules are supposed to be able to correctly extract data. Therefore, it is necessary to select at least one valid regex. This is done by the regex retrieval capacity that evaluates the generated regex calculating the F-measure using a dataset composed by several documents (see equation 3). This measure is based on the weighted *harmonic mean* from classical Information Retrieval *Precision* (*P*) and *Recall* (*R*) values. Of course, other extraction quality measures such as F_β or E are also valid. Based on these quantitative measures, automatic estimation of the best composed regex is possible.

$$F_{measure} = \frac{2PR}{P+R} \tag{9.3}$$

9.5 Experimental Evaluation

The experimental evaluation has been divided into three stages with different goals. The first stage is the setup of the experiment, in which several tests were carried out in order to set the basic GA parameters, necessary for the second stage in which the regex evolution uses a MAS. Finally, some grammatical rules are used with the evolved regex and they are automatically evaluated in order to obtain a final composed regex.

Some initial experiments were carried out to acquire knowledge about the behaviour of the regex evolution. In order to achieve this goal a single GA was used with the same configuration required by the MAS. Due to the lack of space, only the most significant results are enumerated without further discussion.

Despite the differences between phone and URL, both evolved regex have similar behaviours, in this way it is possible to extrapolate experimental results. The best results are achieved with mutation probabilities between 0.01 and 0.02 thus an average mutation probability of 0.015 was fixed to carry out the rest of the experiments. Tournament size has shown to have a remarkable impact to obtain a faster convergence while avoiding local maximum. Experiments have demonstrated that a tournament size of three is a good balance. Variation in the size of the population shows the usual GA behaviour, a population with fifty individuals is a good trade-off between convergence speed and computational resources.

9.5.1 Results: Regex Evolution

The second experimental stage aims to use a MAS with subsets of agents where populations of chromosomes with different lengths are evolving. Six subsets of agents have been used with chromosomes sizes of 6, 9, 12, 15, 18 and 21 bits organized in chunks of three bits. The emigration strategy has been set as described in section 3.1. Agents in the MAS run a GA with the parameters obtained in the experiment setup phase and following the same experimental procedure.

Fig. 9.4 depicts the evolution of the average fitness of each subset of agents for the phone numbers regex evolution. Since the results for the URL regex evolution are analogous, no figure is included. It should be noticed the close relationship between the convergence speed and the chromosome size. The longer is the chromosome, the longer it takes to converge because the chromosome codes a solution in a bigger search space. Another fact that influences the difference in the convergence speed is found in the limited speed of propagation of good chromosomes along the MAS.

Fig. 9.4 shows an interesting fact in relation to the different fitness convergente values in the agents for the phone number evolution. shows is the different fitness convergence values in the agents for the phone number regex evolution. A population with a chromosome length of 6 bits presents a fast convergence to 0.54. With 6 bits only very small phenotypes can be represented, just two symbols, so only part of the examples can be extracted, achieving a maximum fitness of 0.54. This fact is also found in populations of size 9, but with less dramatic effects. The populations with chromosomes of length 12 present a very important growth in the fitness. The reason is that the shortest ideal regex must be coded with at least 12 bits.

9.5.2 Results: Regex Composition

Two basic regex have been selected for the third evaluation stage, where they are composed and its extraction capabilities are measured using precision, recall and F-measure.

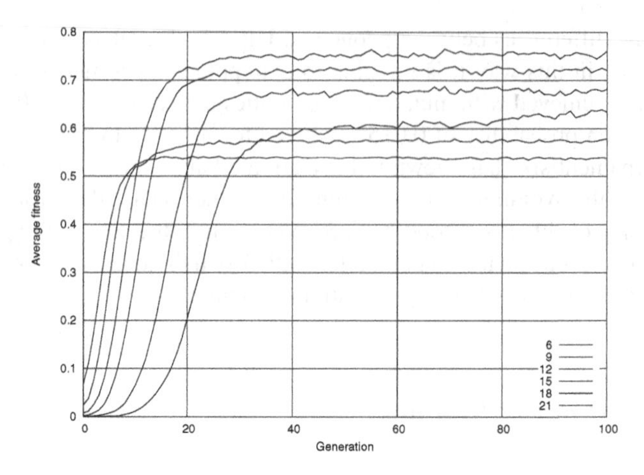

Fig. 9.4: Average fitness for the phone number regex evolution

Table 9.1: Evolved regular expressions

Evolved regex (URL)	Fitness	Evolved regex (Phone)	Fitness
http://-http://http://	0	\w+	0
conwww\.http://com-www\.com	0	\(\d+\)	0.33
/\w+\.	0.55	\(\d+\)\d+	0.58
http://\w+\.\w+\	0.8	\(\d+\)\d+-\d+	1
http://\w+\.\w+\.com	1		

A dataset was needed to measure the precision and recall. The experiment used a dataset composed by ten documents from different origins containing URLs and phone numbers, mixed and alone. Table 9.3 shows basic information about the dataset and its records. Documents one, two and three are composed by examples extracted from the training set. The rest of documents are web pages retrieved from the Web; however documents five and six were transformed to a plain text format in order to remove all the URLs.

The calculus of precision and recall use the total number of records in the document, i.e., the sum of URLs and phone numbers, regardless the evaluated regex. It means that the result is strongly biased against regex that are not able to extract both URLs and phone numbers. It should be noticed also that an extracted string is true if and only if it matches exactly the records, otherwise it has been computed as a false positive.

Results, as can be seen in table 9.2, are quite satisfactory for the pre-processed documents, i.e., documents one to five, but measures get worse for real raw documents. X has a perfect precision; meanwhile Y has a poor average precision of 0.41. It can be explained looking at Y. This regex has the form *http://\w+\.\w+\.com*, which means that it only extracts the protocol and the host name from the URLs, but it cannot extract the path, a common part of URLs found in the Web.

A special case is the regex XY, a direct concatenation of X and Y. This regex extracts URLs followed directly by a phone number; obviously, such situation is not likely to happen. So, it is unable to extract any record (for this reason these results are not shown in the table). After all, the best balance between precision and recall is achieved by the composed regex $(X) \mid (Y)$, with a precision of 0.63 and a recall of 0.62.

Table 9.2: Extraction capacity of basic and composed regex. It is calculated using traditional precision (Prec.) and Recall values. The table shows the Retrieved elements (Retr) and the True Positives (TPos) detected.

	X				Y				(X)|(Y)			
	Retr	**TPos**	**Prec.**	**Recall**	**Retr**	**TPos**	**Prec.**	**Recall**	**Retr**	**TPos**	**Prec.**	**Recall**
Document 1	5	5	1	1	0	0	-	-	5	5	1	1
Document 2	0	0	-	-	5	5	1	1	5	5	1	1
Document 3	5	5	1	0.5	5	5	1	0.5	10	10	1	1
Document 4	99	99	1	1	0	0	-	-	99	99	1	1
Document 5	10	10	1	1	0	0	-	-	10	10	1	1
Document 6	0	0	-	-	43	6	0.14	0.12	43	6	0.14	0.12
Document 7	20	20	1	0.21	773	12	0.16	0.12	97	32	0.33	0.33
Document 8	37	37	1	0.05	668	76	0.11	0.11	705	113	0.16	0.16
Document 9	24	24	1	0.13	88	1	0.01	0.01	112	25	0.22	0.14
Document 10	0	0	-	-	49	23	0.47	0.45	49	23	0.47	0.45
Average			1	0.56			0.41	0.33			0.63	0.62

Precision and recall balance can be quantified with the F-measure (shown in table 9.3). The $(X) \mid (Y)$ regex obtained the best value followed not far by X. As it was expected, the composition agent selectes $(X) \mid (Y)$ based on the F-measure.

Table 9.3: Document record types and F-measure of regexes.

	URL	Phone	**X**	**Y**	**(X)|(Y)**
Document 1	0	5	1	-	1
Document 2	5	0	-	1	1
Document 3	5	5	0.67	0.67	1
Document 4	0	99	1	-	1
Document 5	0	10	1	-	1
Document 6	0	51	-	0.12	0.12
Document 7	77	20	0.35	0.14	0.33
Document 8	436	37	0.09	0.11	0.16
Document 9	241	24	0.23	0.01	0.17
Document 10	51	0	-	0.46	0.46
Average			0.62	0.35	0.69

9.6 Conclusions

An innovative approach for data extraction based on regex evolution and grammatical composition of regex has been presented. We have shown that it is possible to use a GA to evolve regex in a MAS and to apply grammatical rules to the evolved regex in order to generate a composition of regular expressions with the capacity to extract different records of data.

Using a MAS to simulate a variable-length genome population has showed to be a successful way to generate a variable-length chromosome evolution. Each agent is able to evolve a population and the MAS presents a macroevolution that tends to generate regex correctly sized.

However, the experiments carried out show some limitations. The linear nature of the GA codification is not the best option to represent a hierarchical structure such as a regex. The result is a natural difficulty to define a fine-grained fitness function able to evaluate not only all the regex, but also its parts. For these reasons the next step to follow is to use other evolutionary algorithms, such as Genetic Programming and Grammatical Evolution that overcome this limitation.

Finally, grammatical rules offer a simple way to automatically compose basic regex and select the best composed regex measuring its F-measure with a set of documents.

Acknowledgements

The authors wish to thank Mónica Gamboa for her review. This work has been supported by the research projects TIN2007-65989, TIN2007-64718 and Junta de Castilla-La Mancha project PAI07-0054-4397.

References

1. David Camacho, Maria D. R-Moreno, David F. Barrero, and Rajendra Akerkar. Semantic wrappers for semi-structured data extraction. *Computing Letters (COLE)*, 4(1), 2008.
2. Longbing Cao, Chao Luo, and Chengqi Zhang. Agent-mining interaction: An emerging area. In *AIS-ADM*, pages 60–73, 2007.
3. John H. Holland. *Adaptation in Natural and Artificial Systems: An Introductory Analysis with Applications to Biology, Control, and Artificial Intelligence*. The MIT Press, April 1992.
4. Marat Kanteev, Igor Minakov, George Rzevski, Petr Skobelev, and Simon Volman. Multi-agent meta-search engine based on domain ontology. In *AIS-ADM*, pages 269–274, 2007.
5. Nicholas Kushmerick. Wrapper induction: Efficiency and expressiveness. *Artificial Intelligence*, 118:2000, 2000.
6. M. Michalowski, J.L. Ambite, S. Thakkar, R. Tuchinda, C.A. Knoblock, and S. Minton. Retrieving and semantically integrating heterogeneous data from the web. *IEEE Intelligent Systems*, 19(3), 2004.
7. Ken Thompson. Programming techniques: Regular expression search algorithm. *Commun. ACM*, 11(6):419–422, 1968.

Chapter 10
Establishment and Maintenance of a Knowledge Network by Means of Agents and Implicit Data

Jaime Moreno–Llorena, Xavier Alamán and Ruth Cobos

Abstract Semantic KnowCat (SKC) is a groupware system prototype for Knowledge Management on the Web by means of semantic information without supervision. The main aim of SKC is to select the knowledge contained in the system by paying attention to its use. This paper presents the SKC Network Module (NM), which is in charge of discovering other instances of the system on the Internet and establishing contact with them to create a knowledge network on the Web. In order to do this, each instance of the system is represented by a software agent, which is in charge of interacting with Web search engines and collaborating with the agents that represent other system instances, thereby using data mining techniques. As a result, SKC manages to build and maintain a network node to share knowledge.

10.1 Introduction

Information overload is one of the problems of the ICT use extension. The Web is the most general and significant example of this phenomenon. We think network knowledge management systems have the most important characteristics of the systems with this problem, but these systems are more scalable and have more controllable parameters, so they may be used as an experimental model for research. For example, the Web could be seen as a global knowledge management system.

Our general hypothesis is that there are several hidden aspects in the systems affected by the information overload which can contribute positively to the solution of this problem. On one hand, taking advantage of the excess energy of the active elements that are involved in the systems, such as users, services, applications and other entities related to them -e.g. software agents-. On the other hand, using the

Jaime Moreno-Llorena, Xavier Alamán and Ruth Cobos
Department of Computer Science, Universidad Autónoma de Madrid, Spain, 28049 Madrid
Telephone: +34914972212, Fax: +34914972235
e-mail: `(jaime.moreno, xavier.alaman, ruth.cobos)@uam.es`

L. Cao (ed.), Data Mining and Multi-agent Integration, DOI: 10.1007/978-1-4419-0522-2_10,

properties of both the elements and the activities related to the systems affected by the problem, e.g. the network, the active entities mentioned above, both the information and the knowledge involved, or the processes and the interactions of that elements and activities.

In order to investigate this hypothesis we have used, as an experimental platform, a knowledge management system called KnowCat (KC) [2] [6]. This is a groupware system that facilitates the management of a knowledge repository about a knowledge area by means of user community interaction through the Web. This can be done without supervision by using information about user activities and user opinions about the knowledge elements that are part of the knowledge repository.

The knowledge repository is constituted fundamentally by two components: the Documents, that are the basic knowledge units; and the Topics, that are structured hierarchically as a Knowledge Tree. Each document describes the topic where the document is located in the tree. Each topic appears once in the tree, and it could include several documents that describe it and it could include some subtopics too.

The User Community is composed by users that could interact with the knowledge repository and with other users through the system. The users could contribute to the repository with new documents, placing them in one of the knowledge tree topics, and they could review all the published documents by the community. The users could also express their opinion about all the documents of their node repository by means of annotations and votes.

Each system instance is a KC node that deals with a specific subject and has a user community and its own knowledge repository. KC nodes are independent and there may be many of them in active and connected to the Web at the same time, as happens with Web sites. Even though all the nodes are connected to the Web, they ignore one another and they don't have any mechanism to collaborate in order to improve the service that they offer to their respective users. This could change if the nodes became active entities and were integrated in a node community dedicated to stimulate and facilitate the activity of their respective users. This is the approach proposed in this paper by means of software agents and some data mining techniques integration.

In order to corroborate the design hypothesis of the proposed approach, a prototype has been developed on KC, which has been called Semantic KnowCat (SKC) [13]. SKC includes an Analysis Module (AM) [14] that is in charge of analyzing the system knowledge repository with data mining techniques [4] to describe its elements by means of Words Weight Vectors (WWV) [3]. In addition, SKC incorporates a Network Module (NM) that is in charge of finding other existing KC nodes on the Web, establishing contact and collaborating with them, in order to create and maintain a network that integrates the known active nodes in each moment, and allows to determine and preserve a particular knowledge interchange relationship between those that share the same interests. To achieve this, the module uses software agents [10] that act on behalf of the system nodes in a decentralized manner and interact one with the other and with other Web applications to achieve the so mentioned objectives. The nodes' WWVs and the WWVs of their knowledge repository elements are the agents' transaction objects and the comparison of these vectors guides the agents' behaviour. As a result, in every system instance two views are available, one of the nodes accessible through the Web, and the other of the contents linked when the nodes are alike. The agents are able to react to nodes' changes, to their appearance and disappearance, and to their content changes, accommodating the network to the variable circumstances throughout the time.This paper shows how a knowledge network by means of both software agents and data mining techniques could be established and maintained on the Web.

The proposed approach has its antecedents in diverse references, such as: the procedures used in the current Web searcher engines, as Google, for indexing Web sites and relate search queries to Web pages [3]; the techniques used in some recommendation systems [1], as the book shop Amazon; the recent Web 2.0 tendencies [16] that use the Web structure, the social networks, the collective intelligence and the collaboration, in collaborative applications as wikis [12] or folksonomies [8]; as well as the automatic annotation procedures [11] and the ontology mapping techniques [15] -KC knowledge tree may be a light ontology [9]-, that are still open questions and essential in the transition from the current Web to the Semantic Web. Furthermore, this approach is related to data mining field, especially to text mining techniques based on vector model [3], and also to the Semantic Web, regarding the agents' use [10] to provide systems with pro-activity and gregarious capacity.

The problem of mapping between ontologies, has been treated with several processes [5] that always seem to require human validity. The systems and algorithms that use the classified contents in the ontologies to carry out mapping seem to be the most autonomous; among them you can find FCA-Merge, LOM, CTXMATCH [5]. We believe that this strategy is the most appropriate to work with light ontologies in an unattended way within environments that may profit from their mapping, but that don't have this functionality as their main objective, just like some characteristic applications of Web 2.0 -wikis, weblogs, folksonomias, etc.- and of the SKC.

Moreover, the SKC system is a collaborative system that exploits user networks to achieve its objectives. This same strategy is the one that the system instances intend to exploit when being integrated in knowledge node "collaborative" networks, embodying in software agents that cooperate with objectives coinciding with the ones of human users, but at a different level. Software agents are potential users of the Semantic Web and their gregarious capacity may be used in it as the human tendency of grouping together does on the Web 2.0.

Although the central functionalities of SKC node integration could have been achieved with the use of Web Services (WS) or Remote Procedure Calls (RPCs), for instance, with this type of choice the light, flexible, collaborative and network model, what is aimed to be provided in the system, would have been distorted. Web 2.0 applications prefer open and straightforward communication processes, such as the family of Web feed formats RSS, to the most sophisticated mechanisms, such as WS. This same philosophy has led us to Web search engines as alternative means of system instances discovery, instead of more conventional and rigid options, such WS or RPCs.

10.2 The SKC Analysis Module

As explained deeply in a previus paper [14], the Analysis Module (AM) of SKC is in charge of processing the knowledge repository of the system by means of text mining techniques to enable the comparison between the elements of knowledge within and outside the nodes. Thus AM provides the basis for features such as: automatic annotation of documents, which enables its automatic classification on the knowledge tree topics of the node; the clustering of knowledge elements to support recommendation services; or the mapping between knowledge trees of different nodes -that is a case of light ontologies mapping-, which allows the establishment of the knowledge network and exchange of knowledge between nodes, by means of software agents.

In our approach we have considered, initially, four types of knowledge items : nodes that are system instances in charge of the knowledge management about an area with the help of a user community; topics structured in the form of a knowledge tree that develop the different aspects of the main node topic; users that constitute the community that participates in the node; and documents that describe the different topics and are provided by the users, searched by them and which are the object of their consideration.

It is possible to associate a text document with all the knowledge items that are considered in the system that describes them. These contributions may have several backgrounds. First, text associations to knowledge items may come from the nature of these items; for instance the documents used in experiments are textual type. Secondly, these associations may stem from the explicit relations of knowledge items with other items that already have associated texts. This occurs with topics that organize the documents, users that provide documents to the system or the node that contains both. Thirdly, text associations descriptive to the items may be inferred from more dynamic relations, like the one that is established between users and the documents that are more frequently visited by them or the ones they give their opinion about, or like the ones that are shown between documents referenced among them. Lastly, it is always possible to associate texts descriptive to the items that affect some aspects of utility, such as users' curricula vitae, their topics of interest, key words associated with documents or descriptions of topics. This case is completely general and may be applied to non textual documents, such as images, sound, etc. In this paper, the first two cases have been considered.

The explicit relations of knowledge items are established by analysing how these are related to each other in the knowledge repository. In our approach we have considered the way in which the documents and the topics within the knowledge tree are organised, and how users relate to the documents they provide to the system. The analysis follows a process that goes through three stages. First, the AM establishes the knowledge items included in the tree that need to be treated. Next, the AM identifies the users responsible for the valid knowledge items in the knowledge tree. Finally, the AM recovers the textual components that constitute the texts associated with the knowledge items through the Web, e.g. the text associated with a topic shall be made up of texts associated with each document and subtopic included.

Once a descriptive text has been associated with one of the items considered, it is necessary to put it in a way that it can be used as a comparison instrument. This is achieved by converting the text into a descriptor that shall be connected to the aspect it is referring to. For instance, if the text associated with a user describes the topics of interest for the latter, the corresponding descriptor shall refer to the users' preferences; but if the text describes the documents it itself has elaborated, the corresponding descriptor shall refer to its creative job. Like this, items shall be able to have as many descriptors as aspects of them are taken into account.

In our approach descriptors are Words Weight Vectors (WWV) that may be used to determine similarities with other vectors of the same kind and thus relate the corresponding knowledge items. The similarity between two vectors is the measurement of the distance between them. In this approach we have considered the distance between vectors is estimated according to the cosine of the angle they form. The level of similarity between two vectors is a coefficient between zero and one. The closer the value is to the unit, the more similar vectors are, and the closer to zero, the less similar they shall be. The relation of similarity established between knowledge items is described with this coefficient. In our approach the knowledge items that exceed a specific threshold of similarity coefficient of the relation between both are considered to be related. Unfortunately, this threshold may not be established neither in a fixed way nor in a general way for all cases, given that depending on circumstances such as theme nodes or the nature of the documents taken into account, the election of its value may vary greatly. [3] [4]

10.3 The SKC Network Module

The intention of the SKC Network Module (NM) is to increase projection of the nodes in the system, enrich their knowledge repositories and stimulate the corresponding user communities. The aim is to achieve all this by means of popularization of the knowledge available among collective groups interested, on a specific moment and place where it can be relevant, handy and very little intrusive, and with less need for attention by the recipients.

As mentioned above, the NM is based on agents that collaborate in the creation and maintenance of a network formed by all the SKC nodes available and is in charge of the establishment and support of the knowledge interchange relationship between nodes related to the same topics. In order to do this, each agent must develop several tasks in an autonomous and proactive manner: find other nodes on the Web, contact them to check their activity and obtain basic information about them, determine the appropriate type of link to be used with them, obtain the necessary information to establish the corresponding links, periodically visit the identified nodes to check the continuity of their activity, deactivate all the registered nodes that have stopped all contact, activate inactive nodes that are fully functioning again and verify in a reactive manner the link with active nodes according to the development of the node itself and of others. To localize other nodes on the Web agents must not only interact within each other but also with other Web applications, which are considered Web search engines for general use in the prototype development. In addition, each agent must attend to the demands of others and must make use of the interaction of the unknown to identify them and establish a relationship with them. All this activity requires the use of several social skills based on communication, collaboration and the use of appropriate protocols.

Internally, agents have an interface to interact with similar agents (AII) and another interface to interact with other Web applications (W3I). Both interfaces use TCP/IP network as a communication infrastructure. The AII establishes a low-level communication between agents, directly using sockets, in order to support the conversation in FIPA ACL language [7], which encapsulates the messages typically used by the agents into messages and sequences interaction of this language. Each agent establishes its AII on port IP free from the machine where it resides. Only FIPA interaction message sequences indispensable for a basic operation have been implemented in the prototype developed. The W3I behaves like a Web browser controlled by the agent it belongs to, therefore it uses HTTP protocol to access Web servers of interest -that are Web search engines in the prototype-, such as any human being would do. W3I allows sending petitions to the servers and receiving replies in HTML format.

Each agent is an instance of a Perl program, which runs on the machine where its SKC node is located. The two interfaces, W3I and AII, are implemented with three processes that the program creates. The first one corresponds to the W3I and it is an HTTP client that allows the agent to interact with Web search engines. The second one is the IIA input and it is responsible for dealing with the agents of other SKC nodes. Finally the third process is the IIA output and it allows the agent to visit other nodes agents. These processes use a database to exchange data that regulate their behaviour throughout time.

Agents shall pay attention to the communications of other agents at all times. In the prototype only one visit shall be attended to at a time and the concurrent visits shall have to wait until the previous one has ended, although the waiting time is limited for the visitor and the visits shall be considered vain if this time limit is exceeded. In addition, each agent shall follow a visit policy to maintain contact with all the known agents that are considered to be active. In the prototype, the agents that haven not paid attention to a certain number of consecutive visits are considered inactive, and the visit policy shall consist of attempting to access the active agents with which no contact has been attempted for a longer period of time. In a similar way, agents shall maintain a periodic visitation schedule to Web servers used to discover new nodes on the Internet.

In order to interact with the Web applications mentioned before, agents need to have information about them. In the case of the Web search engines taken into consideration, the information needed is the following: searcher URL, content and format of the data required, delivery channels and a reply format. In the prototype these are considered to be Web search engines that return HTML pages that link to the elements being searched. Agents will process these pages to obtain the data required.

The data base for each KnowCat (KC) node includes complete information on all items that constitute its traditional repository of knowledge. In prototypes developed using KC, presented thus far in previous papers [13] [14], the knowledge items taken into consideration were local in each node: documents, users as authors, topics, knowledge trees, relationship that linked ones with others and the node itself.

In addition, for NM support, the system needs to keep information on knowledge items of other nodes in its database: nodes, topics, users and documents, all of them remote. The information about knowledge items alien to a node does not have to be as thorough as their own, and their composition depends on the kind of items it is dealing with, on the existing link with the node to which they belong and the use they are going to be given. In every case, the other nodes knowledge items live in their own nodes and in the databases of the rest only references and indispensable data are kept for manipulation.

Despite the fact that the system takes into consideration the remote condition of the new knowledge items for its process, as it can distinguish the difference between other local item categories, the general principles of their use are the same: The knowledge items are classified into different types -documents, remote documents, topics, remote topics, etc.- each type of item has a particular representation in the database, items may be associated multiple descriptors to describe them, such as WWVs (in term vector model) [3], and some items can be related to others based on their descriptors. In this way, it is possible to analyse the data records and the activity of new items as it was with the already existing records, incorporate these items in pre-existing services and system functionalities [14], and integrate them with the rest into new tools specifically designed for the SKC nodes network.

To find new nodes on the Web the agents use three different strategies. The first one consists of using the references provided during installation, this is the simplest alternative but the least powerful and it aims at exploiting the knowledge of the ones in charge for the installation of new nodes over other nodes on the internet. The second one tries to use conventional Web search engines; this option is more powerful than the first one because it allows exploiting general use and public domain resources available on the Web to recuperate information. These resources are relatively persistent, autonomous and independent from SKC. In order to use them it is necessary to provide information about each of them to the system, basically, the corresponding URLs, ways of invoking them and the formats and data they require. The third strategy for finding nodes on the Web consists of using proportionate references by other nodes with which there is contact, this possibility matches with the two previous ones and dynamizes these, as it allows to renovate the nodes and Web search engines references, making use of the communication and collaboration between nodes, which are two fundamental aspects of the SKC network.

As mentioned above, in order that the Web search engines can find the nodes' references on the Web, these must be published in an appropriate way: in a location tracked by Web crawlers, with access authorized by them and with data of interest for human users, which is what Web indexers expect to discover. In the prototype development, each SKC server is provided with a special page with reviews of all their active nodes and links, easy to identify automatically, in order to access to the data really needed by the agents.

By means of the NM the SKC nodes may be linked in two ways. In both cases, the nodes must have contact to interchange information about the active nodes they know and that constitute the SKC network at every moment. As time goes by the group of active nodes may change, as the new nodes are created and others disappear. With data interchanged in this way, every node in SKC network has a relatively updated image of the active nodes that integrate it at every moment. In the most basic link between nodes, only this type of data is interchanged. When the nodes deal with similar topics, apart from the previous information, they interchange information about theirs knowledge repositories contents. With this information linked nodes may establish a relationship among the items of their respective repositories. This is the strongest link category taken into consideration in the prototype development. As nodes are systems in continuous development, in which the content changes, their links are likely to be altered as time goes by.

The link between nodes is a process that goes through several phases depending upon the circumstances under which it occurs. When the identification occurs through a third node or as a result of the first visit from the node discovered, only a minimum information contact is obtained. After the first visit to the new node, the remaining data is obtained. However, when a new node is found through a Web search engine, all the data is obtained directly. Once you have all the discovered node identification signs, in the following interaction, the discoverer obtains the information to determine the similarity of theirs subjects and to establish the kind of relationship with the other. If the subjects are not alike, nodes are linked in a basic way and periodically will interchange data on the active nodes they know, as well as subject information to revise its link. On the other hand, if the subjects are similar, the link is a content link and additional periodic interactions will be needed to obtain information of the knowledge items alien to the discoverer node and to relate them to its own.

As we have seen, in order to establish the kind of link between the nodes and the relationship between its knowledge items, it is necessary to compare the corresponding knowledge repositories. However, it is not convenient that the nodes interchange their entire repositories to do this, due to the great size they usually have, and the resources and time needed for their repeated process. In the prototype development descriptors that represent the knowledge items of the repositories are used, they are used as tools of comparison and usually have a smaller size than the item they refer to [14]. Specifically, in the experiments carried out descriptors of WWVs (term vector model) have been used [3], obtained from texts associated with the knowledge items they represent. These descriptors are previously created in the nodes that contain the corresponding items, they can easily be sent through the network, and allow the comparison between elements with a reasonable effort.

As a result of the integration of the nodes in the SKC network, each of them has a register of the nodes in it. With the data registered the nodes can establish the similarity among the subjects of the different nodes and fix those which are related to theirs. With this information it is possible to draw a map of the SKC network in which the nodes and the links may be shown. In the developed prototype a view of the network in an interactive graph shape is provided, where all the nodes registered are shown with a visual indication of their subject links. Clicking on the nodes in the graph you may surf through the SKC network, having access to the corresponding nodes.

In addition, each node has a knowledge item register of nodes with subjects similar to its own. With this data nodes can establish the similarity among their knowledge items and related node items. In the developed prototype, this data can help integrate network nodes into a relationviewer between knowledge items and into a recommendation service previously prototyped [14]. This integration states how node item references similar to one in particular are incorporated into their knowledge repository as items of remote type, handled in a similar and compatible way with the local ones, previously taken into consideration.

10.4 Experiments Carried Out

To check the viability of the proposed approach, we have developed a Nodes Network Module prototype (NM) for Semantic KnowCat (SKC), which has been incorporated to a KnowCat system (KC) together with other previously prototyped modules as Analysis Module (AM). The new module has allowed to carry out several experiments with KC nodes, prepared to try SKC prototypes during several academic years, in teaching activities carried out in the Escuela Politécnica Superior (Higher Polytechnic School) of the Universidad Autónoma de Madrid (Spain). In particular, the same KC nodes used in the previous projects have been used [14]: one on Operating Systems (OS) with a two level knowledge tree with 20 topics and 350 documents; two nodes on Automaton Theory and Formal Languages (ATFL), with different trees of two levels, one with 12 topics and 50 documents and the other with 6 topics and 24 documents; and a node on Computer Systems (CS) with a one level tree of 40 topics and 180 documents.

In order to test the establishment and maintenance process of the knowledge nodes network by means of the automatic link two experiments have been carried out in which five nodes have been used. Three of which are the original nodes for ATFL and CS, the other two have been generated from OS, maintaining knowledge trees the same as the first one in both and sharing documents of the original node so that in each couple of comparable topics of the new nodes there is a similar number of different documents and of similar relevance. The first experiment was initiated manually registered the basic identification data of some other nodes in each node, in such a way that none of them get to know more than two of the others. In Figure 15.3 (left) nodes are represented by circles with their names inside. Initially each node knows other nodes to which it points to with arrows labeled with number "1". Next the agents of every node have started up with the Web search function deactivated. The agents have begun to visit each other, first to complete the identification data of the nodes they already know. As a result they have obtained the identification

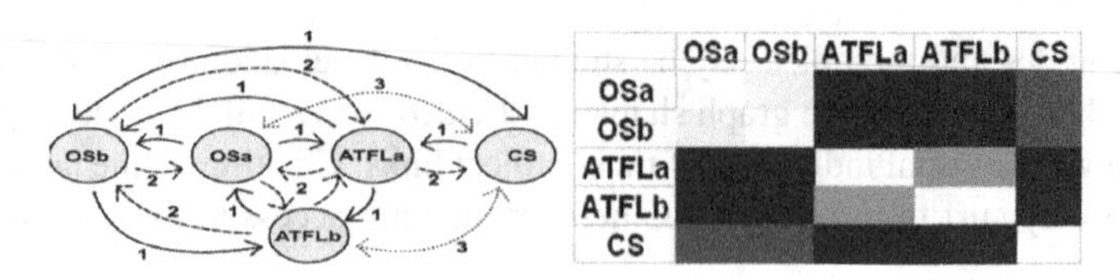

Fig. 10.1: Discovery nodes on the Internet (left) and classification of nodes by subject (right)

data of the nodes visited and have discovered unknown nodes that knew them and had visited them, indicated with arrows labeled with the number "2" in Figure 15.3 (left). When the discovery by direct contact has been exhausted, the agents have established their links with the nodes they knew and are prepared to visit them to obtain the referentes of the nodes registered by these. In this way, the agents that had not been directly discovered will get the references of others or will be visited by the ones that have found theirs, marked with arrows labeled with the number "3" in Figure 15.3 (left). After some activity of the agents, every node knew all the other existing active nodes. The time needed to complete the process depends upon the order of the visits followed by the agents, others availability when trying to interact with them and on the periods established between interactions. At best the process can be completed in three stages. If the experiment is repeated with one of the deactivated agents, every agent that knows the inactive one will end up considering it as such. If later this agent is activated again and has references of active agents, all the nodes will end up recognising its new situation. In this case, the visits initiated by this node will cause a change of consideration in the visited nodes that had believed it was inactive.

The second experiment starts off with a scenario where each node knows all the other network nodes and has completed their registers. In these conditions, every node has the descriptors of others available and is in disposition of establishing the kind of link with them, comparing their own descriptors (WWVs) with the rest. As shown before in this paper and in a previous one [14], the levels of similarity between the node descriptors allow identifying the ones which have similar subjects. This can be appreciated in Figure 10.1 (right), where the lighter colours represent a greater similarity. As explained before, this threshold must be established in an empirical way, as it depends on the knowledge field taken into consideration, although it seems to be at this point in the fields dealt with. Where activating the agents in this situation the agents of nodes which related subjects will establish a content link, whereas the rest will establish a basic one. From this moment, interactions between the nodes with different links shall be different. In the case of nodes with basic links, the interaction will be reduced to interchanging their node descriptors and the active node references that each of them has. In the case of nodes with a content link, the descriptors of every knowledge item will also be interchanged. If one of the linked nodes is deactivated, the other nodes will consider it inactive after some time. Where the node is reactivated and the other nodes detect it, they will mark it as active again, but will withdraw the link previously established, and will have to recover the corresponding descriptor and determine the type of link again. Where a content link node remains inactive, every knowledge item will be excluded and the corresponding descriptors that must be obtained again in the case of reactivation and restoration of the lost link. If during the agents activity the descriptor of any node is changed and the levels of similarity change to higher or lower than 0.3 (the threshold considered) or vice versa, after some time, the type of link between the affected nodes will change and the knowledge repository will adapt to the new situation. As a result of these experiments it may be concluded that the agents are capable of recognising a network node and of adapting to the changes by collaborating, acting over the environment and reacting to its changes in a reactive way. Therefore, it is possible to establish and maintain a SKC network node with selective knowledge interchange using the proposed procedure.

In order to test the discovery process of unknown knowledge nodes on the Web an additional experiment was carried out. To do this, the nodes taken into consideration in previous experiments will be used. These nodes are deployed in a SKC server in a way that none of them know the other nodes. In addition, a Web page is prepared in the Server with data of all the nodes included. Later this page is made accessible for the spider of the Web search engine taken into consideration, in this case Google, and is registered in it, to facilitate its discovery. The details of the node discovery process on the Web are explained in the corresponding previous section. Next, node agents are activated and we must wait until the Google crawler registers the nodes page published by the SKC server in the Web search engine data base. From the moment of the activation, the agents will begin to search active nodes pages in the configured Web search engines. Within several day period, depending on the dynamic of search engine indexation, some of the agents will find the page searched for in Google. In that moment the agent discoverer will have access to the nodes page of the SKC server and shall obtain the data of the nodes included. Immediately after it will contact all of them, and the SKC network node will be created as in previous experiments. The conclusions of the experiment stated that the proposed procedure to discover unknown nodes on the Web is possible and that it is a powerful mechanism for the integration of network node. Therefore, it is viable to establish a SKC network node by discovering nodes dispersed through the Web, without knowing anything about them, using the proposed procedure and conventional Web search engine.

10.5 Conclusions

The experimental results have shown evidence of how to take advantage of every knowledge node in order to enrich the knowledge of each individual node and to try to improve the system performance in different ways: stimulating the users with relevant external references that can mark a contrast with their own knowledge elements, showing alternative viewpoints, and avoiding potential practices of "intellectual endogamy", to which isolated nodes seem to have a tendency. For this purpose, the latent knowledge revealed with the WWVs has been fundamental, as well as the use of the system features, derived from their condition of Web application and active entity, by means of software agents.

All this would not have been possible without the integration of data mining techniques and software agents. WWVs provide support for controlling the activity of software agents. Moreover, WWVs are objects suitable for the exchange between agents, because they are the light, compact and comparable representation of the knowledge elements they symbolize. Agents in turn can work in a distributed and changeable environment in which there is not a pre-established full definition of all the elements involved. Thus, a solution without the combination of these two technologies would not have had adaptability and scalability of the proposed approach.

Acknowledgements This research was partly funded by the Spanish National Plan of R+D, projects TIN2007-64718 and TIN2008-02081/TIN, and by the AECI (Spanish Agency for the International Cooperation) project number A/017436/08.

References

1. Adomavicius, G., Tuzhilin, A.: Toward the next generation of recommender systems: a survey of the state-of-the-art and possible extensions. In: IEEE TKDE, no.17, pp. 734-749 (2005)
2. Alamn, X., Cobos, R.: KnowCat, A Web Application for Knowledge Organization. In: Chen, P.P. et al. (eds.) LNCS, vol. 1727, pp. 348-359. Springer (1999)
3. Baeza, R., Ribeiro, B.: Modern Information Retrieval. Addison Wesley (1999)
4. Chang, G., Healey, M., McHugh, J., Wang, J.: Mining the World Wide Web: An introduction search approach. Kluwer (2001)
5. Choi, N., Song, I., Han, H.: A survey on ontology mapping. In: ACM SIGMOD Record archive, vol. 35, no. 3, pp. 34 - 41 (2006)
6. Cobos, R.: Mechanisms for the Crystallisation of Knowledge, a proposal using a collaborative system. Doctoral dissertation. Universidad Autnoma de Madrid (2003)
7. FIPA, The foundation for intelligent physical agents. `http://www.fipa.org` (2007)
8. Golder, S., Huberman, B.A.: The structure of collaborative tagging systems In Journal of Information Science, vol. 32, no. 2, pp. 198-208 (2006)
9. Gruber, T. R.: A Translation Approach to Portable Ontology Specifications. In Knowledge Acquisition, 5(2), pp. 199-220 (1993)
10. dInverno, M., Luck, M.: Understanding Agent Systemas. Springer (2001)
11. Kiryakov, A., Popov, B., Terziev, I., Manov, and D., Ognyanoff, D.: Semantic Annotation, Indexing, and Retrieval. In: Journal of Web Sematics 2, Issue 1, pp. 49-79, Elsevier (2004)
12. Leuf, B., Cunningham, W.: The Wiki Way. Quick collaboration on the Web. Addison-Wesley (2001)
13. Moreno, J., Alamn, X.: A Proposal of Design for a Collaborative Knowledge Management System by means of Semantic Information. In: Navarro-Prieto, R. et al. (eds.): HCI related papers of Interaccin 2004, pp. 307-319. Springer, Dordrecht, The Netherlands (2005)
14. Moreno, J. and Alamn, X.: SKC: Digestin de Conocimiento. In: Proceedings of the VII Congreso Internacional INTERACCION 2007, pp. 281-290, Zaragoza (2007)
15. Noy ,N. F., Musen, M. A.: Evaluating Ontology-Mapping Tools: Requeriments and Experience. In EKAW02 Workshop (WS1) (2002)
16. O'Reilly, T.: What Is Web 2.0? Design Patterns and Business Models for the Next Generation of Software. In O'Reilly (online publishing) `http:/www.oreillynet.com/pub/a/oreilly/tim/news/2005/09/30/what-is-web-20.html` (2005)

Chapter 11
Equipping Intelligent Agents with Commonsense Knowledge acquired from Search Query Logs: Results from an Exploratory Story

Markus Strohmaier, Mark Kröll, and Peter Prettenhofer

Abstract Access to knowledge about user goals represents a critical component for realizing the vision of intelligent agents acting upon user intent on the web. Yet, the manual acquisition of knowledge about user goals is costly and often infeasible. In a departure from existing approaches, this paper proposes Goal Mining as a novel perspective for knowledge acquisition. The research presented in this chapter makes the following contributions: (a) it presents *Goal Mining* as an emerging field of research and a corresponding automatic method for the acquisition of user goals from web corpora, in the case of this paper search query logs (b) it provides insights into the nature and some characteristics of these goals and (c) it shows that the goals acquired from query logs exhibit traits of a long tail distribution, thereby providing access to a broad range of user goals. Our results suggest that search query logs represent a *viable*, yet largely *untapped* resource for acquiring knowledge about explicit user goals.

11.1 Motivation

To realize the vision of intelligent, goal-oriented agents on the web, agents must have programmatic access to the large set and variety of human goals, in order to reason about them and to provide resources and services that can help satisfy users' needs. In Tim Berner's Lee's vision, an agent aiming to, for example, `"plan a trip to Vienna"` would need to have some means to understand that `"plan a trip"` is likely to involve a set of other goals or services, such as `"contact a travel agency"` and `"book a hotel"`. This form of reasoning would

Markus Strohmaier, Mark Kröll
Knowledge Management Institute, Graz University of Technology and Know-Center, Inffeldgasse 21a/II, A-8010 Graz, Austria, e-mail: `{markus.strohmaier,mkroell}@tugraz.at`

Peter Prettenhofer
Web Technology & Information Systems, Bauhaus University Weimar, Bauhausstrasse 11, D-99423 Weimar, Germany, e-mail: `peter.prettenhofer@uni-weimar.de`

L. Cao (ed.), Data Mining and Multi-agent Integration, DOI: 10.1007/978-1-4419-0522-2_11,

require agents to have explicit representations of users' goals, expressed in description languages such as WSMO, the Web Service Modeling Ontology[1]. The type of knowledge necessary for such intelligent behavior has been characterized as commonsense knowledge [14, 17], i.e. knowledge that humans are generally assumed to possess, but which is not easily accessible to software agents.

Commonsense Enabled Agents: Agent-mining interaction has been introduced as an emerging new research area that aims to combine ideas from research on *Agents* and *Data Mining* [4, 28]. The goal is to enable knowledge transfer between one field to the other leading to the notions of agent-driven data mining [6] and mining-driven agents [24]. One important research challenge related to agents involves the exploration of the human role in agents and data mining approaches, including the question of *How can everyday knowledge - trivial to human beings (commonsense knowledge) - be made accessible to agents?*

Current research projects aiming to capture commonsense knowledge, including knowledge about human goals, are CyC [14] or ConceptNet/Openmind [17, 25], which utilize human knowledge engineering [14], volunteer-based [17], game-based [16] or semi-automatic approaches [8] for knowledge acquisition. In particular, knowledge about potential human goals has been found to be an important kind of knowledge for a range of challenging research problems, such as goal recognition from user actions, reasoning about user goals, or the generation of action sequences that implement goals (planning) [16, 23]. Therefore, the acquisition of user goals in order to augment commonsense knowledge bases appears to be beneficial for the development of commonsense enabled intelligent agents.

Search engines appear to be a particularly interesting source of users' goals as they represent a primary instrument through which a large number of users exercise their intent on the web. During search, users frequently refine, generalize and evolve their daily informational needs by formulating and executing a sequence of search queries, thereby leaving *"traces of intent"*, i.e. implicit traces of users' goals and intentions [26]. Table 11.1 gives some examples of actual queries containing/not containing explicit statements of goals (or *"explicit user goals"* from here on) obtained from a real world query log [19]. While search query logs have been utilized

Table 11.1: Comparison of queries containing/not containing goals.

Queries containing explicit statements of goals	**Queries not containing explicit statements of goals**
"sell my car"	"Mazda dealership"
"play online poker"	"online games"
"passing a drug test"	"drug test"

successfully for knowledge acquisition in different contexts [18], they have not been used to acquire knowledge about explicit user goals, partly because queries pose a number of challenges such as shortness and ambiguity. Yet, recent research suggests

[1] http://www.w3.org/Submission/WSMO/

that search query logs represent a viable resource for goal mining and that the set of goals in search query logs is vast and topically diverse [27]. This would make query logs a particularly useful resource for acquiring knowledge about diverse user goals.

The overall goal of this chapter is to study the feasibility of Goal Mining in a web context - particularly search query logs. Section 2 gives an in-depth overview of Goal Mining in different application domains and provides a discussion of related work in this emerging area. To study Goal Mining in a web context, we present the results of a human subject study that aimed to develop a useful definition for "queries containing explicit user goals" in Section 3. Based on this, we present an automatic classification approach that is capable of acquiring goals from search query logs with reasonable precision/recall scores in Section 4. In Section 5, we present results from experiments applying and evaluating the automatic classification approach resulting in the acquisition of an estimated set of 75.000 explicit user goals. Section 6 concludes that explicit user goals extracted from search query logs have the potential to extend commonsense knowledge bases.

11.2 Goal Mining

In the last decades, the AI and web research community has poured significant efforts into the development of intelligent agent-based systems, leading to a number of pioneering projects in academia as well as in industry. Yet, broad adoption has been slow, which has raised questions about the maturity and recent progress of the field [10]. Agent-based systems still appear to suffer from a number of problems, including - among others - human-agent interaction and a lack of broad knowledge about human intentional structures. If agents were able to hypothesize and reason about the range of human goals, they might be in a better position to help users achieving their specific needs on the web. Yet, making this kind of knowledge accessible to agents has been hindered by what has been coined the *"Knowledge Acquisition Bottleneck"* and the complexity and sheer scale of human goals.

We define *Goal Mining* to focus on the acquisition of goals from textual resources. This emerging research area covers a broad range of interesting problems, including the acquisition of goals from patents, scientific articles [11], organizational policies [20], organizational guidelines and procedures [15] and others.

Search query logs have recently caught the attention to serve as corpora for goal mining. Several different definitions of search goals emerged in previous years [3], [7], [9], covering different aspects of user intent. Broder [3] introduced a high level taxonomy of search intent by categorizing search queries into three categories: navigational, informational and transactional. Evolutions of Broder's taxonomy include collapsing categories, adding categories [2] and/or focusing on subsets only [22, 13]. Overall, Goal Mining offers a novel perspective for making knowledge about user goals accessible to intelligent agents. In contrast to existing knowledge acquisition approaches such as manual knowledge engineering, volunteer-based or game-based approaches, Goal Mining taps into the artifacts produced as a by-product of natural user activity on the web.

11.3 Explicit User Goals: Definition and Agreeability

In order to automatically mine goals from search query logs, a stable and agreeable definition of goals is required. In this section we present a practical definition for user goals in search queries. While a multitude of definitions for user goals exist in related literature, our definition seeks to be applicable to the context of search query logs, i.e. capable of distinguishing search queries that contain explicit user goals from queries which do not. We define queries containing explicit user goals in the following way:
A search query is regarded to contain an explicit user goal (or short: explicit goal) whenever the query 1) contains at least one verb and 2) describes a plausible state of affairs that the user may want to achieve or avoid (cf. [21]) in 3) a recognizable way [27].

"Recognizable" refers to what [12] defines as "trivial to identify" by a subject within a given attention span. *"Plausible"* refers to an external observer's assessment whether the goal contained in a query could likely represent the goal of a user who formulates the given query. *"Queries containing explicit goals"* according to our definition can be related to what other researchers have characterized as “better queries”, or queries that have "more precise goals"[2]. A query does not contain an explicit goal when it is difficult or extremely hard to elicit some specific goal from the query. Examples include blank queries, or queries such as `"car"` or `"travel"`, which embody user goals on a very general, ambiguous and mostly implicit level.

11.3.1 Results of Human Subject Study

We approach the problem of acquiring knowledge about user goals from search query logs as a binary classification problem, aiming to separate queries containing explicit user goals from other queries. This classification problem has been shown to represent an orthogonal problem to existing approaches to search intent categorization [27]. While existing approaches such as Broder et al. [3] perform a taxonomic categorization, we are interested in the acquisition of individual goal instances. In order to gauge the results of an automatic classification approach addressing this problem, we conducted a human subject study aiming to 1) define our constructs more rigorously and 2) to learn about their principal agreeability.

Questionnaire Design: In the study, the subjects were required to independently answer a single question for each of 3000 queries randomly obtained from a real world search query log after sanitization and pre-processing steps were performed (more details on our dataset are presented in Section 11.3). The question for each

[2] R. Baeza-Yates at the "Future of Web Search" Workshop 2006, Barcelona

query followed this schema: *"Given a query X, Do you think that Y (with Y being the first verb in X, plus the remainder of X) is a plausible goal of a searcher who is performing the query X?"*. To give two examples:

Given query: "how to increase virtual memory" Question: *Do you think that "increase virtual memory" is a plausible goal of a searcher who is performing the query "how to increase virtual memory"?* Potential Answer: Yes

Given query: "boys kissing girls" Question: *Do you think that "kissing girls" is a plausible goal of a searcher who is performing the query "boys kissing girls"* Potential Answer: No

Agreeability of Constructs: 243 queries out of 3000 have been labeled as containing an explicit goal by all 4 subjects (8.1%), and 134 queries have been labeled as containing an explicit goal by 3 out of 4 subjects. This corroborates the assumption that query logs contain a small number of queries that contain explicit goals. The majority of queries (79.2%) has been labeled as not containing an explicit goal unanimously by all subjects. To further explore agreeability, we calculated the inter-rater agreement κ [5] between all pairs of human subjects A, B, C and D. The κ values in our human subject study range from 0.65 to 0.76 (0.72 on average) hinting towards a principle (yet not optimal) agreeability of our definition. In the next section, we use these results to inform the development of an automatic classification approach.

11.4 Case Study

In the previous section, we approached the problem of acquiring knowledge about user goals from search query logs as a binary classification problem - either a query contains an explicit user goal or not - and presented the results of a human subject study showing the principal agreeability of such an approach. Based on these findings, we conducted an experiment to develop and evaluate an inductive classification approach that aims to perform the task of classifying queries into one of the two categories. In the following we describe the experimental setup, introduce a classification approach based on lexical and syntactic analysis and present the results of an evaluation of the approach on a manually labeled dataset.

Acquiring Explicit User Goals: Our goal mining experiment is based on a large search query log recorded by AOL in early 2006 [19]. It contains $\sim$ 20 million search queries collected from 657,426 unique user IDs between March 1, 2006 and May 31, 2006 by AOL. To our knowledge, the AOL search query log is the most recent, very large corpus of search queries publicly available (2006). We now present an inductive classification approach that aims to perform the task of classifying queries into one of the two categories (containing/not containing an explicit goal) automatically to aid goal mining.

Our approach used a linear Support Vector Machine to classify queries. As features, we utilized a combination of Part-of-Speech trigrams and stemmed unigrams. For further details regarding the pre-preprocessing of the data and the analysis of the binary classification task we refer to [27]. We choose F1, the harmonic mean of precision and recall, for evaluation that are summarized in Table 11.4.

Table 11.2: Precision, Recall and F1-measure for our approach on the manually labeled data set. All values refer to the class that represents queries containing goals. A precision of 77% means that in 77% of cases, our classification approach agrees with the majority of human subjects.

Precision	Recall	F1 measure
0.77	0.63	0.69

A simple baseline approach, where we would guess that a query containing a verb always contains an explicit goal, would perform significantly worse. While the baseline would excel on recall (Recall 1.0, due to our definition of explicit goals that requires a query to contain a verb), it would perform worse on precision (Precision 0.13, based on the data for the manually labeled dataset) and F1 (0.23) scores. While there are differences in the evaluation procedure and the type of knowledge captured, the precision of explicit goals in our result set is roughly comparable to precision scores reported for the ConceptNet commonsense knowledge database (75%) [25]. These encouraging results suggest the relevance of query logs for this task.

11.5 Results

In this section, we present selected statistics of the result set and give some qualitative insights into the nature of acquired goals. Applying our automatic classification method to the AOL search query log yielded a result set comprising 118.420 queries, 97.454 of them unique. With a precision of 77%, the result set comprises an estimated *75.039 queries containing explicit goals*, which might appear small in the light of $\sim$ 20 million queries contained in the original AOL search query log. However, considering that the 20 million queries reportedly represent only 0.33% of the total number of queries served by AOL during that time, the approach would be able to acquire a much larger set of explicit goals on larger datasets. The 10 most frequent queries from the result set are presented in Table 11.3. Each example is accompanied by the rank and the number of different users who submitted the query (frequency). The information in this table could be expected to reflect -

Table 11.3: The 10 most frequent queries in the result set. Queries containing the token "http" were filtered out.

Nr.	Query	#Users	Nr.	Query	#Users
1	add screen name	205	6	pimp my ride	97
2	create screen name	137	7	assist to sell	93
3	rent to own	120	8	wedding cake toppers	64
4	listen to music	108	9	skating with celebrities	58
5	pimp my space	102	10	lose weight fast	56

to some extent - an excerpt of the needs and motivations of the study population.

Some of the most frequent queries containing goals relate to commonsense knowledge goals, such as "lose weight" or "listen to music", which provides some evidence of the suitability of search query logs for the knowledge acquisition task. Queries such as "add screen name", "wedding cake toppers" and "skating with celebrities" are representatives for the bias that is introduced by the corpus (search queries) and the population (i.e. AOL users).

Table 11.4: The 10 most frequent verbs and nouns in the result set and corresponding co-occurrences in queries containing goals.

	home (2512)	card (2188)	name (1844)	screen (1561)	credit (1433)	music (1398)	money (1371)	weight (1338)	school (1221)	car (1189)
make (8763)	210	208	96	96	5	58	631	19	19	32
buy (8557)	237	117	12	10	66	58	43	6	17	224
find (8545)	169	25	192	30	20	57	60	17	104	94
get (6562)	65	103	41	26	130	33	68	13	55	54
do (6391)	70	62	72	69	40	51	52	52	44	25
listen (2485)	18	0	0	0	0	477	0	0	27	2
learn (2014)	12	16	3	1	6	34	10	3	28	5
sell (1962)	141	38	8	1	2	8	15	1	1	90
use (1688)	15	22	5	5	15	3	3	10	9	15
play (1598)	8	63	0	1	1	13	3	1	4	4

The most popular verb/noun co-occurrences in Table 11.4 seem to be indicative of typical user goals on the web, such as "make money", "listen music" or "buy home". Overall, the top 10 verb/noun correlations identified were "lose weight"* (688)[3], "make money"* (631), "listen music"* (477), "find number"* (457), "find address"* (441), "add name" (399), "add screen" (380), "buy online" (339), "find phone"* (333), "find people" (332). Evaluations of the top verb/noun correlations have revealed that many of these goals are also contained in the ConceptNet commonsense knowledge base v2.1 (marked with an *), a knowledge base generated to augment the development of commonsense-enabled intelligent agents.

These findings suggest that search query logs might be useful to automatically expand the knowledge contained in existing commonsense knowledge bases. Extracted user goals could add to the commonsense knowledge that is accessed by commonsense-enabled intelligent agents.

If search query logs would be utilized for such a purpose, a relevant question to ask is: how diverse is the set of goals contained in search query logs? The diversity of goals would ultimately constrain the utility of a given dataset for expanding existing knowledge bases. In order to explore this question, we present a rank/frequency plot of the data depicted in Table 11.3. In Fig. 11.1, goals are plotted according to their rank and the set of different users who share them. The distribution in Fig. 11.1

[3] not listed in Table 11.4

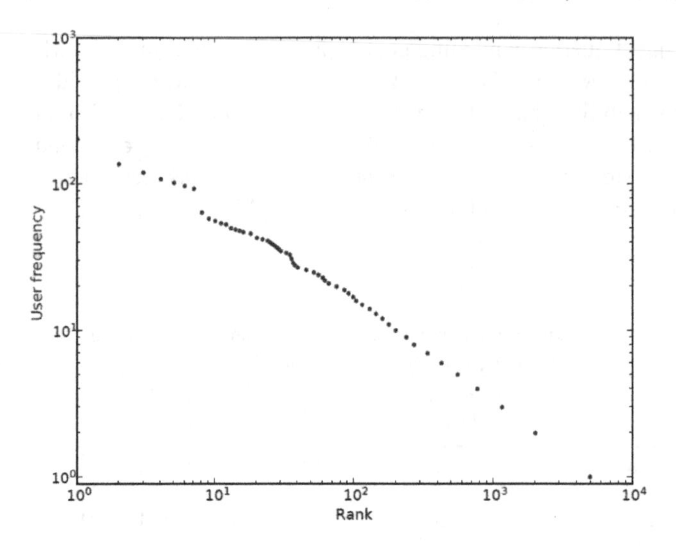

Fig. 11.1: Rank-Frequency Plot of Queries containing unique user goals.

shows that while there are very few popular goals, a majority of goals is shared by only a few users. The curve approximates a power-law distribution, implying the existence of a long tail effect of user goals (similar to the long tail of products [1]). This means that the explicit goals in the result set are extremely diverse; making search query logs a particularly promising resource for the acquisition of *broad* knowledge about user goals in the context of the web.

11.6 Conclusions

The extent and diversity of explicit user goals identified in search query logs suggest that Goal Mining of textual corpora, in particular search query logs, represents a promising alternative to existing approaches to knowledge acquisition. This contribution focused on presenting the feasibility of an automatic method for mining goals from search query logs with useful precision/recall scores. In addition, the overlap with e.g. ConceptNet shows relevance for the development of agent-based systems. Yet, the integration of these goals with actual agent-based systems was beyond the scope of this work. Current work focuses on developing agent-based systems that can produce probabilistic estimates of the explicit user goals behind short queries and on goal-oriented search applications. Future work focuses on the problem of constructing large-scale goal association graphs and on comparing goals mined from search query logs with goals in other, existing knowledge bases.

Acknowledgements This work is funded by the FWF Austrian Science Fund Grant P20269 TransAgere. The Know-Center is funded within the Austrian COMET Program under the auspices of the Austrian Ministry of Transport, Innovation and Technology, the Austrian Ministry of

Economics and Labor and by the State of Styria. COMET is managed by the Austrian Research Promotion Agency FFG.

References

1. Anderson C.: The Long Tail. Wired Magazine, (12)10, 2004.
2. Baeza-Yates, R., Calderon-Benavides, L., Gonzalez-Caro, C.: The intention behind web queries. In String Processing and Information Retrieval (SPIRE'06), 98–109, 2006.
3. Broder, A.: A taxonomy of web search. In SIGIR Forum, vol. 36, no. 2, pp. 3–10, 2002.
4. Cao, L., Luo, C., Zhang, C.: Agent-Mining Interaction: An Emerging Area. In Vladimir Gorodetsky; Chengqi Zhang; Victor A. Skormin & Longbing Cao, Springer, pp. 60-73, 2007.
5. Cohen J.: A coefficient of agreement for nominal scales. In Educational and Psychological Measurement, (20)1–37, 1960.
6. Davies, W.: ANIMALS: A Distributed, Heterogeneous Multi-Agent Learning System. MSc Thesis, University of Aberdeen, 1993.
7. Downey, D., Dumais, S., Liebling, D., Horvitz, E.: Understanding the relationship between searchers' queries and information goals. In Proceedings of the International Conference on Information and Knowledge Management, ACM, New York, NY, USA, pp. 449–458, 2008.
8. Eslick, I.S.: Searching for commonsense, Master's thesis, MIT, 2006.
9. He, K., Chang, Y., Lu, W.: Improving identification of latent user goals through search-result snippet classification. In Proceedings of the IEEE/WIC/ACM International Conference on Web Intelligence, IEEE Computer Society, Washington, DC, USA, pp. 683–686, 2007.
10. Hendler, J.: Where Are All the Intelligent Agents?. In Intelligent Systems, pp. 2–3, 2007.
11. Hui B., Yu E.: Extracting conceptual relationships from specialized documents. In Data And Knowledge Engineering, (54)1:29–55, Elsevier, 2005.
12. Kirsh D.: When is information explicitly represented? In Information, Language and Cognition - The Vancouver Studies in Cognitive Science., 340–365, UBC Press, 1990.
13. Lee, U., Liu, Z., Cho, J.: Automatic identification of user goals in Web search. In Proceedings of the 14th international conference on the World Wide Web (WWW '05), ACM, New York, NY, USA, pp. 391–400, 2005.
14. Lenat, D. B.: CYC: a large-scale investment in knowledge infrastructure. In Communications of the ACM 38(11), 33–38, 1995.
15. Liaskos S., Lapouchnian A., Yu, Y., Yu E., Mylopoulos J.: On goal-based variability acquisition and analysis. In Proceedings of the 14th IEEE International Requirements Engineering Conference (RE'06), Minneapolis, USA, 2006.
16. Lieberman, H., Smith, D., Teeters, A.: Common Consensus: a web-based game for collecting commonsense goals. In Proceedings of the Workshop on Common Sense and Intelligent User Interfaces held in conjunction with the 2007 International Conference on Intelligent User Interfaces (IUI 2007), 2007.
17. Liu, H., Singh, P.: ConceptNet - a practical commonsense reasoning tool-kit. In BT Technology Journal 22(4), 211–226, 2004.
18. Pasca M., Van Durme B., Garera N.: The role of documents vs. queries in extracting class attributes from text. In Proceedings of the International Conference on Information and Knowledge Management (CIKM'07), 485–494, ACM, New York, NY, USA, 2007.
19. Pass G., Chowdhury A., Torgeson C.: A picture of search. In Proceedings of the 1st International Conference on Scalable Information Systems, ACM Press New York, NY, USA, 2006.
20. Potts C., Takahashi K., Anton A. I.: Inquiry-based requirements analysis, 1994.
21. Regev G., Wegmann A.: Where do goals Come from: the underlying principles of goal-oriented requirements engineering. In Proceedings of the 13th International Conference on Requirements Engineering , 253–362, IEEE Computer Society, Washington DC, USA, 2005.
22. Rose, D. E., Levinson, D.: Understanding user goals in web search. In Proceedings of the 13th WWW, ACM, NY, USA, pp. 13–19, 2004.

23. Schank R.C., Abelson R.P.: Scripts, plans, goals, and understanding: an inquiry into human knowledge structures. Lawrence Erlbaum Associates, 1977.
24. Sian, S.: Extending Learning to Multiple Agents: Issues and a Model for Multi-Agent Machine Learning. In Proceedings of the European Workshop Sessions on Learning, Porto, Portugal, pp. 458–472, 1991.
25. Singh, P., Lin, T., Mueller, E., Lim, G., Perkins, T., Zhu, W.: Open mind common sense: knowledge acquisition from the general public. In Proceedings of the First International Conference on Ontologies, Databases, and Applications of Semantics for Large Scale Information Systems, Springer-Verlag London, UK, , pp. 1223–1237, 2002.
26. Strohmaier, M., Lux, M., Granitzer, M., Scheir, P., Liaskos, S., Yu, E.: How do users express goals on the web? - an exploration of intentional structures in web search. In International Workshop on Collaborative Knowledge Management for Web Information Systems We Know'07, in conjunction with WISE'07, Nancy, France, 2007.
27. Strohmaier, M., Prettenhofer, P., Kröll, M.: Acquiring explicit user goals from search query logs. In International Workshop on Agents and Data Mining Interaction ADMI'08, in conjunction with WI '08, 2008.
28. Zhang, C., Zhang, Z., Cao, L.: Agents and Data Mining: Mutual Enhancement by Integration. In Autonomous Intelligent Systems: Agents and Data Mining, 50–61, 2005.

Chapter 12
A Multi-Agent Learning Paradigm for Medical Data Mining Diagnostic Workbench

Sam Chao and Fai Wong

Abstract This chapter presents a model of practical data mining diagnostic workbench that intends to support real medical diagnosis by employing two cutting edge technologies - data mining and multi-agent. To fulfill this effort, i^+DiaMAS - an intelligent and interactive diagnostic workbench with multi-agent strategy, has been designed and partially implemented, where multi-agent approach can perfectly compensate most data mining methods that are only capable of dealing with homogeneous and centralized data. i^+DiaMAS provides an integrated environment that incorporates a variety of preprocessing agents as well as learning agents through interactive interface agent. The philosophy behind i^+DiaMAS is to assist physicians in diagnosing new case objectively and reliably by providing practical diagnostic rules that acquired from heterogeneous and distributed historical medical data, where their new discoveries and correct diagnostic results can benefit other physicians.

12.1 Introduction

Decision tree is one kind of data mining algorithms, which offers a highly practical method for generalizing classification rules from the concrete cases that already solved by domain experts, while its goal is to discover knowledge that not only has a high predictive accuracy but also comprehensible to users. It is an intelligent and powerful tool in supporting decision making, where its technology is well suited for analyzing medical data and assisting physicians to improve the diagnostic perfor-

Sam Chao
Faculty of Science and Technology, University of Macau, Av. Padre Tomás Pereira Taipa, Macau, China, e-mail: lidiasc@umac.mo

Fai Wong
Faculty of Science and Technology, University of Macau, Av. Padre Tomás Pereira Taipa, Macau, China, e-mail: derekfw@umac.mo

L. Cao (ed.), Data Mining and Multi-agent Integration, DOI: 10.1007/978-1-4419-0522-2_12,

mance and reliability. However, a single data mining technique is inappropriate for diverse domains and distributed datasets. The use of agent-based approach to support decision-making is important within the medical industry because they allow medical practitioners to quickly gather information and process it in various ways in order to assist with making diagnosis and treatment decisions [6].

Agents (adaptive or intelligent agents and multi-agent systems) constitute one of the most prominent and attractive technologies in Computer Science at the beginning of this new century, and multi-agent techniques currently play an important role in distributed/heterogeneity data mining [22]. Multi Agent Systems (MAS) are systems composed of multiple interacting agents, where each agent works in a rather focalized way, under specific goals and assumptions. It enables the distributed data mining tasks among different agents work together to find solutions to complex problems with the ability to communicate, cooperate, coordinate and negotiate with each other. Besides, having multiple agents could speed up a system's operation by providing a method for parallel computation [17]. In addition, incorporation of intelligent agents into a data mining system can perfectly compensate the most of existing data mining methods that are only capable of dealing with homogeneous [8] and centralized data. A survey of recent research on the use of agent-based intelligent decision support systems to support clinical management is presented in [6]; while a review of the most representative agent-based distributed data mining systems is provided in [12].

This chapter presents a data mining model of practical diagnostic workbench to support real medical diagnosis by employing multi-agent techniques in accomplishing incremental learning strategy, which is able to automatically extract useful knowledge from heterogeneous and distributed medical data efficaciously. To fulfill this effort, i^+DiaMAS - an intelligent and interactive diagnostic workbench from multi-agent point of view, has been designed and partially implemented. i^+DiaMAS provides an integrated environment that incorporates a variety of preprocessing agents as well as learning agents through an interactive interface agent. The philosophy behind i^+DiaMAS is to assist physicians in diagnosing new case objectively and reliably by providing practical diagnostic knowledge that acquired from heterogeneous and distributed historical medical data, where their new discoveries and correct diagnostic results can benefit other physicians.

The rest of this chapter is organized as follows: section 2 illustrates a general overview of our agent-based diagnostic workbench. Then the functionality of two main agents, data preprocessing agent and data mining agent, are presented in detail in section 3 and section 4 respectively; while how the agent approach can facilitate these data mining tasks are evaluated in section 5. Finally, section 6 draws the conclusions of this chapter as well.

12.2 Multi-Agent Learning Paradigm of i^+DiaMAS

i^+DiaMAS takes advantage of the power of multi-agent architecture, which is the amalgamation of various types of intelligent data mining agents, each responsible for a specific and independent task. It intends to constitute a multi-agent infrastructure aimed at supporting diagnostic and therapeutic decision-making by employing a variety of state-of-the-art data mining methods. Fig. 12.1 briefly describes our proposition.

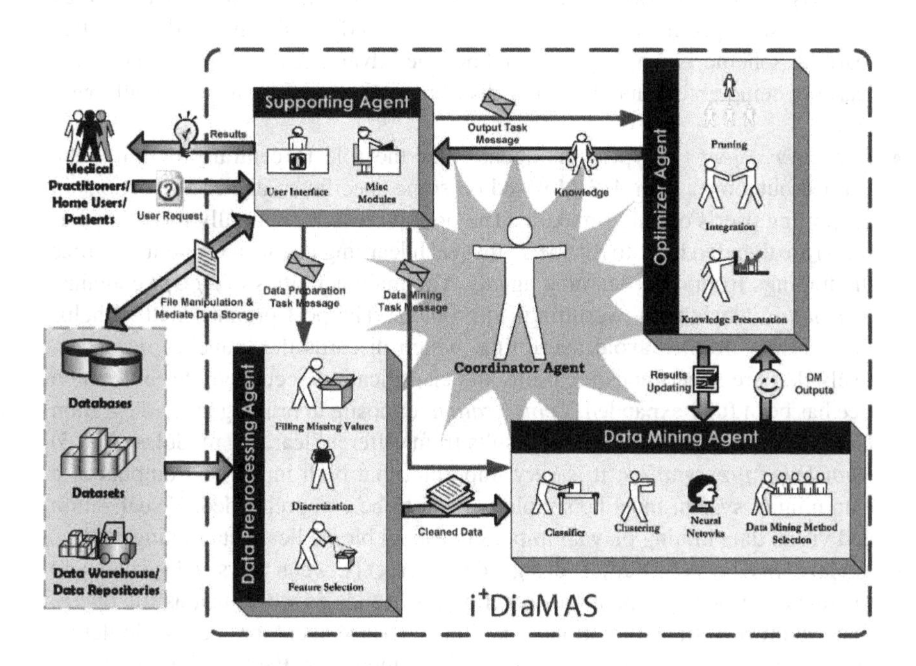

Fig. 12.1: A general overview of agent-based data mining diagnostic workbench.

The principle is to distribute a complex problem to several agents that have the capacity to process in parallel and communicate by sending messages. The main focus of this chapter is on the preprocessing and learning capabilities, which are described in detail in the following two sections. There are altogether five essential types of agents in coping with various data mining tasks. Here we abstract each of these agents concisely.

- *Coordinator Agent* (AgCor): forms the core of i^+DiaMAS, which plays a role to coordinate various agents' activities. It is able to break a task into sub-tasks that are directed toward the appropriate agents, and then combines results from separated agents for the overall response to a task. In addition, *AgCor* is also responsible for the scalability problem of processing agents, such as create or add new agents, delete or suspend out-of-date techniques at run-time dynamically.

Besides, the interactions and communications between agents should be handled by *AgCor* also.

- *Data Preprocessing Agent* (AgPre): the primary key factor of a successful data mining tool is using the right data [18]. This agent sifts the useful data from the raw data by applying our innovative data preprocessing methodology, which consists of filling in missing attribute values, discretization, and feature selection, etc.
- *Data Mining Agent* (AgDM): this is the core of i^+ DiaMAS workbench, which comprises a set of learning agents that each of which implements a methodology to diagnose a specific disease. In which, our novel dynamic on-line decision tree learning scheme is employed. By taking the advantage of MAS, it makes the learning being able to handle a new distributed medical case incrementally with ease.
- *Optimizer Agent* (AgOpt): this agent takes the role to centrally optimize the learned outcomes from *AgDM* based on some specific strategies and actions in supporting user's decision-making. The optimal results (normally formulated as rules) are then fed back to *AgDM* so that each learning agent can benefit to utilize the findings from other learning agents. The main activities of *AgOpt* contains: 1) *Pruning*: deals with overfitting [5] problem. The post-pruning method helps the decision trees to avoid overfitting, which discards the branches retrospectively that are superfluous regarding the classification accuracy after a decision tree has been fully expanded. 2) *Integration*: opposite to pruning task, integration task is necessary to assemble the results from different learning modules. And 3) *Knowledge presentation*: it is very important that both inputs and outputs of a data mining system must be simple enough to be comprehended. Visualization and visual data mining play an important role in biomedical data mining [10].
- *Supporting Agent* (AgSup): is designed to interact between users and system, and to handle various manipulations regarding data files as well, such as file operations and format transformations, etc. Thus interface agent, project manipulation agent and format conversion agent are all its members. Each of them is implemented to deal with a specific task, while communications amongst them are managed by *AgSup*. In which, interface agent receives the requests from users for a data mining service via an interactive Graphical User Interface (GUI). Then delivers the request that reformulated as a set of message scripts to the appropriated agent for processing. After that, the result is generated and demonstrated in a specific form to the users as desired.

12.3 Data Preprocessing Agent (*AgPre*)

Data preprocessing can present equal challenges as a data mining process although it is time consuming. Thus considerable manipulation of a dataset is invariably necessary before a mining process. In our workbench, this effort is concentrated on the feature selection and discretization. As *AgPre* is responsible for sifting the useful data from the raw data, these two tasks are defined under *AgPre* to cope with various real, dirty and uncertain data.

- *Discretization. AgPre* is equipped with various discretization strategies, which are categorized into two major directions: univariate and multivariate. Each of them further contains binary and multi-interval methods. The dominant power of the discretization functionality is so called MIDCA (Multivariate Interdependent Discretization for Continuous Attributes) [3], which is an innovative algorithm that interested mainly in discovering a best interdependent attribute a_{int} relative to a continuous-valued attribute a_i being discretized, and use it as the interdependent attribute to carry out the supervised multivariate discretization regarding the class attribute. MIDCA incorporates the information [13] and relief [11] theories to find out a such perfect interdependent attribute a_{int}. Equation 12.1 formulates the interdependent weight W_{mvi} between a_{int} and a_i, where symmetric and normalized measurements are employed to reduce the bias.

$$W_{mvi}(a_i, a_{int}) = \frac{\left[\frac{SymGain(a_i,a_{int})}{\sqrt{\sum_{M\neq i}^{A} SymGain(a_i,a_M)^2}} + \frac{SymRelief(a_i,a_{int})}{\sqrt{\sum_{M\neq i}^{A} SymRelief(a_i,a_M)^2}}\right]}{2} \quad (12.1)$$

The best interdependent attribute against a continuous-valued attribute being discretized is the one with the highest W_{mvi} amongst all candidate interdependent attributes. MIDCA generates multi-interval discretization and ensures at least binary discretization. This feature is important, since if a continuous-valued attribute generates null cutting point means that such attribute is useless, hence increases the classification uncertainty. Moreover, MIDCA generates one interdependent attribute for each continuous-valued attribute rather than using one for all continuous-valued attributes. This strategy is able to concern each continuous-valued attribute differently and humanly, while handles it with maximum intelligence. It becomes a main factor for improving the classification accuracy accordingly.
- *Feature Selection.* Feature selection is another essential component of *AgPre*, where LUIFS (Latent Utility of Irrelevant Feature Selection) [4] is another breakthrough algorithm in our workbench. LUIFS claims that a true irrelevant attribute in medical domain is the one that provides neither explicit information nor supportive or implicit information. It takes the inter-correlation between a false irrelevant attribute and another attribute into consideration, to discover the potential usefulness of the false irrelevant attribute rather than ignoring it. Such strategy fits the fact in medical diagnosis that a single symptom seems useless regarding the diagnosis, may be potentially important when combined with other symptoms. An attribute that is completely useless by itself can provide a significant performance improvement when taken with others; while two attributes that are useless by themselves can be useful together [9][2].

  LUIFS generates an optimal feature subset in two phases: (1) Relevant Attributes Seeking (RAS): for each attribute in a dataset, work out its relevant weight W_{rel} regarding the target class by employing the information gain theory; selects the ones whose weights are greater than a pre-defined threshold ϖ into the optimal feature subset; (2) Latent Supportive of Irrelevant Attribute (LSIA) Discovery: for each unselected (false irrelevant) attribute a_{irr} that filtered from RAS phase, determine its supportive importance by performing a multivariate interdependent measure I_{mim} with other attribute a_i, equation 12.2 gives the measure. Where combinative relief theory is employed, which is able to discover the interdependences between attributes. Then the false irrelevant attribute becomes potentially relevant and is included in the optimal feature subset if its I_{mim} meets the minimum requirements; otherwise it is a true irrelevant attribute and discarded accordingly.

$$\begin{aligned} I_{mim}(a_{irr}, a_i) = & P(DifferentValueOf < a_{irr}, a_i > |DifferentClass) \\ & - P(DifferentValueOf < a_{irr}, a_i > |SameClass) \end{aligned} \quad (12.2)$$

where P is a probability function for calculating how well the values of $< a_{irr}, a_i >$ distinguish among the instances from the same and the different classes. It measures the level of hidden supportive importance for a false irrelevant attribute a_{irr} to another attribute a_i, hence the higher the weighting, the more information it will provide, such that the better the diagnostic result. LUIFS overcomes the drawback of the ordinary feature selection methods that select only the relevant attributes regarding the learned task, whereas the irrelevant attributes are ignored and discarded. Though the complexity of such method is little higher, it is still worth adopting it in the diagnostic process, as such hypothesis conforms to the real case in medical diagnosis where a useless symptom by itself may sometimes become indispensable when combined with other symptoms; while compound symptoms always could reveal more accurate diagnostic results.

12.4 Data Mining Agent (*AgDM*)

AgDM takes charge of the management of entire data mining processes. An ideal design of *AgDM* may contain: classifier agent, clustering agent, and neural networks agent. Each learning agent contains a set of corresponding learning modules, in which each implements an independent algorithm with mining capabilities working on a prescribed disease. Receiving a task from interface agent, *AgDM* autonomously selects the most appropriate learning module and coordinates all its processing requirements. Currently the learning algorithms of classifier agent are under developing, while others are still empty. Here we briefly discuss several learning modules for classifier agent in handling some critical learning barriers.

- *Ordinary off-line learning module.* Off-line learning means that all examples are available at once, at the beginning of a learning algorithm starts. In i^+DiaMAS, we have implemented several common off-line leaning algorithms, such as ID3 [14][15] family and C4.5 [16]. The purpose is to establish a complete environment for better comparisons and evaluations amongst different data mining algorithms. ID3 family includes ID3 and ID3 Extension, which are the most basic decision tree algorithms. C4.5 is the successor of ID3 algorithm that is able to handle continuous-valued attributes as well as addressing of over-fitting problem by tree pruning, whereas ID3 family is incapable of handling such tasks.

- *Incremental on-line learning module*. Incremental decision tree induction is able to address the learning task with a stream of training instances. When given new training data, the decision tree acquired from previous training data has been restructured, rather than relearning a new tree from scratch. Since the learning environment is prone to dynamic instead of static, revising a learned structure may be more economical than building a new one from scratch, especially in real-time applications. Our novel learning theory, i^+Learning (intelligent and incremental learning), is a new attempt that contributes intelligent and incremental learning architecture. Such learning strategy offers the promise of making decision trees a more powerful, flexible, accurate, and widely accepted paradigm.
- *i^+Learning module*. This module implements i^+Learning algorithm that grows an on-going decision tree with respect to the new available data in two phases: (1) Primary Off-line Construction of Decision Tree (POFC-DT) and (2) Incremental On-line Revision of Decision Tree (IONR-DT).

  - POFC-DT phase constructs a binary decision tree top-down, using a splitting criterion to divide classes as "pure" as possible until a stopping criterion is met. To build a primitive binary tree, we start from a root node d derived from whichever attribute a_i in an attribute space A that minimizes the impurity measure. We employ Kolmogorov-Smirnoff (KS) distance [7][20] as the measurement of impurity at node d, which is denoted by $I_{KS}(d)$ and is shown below:

$$I_{KS}(d) = \max_{v \in val(d)} (|F_L(v) - F_R(v)|) \tag{12.3}$$

    where v denotes either the various values of a nominal attribute a_i with the test criterion $a_i = v$, or a cutting-point of a continuous-valued attribute a_i with the test criterion $a_i < v$; $F_L(v)$ and $F_R(v)$ are two class-conditional cumulative distribution functions that count the number of instances in the left and right classes respectively, which partitioned by a value v of an attribute a_i on decision node d.
  - IONR-DT phase plays a central role in i^+Learning module, it embraces the faith that whenever the new data is coming, this phase dynamically revises the fundamental tree constructed in POFC-DT phase without sacrificing too much the classification accuracy, and eventually produces a decision tree as same as possible to those off-line algorithms. IONR-DT phase adopts the tree transposition mechanism that in ITI [21] as a basis to grow and revise the initial tree. Furthermore, our i^+Learning algorithm remembers the statistical information of instances regarding the respective possible values as well as class labels, in order to process the transposition without rescanning the entire dataset repeatedly. Once a (set of) new instance(s) is ready to be incorporated

with an existing tree, IONR-DT phase first updates the statistical information on each node that the instance traversed, and merges the new instance into an existing leaf or grows the tree one level under a leaf. Then, evaluates the qualification for the test on each node downwards, starting from the root node; for any attribute test that is no longer best to be on a node, the pull-up tree transposition process is called recursively to restructure the existing decision tree.

12.5 Evaluation

To demonstrate the goodness of integrating the agents' capability into a data mining system, let's demonstrate the following situation. Once a new medical case (raw) is received by interface agent, *AgCor* determines first which preprocessing tasks the new raw case should be taken, and then passes it to the corresponding preprocessing modules of *AgPre*, such as handling missing attribute values, continuous-valued attribute discretization and feature selection, etc. After the individual module completes the assigned task, *AgCor* collects the results and assembles them into a new pre-processed case. Then, this new pre-processed case will be delivered to the appropriate learning module in *AgDM*. Finally, the learned result is delivered to *AgOpt* for optimization or directly feedback to the users through interface agent. Such strategy makes the processes in the same level to be performed in parallel; while in the normal non-agent-based structure, process should be carried out one at a time serially. Thus agent-based strategy may speed up the system's operation and cut down the problem's complexity.

In addition, our agent-based learning approach is experimented in terms of classification accuracy over sixteen bench-mark datasets drawn from UCI repository [1] against two prevalent and comparable learning algorithms: C4.5 and ITI (Incremental Tree Induction) [19]. Table 1 lists the empirical results.

In the experiment, ITI and i^{+}Learning algorithms are evaluated under the agent-based architecture; while each dataset is divided into two portions and located distributed. The evaluation of C4.5 is designed as usual without agent-based structure. The results in Table 12.1 revealed that i^{+}Learning is benefit from intelligent agents in accomplishing the incremental learning over distributed data without sacrificing the diagnostic accuracy; where it is the best algorithm amongst three according to the average accuracy in the last row. In which, ITI failed to build a decision tree from dataset *Auto* (indicated as Error), since it has numerical data in the class attribute. Besides, parallel performance is capable of accelerating the learning performance and decreasing the computational complexity since a complicated problem is decomposed into smaller simple ones that to be solved by individual specialized agents.

Table 12.1: Performance comparison in classification accuracy between various algorithms.

Dataset	C4.5	ITI	i^+Learning
Cleve	75.2	65.347	81.188
Hepatitis	82.7	69.231	78.846
Hypothyroid	99.1	98.578	98.01
Heart	82.2	76.667	86.667
Sick-euthyroid	96.9	96.682	94.408
Auto	71	Error	57.971
Breast	94.4	94.421	95.708
Diabetes	69.1	66.797	73.438
Iris	92	94	94
Crx	82.5	76	84
Australian	86.1	77.391	88.261
Horse-colic	80.9	73.529	80.882
Mushroom	100	100	100
Parity5+5	50	50.586	51.563
Corral	90.6	100	90.625
Led7	67.4	63.367	64.767
Average	82.506	75.162	82.521

12.6 Conclusion

This chapter introduces a practical data mining diagnostic workbench - i^+DiaMAS that constitutes from multi-agent perspectives, which is able to extract the useful medical knowledge from distributed and heterogeneous medical data automatically and autonomously. Currently, a prototype of i^+DiaMAS is under development, while the main focus is on the exploration of methodologies for agent cooperative and collaboration.

i^+DiaMAS intends to resolve the incremental learning problem that pervasively exist in medical domain, through the adoption of multi-agent strategy and novel data mining methods. Moreover, it is designed to provide assistance in revolutionizing the traditional medical services from passive, subjective and human-oriented to active, objective and facts-oriented.

References

1. Blake, C.L., Merz, C.J.: UCI Repository of Machine Learning Databases. Department of Information and Computer Science, University of California, Irvine, CA (1998)
2. Caruana, R., Sa, V.R.: Benefiting from the Variables that Variable Selection Discards. J. Mach. Learn. Res. **3**, 1245-1264 (2003)
3. Chao, S., Wong, F., Li, Y.P.: MIDCA — A Discretization Model for Data Preprocessing. Journal of Communication and Computer, **3**, 1–7 (2006)
4. Chao, S., Li, Y.P., Dong, M.C.: Supportive Utility of Irrelevant Features in Data Preprocessing. In: Proceedings of the 11th Pacific-Asia Conference on Knowledge Discovery and Data

Mining, 425–432 (2007)
5. Cohen, P.R., Jensen, D.: Overfitting Explained. In: Proceedings of the Sixth International Workshop on Artificial Intelligence and Statistics, 309–338 (1997)
6. Foster, D., McGregor, C., El-Masri, S.: A Survey of Agent-Based Intelligent Decision Support Systems to Support Clinical Management and Research. In: Proceedings of the 2nd International Workshop on Multi-Agent Systems for Medicine, Computational Biology, and Bioinformatics, 16–34 (2005)
7. Friedman, J.H.: A Recursive Partitioning Decision Rule for Nonparametric Classification. IEEE Trans. Comput. **C-26**, 404–408 (1977)
8. Gao, J., Denzinger, J., James, R.C.: A Cooperative Multi-agent Data Mining Model and Its Application to Medical Data on Diabetes. In: Gorodetsky, V., Liu, J., Skormin, V.A. (eds.) Autonomous Intelligent Systems: Agents and Data Mining, pp. 93-107. Springer, Heidelberg (2005)
9. Guyon, I., Elisseeff, A.: An Introduction to Variable and Feature Selection. J. Mach. Learn. Res. **3**, 1157-1182 (2003)
10. Han, J.: How Can Data Mining Help Bio-Data Analysis. In Zaki, M.J., Wang, J.T.L., Toivonen, H.T.T. (eds.): Proceedings of the Second Workshop on Data Mining in Bioinformatics (BIOKDD02), 1–2 (2002)
11. Kira, K., Rendell, L.: The Feature Selection Problem — Traditional Methods and New Algorithm. In: Proceedings of AAAI'92, 129–134 (1992)
12. Klusch, M., Lodi, S., Moro, G.: The Role of Agents in Distributed Data Mining — Issues and Benefits. In: Proceedings of IEEE/WIC International Conference on Intelligent Agent Technology (IAT'03), 211–217 (2003)
13. Mitchell, T.M.: Machine Learning. McGraw-Hill (1997)
14. Quinlan, J.R.: Learning Efficient Classification Procedures and Their Application to Chess End Games. In: Michalski, R.S., Carbonell, J.G., Mitchell, T.M. (eds.) Machine Learning: An Artificial Intelligence Approach, pp.463-482. Springer, Heidelberg (1984)
15. Quinlan, J.R.: Induction of Decision Trees. Mach. Learn. **1**(1), 81–106. Kluwer Academic Publishers, Hingham, MA, USA (1986)
16. Quinlan, J.R.: C4.5 — Programs for Machine Learning. Morgan Kaufmann, San Mateo, CA (1993)
17. Stone, P., Veloso, M.: Multiagent Systems — A Survey from a Machine Learning Perspective. Autonomous Robots, **8**, 345–383 (2000)
18. Introduction to Data Ming and Knowledge Discovery. Third editions, Two Crows Corporation (1999)
19. Utgoff, P.E.: An Improved Algorithm for Incremental Induction of Decision Trees. In: Proceedings of the 11th International Conference on Machine Learning, 318-325 (1994).
20. Utgoff, P.E., Clouse, J.A.: A Kolmogorov-Smirnoff Metric for Decision Tree Induction. Technical report (UM-CS-1996-003). University of Massachusetts, Amherst, MA (1996)
21. Utgoff, P.E., Berkman, N.C., Clouse, J.A.: Decision Tree Induction Based on Efficient Tree Restructuring. Mach. Learn. **29**, 5–44 (1997)
22. Zhang, C., Zhang, Z., Cao, L.: Agents and Data Mining: Mutual Enhancement by Integration. Autonomous In: Gorodetsky, V., Liu, J., Skormin, V.A. (eds.) Autonomous Intelligent Systems: Agents and Data Mining, pp. 50-61. Springer, Heidelberg (2005)

Part III
Agent Driven Data Mining

Chapter 13
The EMADS Extendible Multi-Agent Data Mining Framework

Kamal Ali Albashiri, and Frans Coenen

Abstract In this chapter we describe EMADS, an Extendible Multi-Agent Data mining System. The EMADS vision is that of a community of data mining agents, contributed by many individuals and interacting under decentralized control, to address data mining requests. EMADS is seen both as an end user platform and a research tool. This chapter details the EMADS vision, the associated conceptual framework and the current implementation. Although EMADS may be applied to many data mining tasks; the study described here, for the sake of brevity, concentrates on agent based Association Rule Mining and agent based classification. A full description of EMADS is presented.

13.1 Introduction

Agent-Driven Data Mining (ADDM), also known as multi-agent data mining, seeks to harness the general advantageous of MAS to the application domain of DM. It is clear that MAS technology has much to offer DM, particularly in the context of various forms of distributed and cooperative DM. MAS have a clear role in both these areas. MAS technology also offers some further advantageous for ADDM, namely:

- extendibility of DM frameworks
- resource and experience sharing,
- greater end-user accessibility, and
- the addressing of privacy and security issues.

Kamal Ali Albashiri, and Frans Coenen
Department of Computer Science, The University of Liverpool,
Ashton Building, Ashton Street, Liverpool L69 3BX, United Kingdom
e-mail: \{ali, frans\}@csc.liv.ac.uk

L. Cao (ed.), Data Mining and Multi-agent Integration, DOI: 10.1007/978-1-4419-0522-2_13,

Consequently the MAS approach would support greater end user access to DM techniques. In the context of privacy and (to an extent) security, by its nature data mining is often applied to sensitive data; the MAS approach would allow data to be mined remotely. Similarly, with respect to DM algorithms, MAS can make use of algorithms without necessitating their transfer to users, thus contributing to the preservation of intellectual property rights.

It is suggested in this chapter that a method of addressing the communication requirement of ADDM is to use a system of *mediators* and *wrappers* coupled with an ACL such as FIPA ACL, and that this can more readily address the issues concerned with the variety and range of contexts to which any ADDM system should be applicable.

To investigate and evaluate the expected advantageous of wrappers and mediators, in the context of the disparate nature of ADDM, the authors have developed and implemented (in JADE) an ADDM platform, EMADS (the Extendible Multi-Agent Data mining System). Extendibility is seen as an essential feature of ADDM primarily because it allows the functionality of EMADS to grow in an incremental manner. The vision is of an anarchic collection of agents, contributed to by a community of EMADS users, that exist across an *internet space*, that can negotiate with each other to attempt to perform a variety of data mining tasks (or not if no suitable collection of agents can come together) as proposed by other (or the same) EMADS users. An EMADS demonstrator is currently in operation.

In the context of EMADS three categories of data mining tasks are considered to exist: classification, clustering and Association Rule Mining (ARM). The current EMADS demonstrator includes ARM and classification agents. To evaluate the operation of EMADS two data mining scenarios are considered in this chapter. The first (Sub-Section 13.5.1) is a distributed merge-mining scenario where EMADS agents are used to merge the results of a number of ARM operations, a process referred to as meta-ARM, to produce a global set of Association Rules (ARs). The challenge here is to minimise the communication overhead, a significant issue in distributed DM (regardless of whether it is implemented in an agent framework or not); this is also an issue in parallel DM. The second scenario (reported in Sub-Section 13.5.2) is a classification scenario where the objective is to generate a classifier (predictor) fitted to a, EMADS user, specified dataset. The aim of this second scenario is to identify a "best" classifier given a particular dataset.

In summary the chapter describes an operational ADDM framework, EMADS. The framework is currently in use and is providing a useful facility, not only to achieve ADDM, but as a platform for conducting ADDM research.

The rest of this chapter is organized as follows. A brief review of some related work on ADDM is presented in Section 13.2. The conceptual framework for EMADS is presented in Section 13.3. The current implementation of EMADS, together with an overview of the wrapper principle is given in 13.4. The operations of EMADS are illustrated in Section 13.5 with a Meta ARM and classification scenarios. Some conclusions are presented in Section 13.6.

13.2 Related Work

Agent-based systems have shown much promise for flexible, fault-tolerant, distributed problem solving. Much of the foundational work on agent technology has focused on inter-agent communication protocols, patterns of conversation for agent interactions, and basic facilitation capabilities. Some ADDM frameworks consider agents to be relatively trivial models for single platform execution. Others focus on developing complex features for specific DM task, while providing little support in the context of usability or extendibility. The success of peer-to-peer systems and negotiating agents has engendered a demand for more generic, flexible, robust frameworks.

There have been only few ADDM systems directed at such a generic framework. An early example was IDM [7], a multi-agent architecture for direct DM to help businesses gather intelligence about their internal commerce agent heuristics and architectures for KDD. In [4] a generic task framework was introduced, but designed to work only with spatial data. The most recent system was introduced in [11] where the authors proposed a multi-agent system to provide a general framework for distributed DM applications. In this system the effort to embed the logic of a specific domain has been minimized and is limited to the customization of the user. However, although its customizable feature is of a considerable benefit, it still requires users to have very good DM knowledge.

13.3 The EMADS Conceptual Framework

Conceptually EMADS is a hybrid peer to peer agent based system comprising a collection of collaborating agents that exist in a set of containers. Agents may be created and contributed by any EMADS user/contributor. The implementation includes a “main container” that houses a number of housekeeping agents that have no particular connection with ADDM, but provide various facilities to maintain the operation of EMADS. In particular the main container holds an Agent Management System (AMS) agent and a Directory Facilitator (DF) agent. The terminology used is taken from the JADE (Java Agent Development) framework [5] in which EMADS is implemented. Briefly the AMS agent is used to control the life cycles of other agents in the platform, and the DF agent provides an agent lookup service. Both the main container and the remaining containers can hold various DM agents. Note that the EMADS main container is located on the EMADS host organisation site (currently the University of Liverpool in the UK), while other containers may be held at any other sites worldwide.

EMADS agents are responsible for accessing local data sources and for collaborative data analysis. EMADS includes: (i) data mining agents, (ii) data agents, (iii) task agents, (iv) user agents, and (v) mediators (JADE agents) for agents coordination. The data and mining agents are responsible for data accessing and carrying through the data mining process; these agents work in parallel and share information through the task agent. The task agent co-ordinates the data mining operations, and presents results to the user agent. Data mining is carried out by means of local data mining agents (for reasons of privacy preservation).

13.3.1 EMADS End User Categories

EMADS has several different modes of operation according to the nature of the participant. Each mode of operation (participant) has a corresponding category of user agent. Broadly, the supported categories are as follows:

- **EMADS Users**: Participants, with restricted access to EMADS, who may pose data mining requests.
- **EMADS Data Contributors**: Participants, again with restricted access, who are prepared to make data available to be used by EMADS mining agents.
- **EMADS Developers**: Developers are EMADS participants, who have full access and may contribute data mining algorithms.

Note that in each case, before interaction with EMADS can commence, appropriate software needs to be downloaded and launched by the participant. Note also that any individual participant may be a user as well as a contributor and/or developer at the same time.

Conceptually the nature of EMADS data mining requests, that may be posted by EMADS users, is extensive. In the current implementation, the following types of generic request are supported:

- Find the "best" classifier (to be used by the requester at some later date in off line mode) for a data set provided by the user.
- Find the "best" classifier for the indicated data set (i.e. provided by some other EMADS participant).
- Find a set of Association Rules (ARs) contained within the data set(s) provided by the user.
- Find a set of Association Rules (ARs) contained within the indicated type of data set(s) (i.e. provided by other EMADS participants).

The Association Rule Mining (ARM) style of request is discussed further in Sub-Section 13.5.1. The idea was that an agent framework could be used to implement a form of Meta-ARM where the results of the parallel application of ARM to a collection of data sets, with not necessarily the same schema but conforming to a global schema, are combined. Details of this process can be found in Albashiri et al. [2, 3]. A "best" classifier is defined as a classifier that will produce the highest accuracy on a given test set (identified by the mining agent) according to the detail of the request. To obtain the "best" classifier EMADS will attempt to access and communicate with as many classifier generator DM agents as possible and select the best result. The classification style of user request will be discussed further in Sub-Section 13.5.2 to illustrate the operation of EMADS in more detail.

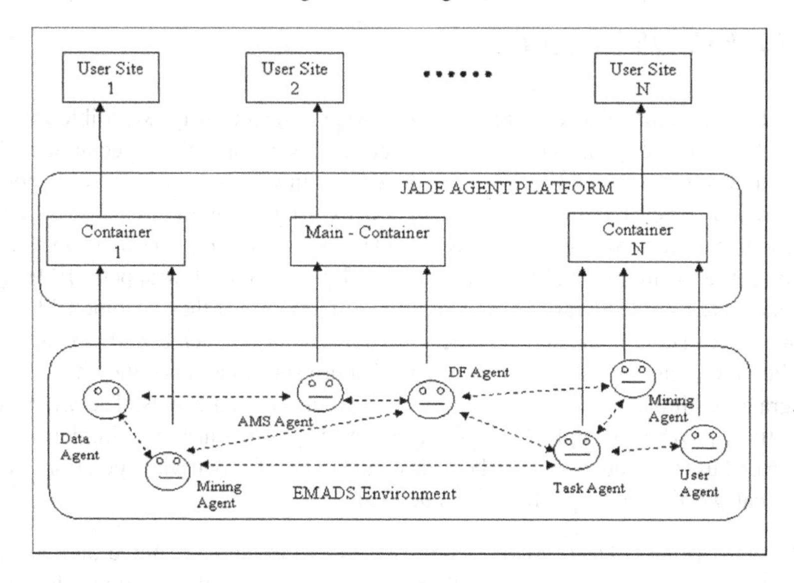

Fig. 13.1: EMADS Architecture as Implemented in Jade

13.4 The EMADS Implementation

EMADS is implemented using the JADE framework. JADE is FIPA (Foundation for Intelligent Physical Agents) [10] compliant middleware that enables development of peer to peer applications based on the agent paradigm. JADE defines an agent platform that comprises a set of containers, which may be distributed across a network as in the case of EMADS. A JADE platform includes a main container in which is held a number of mandatory agent services. These include the AMS and DF agents whose functionality has already been described in Section 13.3. Recall that the AMS agent is used to control the lifecycles of other agents in the platform, while the DF agent provides a lookup service by means of which agents can find other agents. When a mining or data agent is created, upon entry into the system, it announces itself to the DF agent after which it can be recognized and found by other agents.

Fig. 13.1 gives an overview of the implementation of EMADS using JADE. The figure is divided into three parts: at the top are listed N user sites. In the middle is the JADE platform holding the main container and N other containers. At the bottom a sample collection of agents is included. The solid arrows indicates a "belongs to" (or "is held by") relationship while the dotted arrows indicate a "communicates with" relationship. Thus the data agent at the bottom left belongs to container 1 which in turn belongs to User Site 1; and communicates with the AMS agent and (in this example) a single mining agent. The principal advantage of this JADE architecture is that it does not overload a single host machine, but distributes the processing load among multiple machines.

13.4.1 EMADS Wrappers

One of the principal objectives of EMADS is to provide an easily extendible framework that can readily accept new data sources and new data mining techniques. In general, extendibility can be defined as the ease with which software can be modified to adapt to new requirements or changes in existing requirements. Adding a new data source or data mining techniques should be as easy as adding new agents to the system. The desired extendibility is achieved by a system of wrappers. EMADS wrappers are used to "wrap" up data mining artifacts so that they become EMADS agents and can communicate with other EMADS agents. As such EMADS wrappers can be viewed as agents in their own right that are subsumed once they have been integrated with data or tools to become data or data mining agents. The wrappers essentially provide an application interface to EMADS that has to be implemented by the end user, although this has been designed to be a fairly trivial operation. Two broad categories of wrapper have been defined:

- **Data wrappers:** Data wrappers are used to "wrap" a data source and consequently create a data agent. Broadly a data wrapper holds the location (file path) of a data source, so that it can be accessed by other agents; and meta information about the data. To assist end users in the application of data wrappers a data wrapper GUI is available. Once created, the data agent announces itself to the DF agent as consequence of which it becomes available to all EMADS users.
- **Tool wrappers:** Tool wrappers are used to "wrap" up data mining software systems and thus create mining agents. Generally the software systems will be data mining tools of various kinds (classifiers, clusters, association rule miners, etc.) although they could also be (say) data normalization/discretization or visualization tools. It is intended that EMADS will incorporate a substantial number of different tool wrappers each defined by the nature of the desired I/O which in turn will be informed by the nature of the generic data mining tasks that it is desirable for EMADS to be able to perform.

Currently the research team has implemented two tool wrappers: the binary valued data, single label, classifier generator and the data normalization/discretization wrapper. However, many more categories of tool wrapper can be envisaged. Mining tool wrappers are more complex than data wrappers because of the different kinds of information that needs to be exchanged.

In the case of a *binary valued, single label, classifier generator* wrapper the input is a binary valued data set together with meta information about the number of classes and a number slots to allow for the (optional) inclusion of threshold values. The output is then a classifier expressed as a set of Classification Rules (CRs). As with data agents, once created, the data mining agent announce themselves to the DF agent after which they will becomes available for use to EMADS users.

In the case of the data normalization/discretization wrapper, the LUCS-KDD (Liverpool University Computer Science - Knowledge Discovery in Data) ARM

DN (Discretization/ Normalization) software [1] is used to convert data files, such as those available in the UCI data repository [6], into a binary format suitable for use with Association Rule Mining (ARM) applications. This tool has been "wrapped" using the data normalization/discretization wrapper.

13.5 EMADS Operations

In the following two sub-sections the operation of EMADS is illustrated using two DM scenarios: Meta Association Rule Mining and Classification.

13.5.1 Meta ARM (Association Rule Mining) Scenario

The term Meta Mining is defined, in the context of EMADS, as the process of combining individually obtained results of N applications of a DM activity. The motivation behind the scenario is that data relevant to a particular DM application may be owned and maintained by different, geographically dispersed, organizations. Information gathering and knowledge discovery from such distributed data sources typically entails a significant computational overhead; computational efficiency and scalability are both well established critical issue in data mining [12]. One approach to addressing problems, such as the Meta ARM problem, is to adopt a distributed approach. However this entails expensive computation and communication costs. In distributed data mining, there is a fundamental tradeoff between accuracy and computation cost. If we wish to improve the computation and communication costs, we can process all the data locally obtaining local results, and combine these results centrally to obtain the final result. If our interest is in the accuracy of the result, we can ship all the data to a single node (and apply an appropriate algorithm to produce this desired result). In general the latter is more expensive while the former is less accurate. The distributed approach also entails a critical security problem in that it reveals private information; privacy preserving issues [1] are of major concerns in inter enterprise data mining when dealing with private databases located at different sites.

13.5.1.1 Dynamic Behaviour of EMADS for Meta ARM Operations

The meta ARM scenario comprises a set of N data agents, N ARM mining agents and a meta ARM agent. Note that each ARM mining agent could have a different ARM algorithm associated with it, however, it is assumed that a common data structure is used to facilitate data interchange. For the scenario described here the

[1] *http://www.csc.liv.ac.uk/~frans/KDD/Software/*

common data structure is a T-tree [8], a *set enumeration* tree structure for storing item sets. Once generated the N local T-trees are passed to the Meta ARM agent which creates a global T-tree. Each of the Meta ARM algorithms makes use of *return to data* (RTD) lists, one per data set, to contain lists of itemsets whose support was not included in the current T-tree and for which the count is to be obtained by a return to the raw data. During the global T-tree generation process the Meta ARM agent interacts with the various ARM agents in the form of the exchange of RTD lists. There are a number of strategies that can be adopted with respect to when in the process the RTD lists should be exchanged; the research team identified four distinct strategies (Apriori, Brute Force, Hybrid 1 and Hybrid 2). A full description of the algorithms can be found in [2].

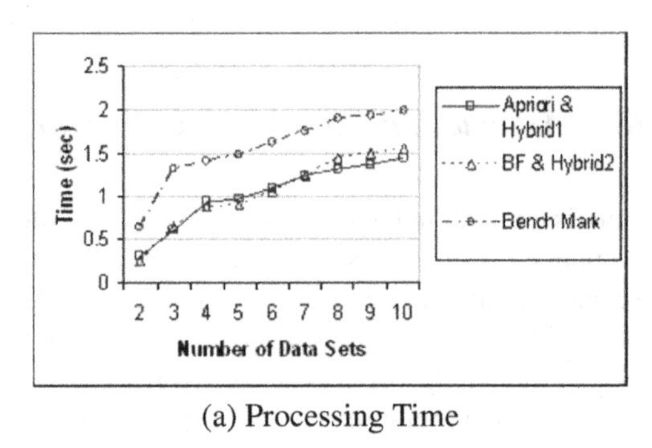

(a) Processing Time

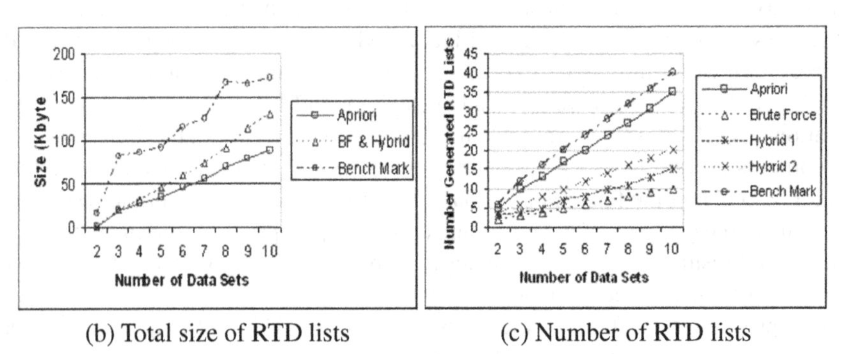

(b) Total size of RTD lists (c) Number of RTD lists

Fig. 13.2: Effect of number of data sources.

13.5.1.2 Experimentation and Analysis

To evaluate the five Meta ARM algorithms, in the context of EMADS, a number of experiments were conducted. The experiments were designed to analyze the effect of the following: (i) the number of data sources (data agents), (ii) the size of the datasets (held at data agents) in terms of number of records, and (iii) the size of the datasets (held at data agents) in terms of number of attributes.

Experiments were run using two Intel Core 2 Duo E6400 CPU (2.13GHz) computers with 3GB of main memory (DDR2 800MHz), Fedora Core 6, Kernel version 2.6.18 running under Linux except for the first experiment where two further computers running under Windows XP were added. For each of the experiments we measured: (i) processing time (seconds/mseconds), (ii) the size of the RTD lists (Kbytes), and (iii) the number of RTD lists generated.

Fig. 13.2 shows the effect of adding additional data sources using the four Meta ARM algorithms and the bench mark algorithm. For this experiment ten different artificial data sets were generated and distributed among four machines using T = 4 (average number of items per transactions), N = 20 (Number of attributes), D=100k (Number of transactions). Note that the slight oscillations in the graphs result simply from a vagary of the random nature of the test data generation. For other experiments results readers are referred to [3].

Fig. 13.2 also indicates, at least with respect to Meta ARM, that EMADS offers positive advantages in that all the Meta ARM algorithms were more computationally efficient than the bench mark algorithm. The results of the analysis also indicated that the Apriori Meta ARM approach coped best with a large number of data sources, while the Brute Force and Hybrid 1 approaches coped best with increased data sizes (in terms of column/rows).

13.5.2 Classifier Generation Scenario

In this section the operation of EMADS is further illustrated in the context of a classifier generation task; however much of the discussion is equally applicable to other generic data mining tasks such as clustering and ARM. The scenario is that of an end user who wishes to obtain a "best" classifier founded on a given, pre-labeled, data set; which can then be applied to further unlabelled data. The assumption is that the given data set is binary valued and that the user requires a single-label, as opposed to a multi-labeled, classifier. The request is made using the individual's user agent which in turn will spawn an appropriate task agent. For this scenario the task agent interacts with mining agents that hold *single labeled classifier* generators that take binary valued data as input. Each of these mining agents is then accessed and a classifier, together with an accuracy estimate, requested. Once received the task agent selects the classifier with the best accuracy and returns this to the user agent. The data mining agent wrapper in this case provides the interface that allows input of: (i) the identifier for the data set to be classified, and (ii) the number of class attributes (a value that the mining agent cannot currently deduce for itself); while the user agent interface allows input for threshold values (such as support and confidence values). The output is a classifier together with an accuracy measure. To obtain the accuracy measures the classifier generators (data mining agents) build their classifiers using the first half of the input data as the "training" set and the second half of the data as the "test" set.

From the literature there are many reported techniques available for generating classifiers. For the scenario reported here the authors used implementations of eight different algorithms[2]:

1. FOIL (First Order Inductive Learner) [15]: The well established inductive learning algorithm for the generation of Classification Association Rules (CARs).
2. TFPC (Total From Partial Classification): A CAR generator [9] founded on the P and T-tree set enumeration tree data structures.
3. PRM (Predictive Rule Mining) [16]: An extension of FOIL.
4. CPAR (Classification based on Predictive Association Rules) [16]: A further development from FOIL and PRM.
5. IGDT (Information Gain Decision Tree) classifier: An implementation of the well established C4.5 style of decision tree based classifier using information gain as the "splitting criteria".
6. RDT (Random Decision Tree) classifier: A decision tree based classifier that uses most frequent current attribute as the "splitting criteria".
7. CMAR (Classification based on Multiple Association Rules): A well established Classification Association Rule Mining (CARM) algorithm [13].
8. CBA (Classification Based on Associations): Another well established CARM algorithm [14].

These were placed within an appropriately defined tool wrapper to produce eight (single label binary data classifier generator) data mining agents. This was a trivial operation indicating the versatility of the wrapper concept.

Thus each mining agent's basic function is to generate a classification model using its own classifier and provide this to the task agent. The task agent then evaluates all the classifier models and chooses the most accurate model to be returned to the user agent to be presented to the user.

13.5.2.1 Experimentation and Analysis

To evaluate the classification scenario, as described above, a sequence of data sets taken from the UCI machine learning data repository [6] were used (preprocessed by data agents so that they were discretized/normalized into a binary valued format). The results are presented in Table 13.5.2 Each row in the table represents a particular request and gives the name of the data set, the selected best algorithm as identified from the interaction between the EMADS agents, the resulting best accuracy and the total EMADS execution time from creation of the initial task agent to the final "best" classifier being returned to the user agent. The naming convention used in the Table is that: D equals the number of attributes (after discretization/normalization),

[2] Taken from the LUCS-KDD software repository at *http : //www.csc.liv.ac.uk/~frans/KDD/Software/*

Data Set	Classifier	Accuracy	Generation Time (sec)
connect4.D129.N67557.C3	RDT	79.76	502.65
adult.D97.N48842.C2	IGDT	86.05	86.17
letRecog.D106.N20000.C26	RDT	91.79	31.52
anneal.D73.N898.C6	FOIL	98.44	5.82
breast.D20.N699.C2	IGDT	93.98	1.28
congres.D34.N435.C2	RDT	100	3.69
cylBands.D124.N540.C2	RDT	97.78	41.9
dematology.D49.N366.C6	RDT	96.17	11.28
heart.D52.N303.C5	RDT	96.02	3.04
auto.D137.N205.C7	IGDT	76.47	12.17
penDigits.D89.N10992.C10	RDT	99.18	13.77
soybean-large.D118.N683.C19	RDT	98.83	13.22
waveform.D101.N5000.C3	RDT	96.81	11.97

Table 13.1: Classification Results

N the number of records and C the number of classes (although EMADS has no requirement for the adoption of this convention).

The results demonstrate firstly that EMADS can usefully be adopted to produces a best classifier from a selection of classifiers. Secondly that operation of EMADS is not significantly hindered by agent communication overheads, although this has some effect. Generation time, in most cases does not seem to be an issue, so further classifier generator mining agents could easily be added. The results also reinforce the often observed phenomena that there is no single best classifier generator suited to all kinds of data set.

13.6 Conclusions

This chapter describes EMADS, a multi-agent framework for data mining. The principal advantages offered are that of experience and resource sharing, flexibility and extendibility, and (to an extent) protection of privacy and intellectual property rights.

This chapter presented the EMADS vision, the associated conceptualization and the JADE implementation. Of particalar note is the use of wrappers to incorporate existing software into EMADS. Experience indicates that, given an appropriate wrapper, existing data mining software can be very easily packaged to become an EMADS data mining agent. The EMADS operation was illustrated using Meta ARM and classification scenarios.

A good foundation has been established for both data mining research and genuine application based data mining. It is acknowledged that the current functionality of EMADS is limited to classification and Meta-ARM. The research team is at present working towards increasing the diversity of mining tasks that EMADS can address. There are many directions in which the work can (and is being) taken forward. One interesting direction is to build on the wealth of distributed data mining research that is currently available and progress this in an MAS context. The research team is also enhancing the system's robustness so as to make it publicly available. It is hoped that once the system is live other interested data mining practitioners will be prepared to contribute algorithms and data.

References

1. Aggarwal, C. and Yu, P., "A Condensation Approach to Privacy Preserving DataMining". Lecture Notes in Computer Science, Vol. 2992, pp. 183-199, (2004).
2. Albashiri, K., Coenen, F., Sanderson, R. and Leng. P., "Frequent Set Meta Mining: Towards Multi-Agent Data Mining". In Bramer, M., Coenen, F.P. and Petridis,M. (Eds.), Research and Development, Springer, London, pp139-151, (2007).
3. Albashiri, K., Coenen, F., and Leng. P., "Agent Based Frequent Set Meta Mining: Introducing EMADS". AI in Theory and Practice II, Proceedings IFIP, Springer, pp23-32, (2007).
4. Baazaoui H., Faiz S., Hamed R., and Ghezala H., "A Framework for data mining based multi-agent: an application to spatial data". 3rd World Enformatika Conference, Istanbul, (2005).
5. Bellifemine, F. Poggi, A. and Rimassi, G., "JADE: A FIPA-Compliant agent framework". Proceedings Practical Applications of Intelligent Agents and Multi-Agents, pg 97-108, (1999). (See http://sharon.cselt.it/projects/jade for latest information).
6. Blake, C. and Merz, C. "UCI Repository of machine learning databases". (1998). Irvine, CA: University of California, Department of Information and Computer Science. http://www.ics.uci.edu/mlearn/MLRepository.html
7. Bose, R. and V. Sugumaran, "IDM: An Intelligent Software Agent Based Data Mining Environment". In Proceedings of the IEEE International Conference on Systems, Man, and Cybernetics, 2888-2893 San Diego, CA: IEEE Press, (1998).
8. Coenen, F., Leng, P., and Goulbourne, G., "Tree Structures for Mining Association Rules". Journal of Data Mining and Knowledge Discovery, Vol 8, No 1, pp25-51, (2004).
9. Coenen, F., Leng, P. and Zhang, L. "Threshold Tuning for Improved Classification Association Rule Mining". Proceeding PAKDD, LNAI3158, Springer, pp216-225, (2005).
10. Foundation for Intelligent Physical Agents, FIPA 2002 Specification. Geneva, Switzerland. (See http://www.fipa.org/specifications/index.html).
11. Giuseppe, D., Giancarlo, F., "A customizable multi-agent system for distributed data mining". Proceedings of the 2007 ACM symposium on applied computing, Pages: 42 47, (2007).
12. Kamber, M.,Winstone, L.,Wan, G., Shan, S. and Jiawei, H., "Generalization and Decision Tree Induction: Efficient Classification in Data Mining". Proceedings of the Seventh International Workshop on Research Issues in Data Engineering, pp.111-120, (1997).
13. Li W., Han, J. and Pei, J., "CMAR: Accurate and Efficient Classification Based on Multiple Class-Association Rules". Proceedings ICDM, pp369-376, (2001).
14. Liu, B. Hsu, W. and Ma, Y., "Integrating Classification and Assocoiation Rule Mining". Proceedings KDD-98, New York, 27-31 August. AAAI. pp80-86, (1998).
15. Quinlan, J. R. and Cameron-Jones, R. M., "FOIL: A Midterm Report". Proceedings ECML, Vienna, Austria, pp3-20. (1993).
16. Yin, X. and Han, J., "PAR: Classification based on Predictive Association Rules". Proceedings SIAM Int. Conf. on Data Mining (SDM'03), San Fransisco, CA, pp. 331-335, (2003).

Chapter 14
A Multiagent Approach to Adaptive Continuous Analysis of Streaming Data in Complex Uncertain Environments

Igor Kiselev and Reda Alhajj

Abstract The data mining task of *online unsupervised learning* of streaming data continually arriving at the system in complex dynamic environments under conditions of uncertainty is an $\mathcal{NP}$-hard optimization problem for general metric spaces and is computationally intractable for real-world problems of practical interest. The primary contribution of this work is a multi-agent method for continuous agglomerative hierarchical clustering of streaming data, and a knowledge-based self-organizing competitive *multi-agent system* for implementing it. The reported experimental results demonstrate the applicability and efficiency of the implemented adaptive multi-agent learning system for continuous online clustering of both synthetic datasets and datasets from the following real-world domains: the RoboCup Soccer competition, and gene expression datasets from a bioinformatics test bed.

14.1 Introduction

14.1.1 Problem Definition

Continuous decision-making and *anytime data analysis* in dynamic uncertain environments represent one of the most challenging problems for developing robust intelligent systems. It is indispensable for intelligent applications working in such

Igor Kisev
The David R. Cheriton School of Computer Science, University of Waterloo,
200 University Avenue West, Waterloo, Canada, ON N2L 3G1, e-mail: ipkiselev@cs.uwaterloo.ca

Reda Alhajj
Department of Computer Science, University of Calgary,
2500 University Drive NW, Calgary, Canada, AB T2N 1N4, e-mail: alhajj@cpsc.ucalgary.ca

* Supplementary materials, software demonstration, and video recordings of the developed multi-agent system are available at: \sf\footnotesize\ttfamilyhttp://www.multiagent.org/mining

L. Cao (ed.), Data Mining and Multi-agent Integration, DOI: 10.1007/978-1-4419-0522-2_14,

complex environments as autonomous robotic systems, dynamic manufacturing and production processes, and distributed sensor networks to be capable for successfully responding to environmental dynamics and making time critical decisions online under conditions of uncertainty. To develop a probabilistic theory of the operating environment (a representation of the world), an intelligent system solves the problem of *unsupervised learning*, that is the process of discovering significant patterns or features in the input data when no certain output or response categories (classes) are specified. The task of unsupervised clustering in statistical learning requires the maximizing (or minimizing) of a certain similarity-based objective function defining an optimal segmentation of the input data set into clusters [7].

There are two types of unsupervised clustering algorithms: *partitional* and *hierarchical*. The main goal of a partitional optimization algorithm can be defined by finding such assignments $\mathcal{M}^*$ of observations $\mathcal{X}$ to output subsets $\mathcal{S}$ that minimize a mathematical energy function, which characterizes the degree to which the clustering goal is not met and is the sum of the cost of the clusters: $\mathcal{W}(\mathcal{M}^*) = \sum_{i=1}^{k} c(\mathcal{S}_i)$. The problem of partitional clustering is known to be computationally challenging ($\mathcal{NP}$-hard) for general metric spaces and is computationally intractable for real-world problems of practical interest. In comparison to partitional clustering algorithms, the goal of a hierarchical optimization algorithm is to extract an optimal multi-level partitioning of data by producing a hierarchical tree $\mathcal{T}(\mathcal{X})$ in which the nodes represent subsets $\mathcal{S}_i$ of $\mathcal{X}$. The time and space complexities of the hierarchical clustering are higher than partitional one: in standard cases a typical implementation of the hierarchical clustering algorithm requires $\mathcal{O}(\mathcal{N}^2 \log \mathcal{N})$ computations.

The task of online learning in complex dynamic environments assumes near real-time mining of streaming data continually arriving at the system, which imposes additional requirements for continuous data mining algorithms of being sensitive to environmental variations to provide a *fast dynamic response to changes* with an event-driven incremental improvement of mining results (cf. Table 14.1).

Table 14.1: Required properties of online unsupervised learning methods

Feature	Classical methods	Required functionality
Model:	**Static:**	**Dynamic:**
• Data set	• ***Static*** input data sets	• ***Dynamic*** and ***Streaming***
• Decision criteria	• ***Fixed*** and ***Single-objective*** criteria for learning	• ***Dynamic*** and ***Multi-objective*** quality metrics with trade-off balancing
• Learning parameters	• ***Invariable*** (cannot be changed at run-time)	• ***Adjustable*** at run-time during algorithm execution
Method:	**Batch-oriented:**	**Continuous:**
• Learning mode	• ***Batch-oriented*** processing of a static data set	• ***Near-real time learning*** of streaming data continually arriving at the system
• Availability of results	• ***Only after full completion*** (needs time to get a result)	• ***At anytime*** during algorithm execution (always see a result and its improvement)
• Reaction to changes	• ***Must be restarted again*** from scratch with full retraining of models or extra repair methods	• ***Without restarting***, and with event-driven ***incremental improvement*** of results, trading off operating time and result quality
Environment:	**Centralized and Deterministic:**	**Distributed and Stochastic:**
• Uncertainty of learning	• Assume predictable outcome (***Rigid***)	• Consider random effects (***Resilient***)
• Data location	• ***Single or Decentralized***, but with an additional centralized algorithm of aggregating partial results	• ***Both centralized and decentralized*** without any additional aggregation procedures

Additionally, a dynamic data mining algorithm, operating in complex uncertain environments with incomplete knowledge about parameters of the learning problem, should be suitable for *online exploratory data analysis* using different measures of similarity in order to be able to continually operate on the basis of various dynamic learning criteria.

The application of the proposed multi-agent solution to continuous online learning is appropriate for various scenarios of near real-time data processing: online intrusion detection, emergency response in hazardous situations (e.g. forest fires, chemical contaminants in drinking water), control of military operations with time critical targets, online learning of distributed robotic systems (e.g. in the RoboCup Soccer and Rescue domains), and run-time detection of previously unknown dispatching rules and effective scheduling policies in transportation logistics.

14.1.2 Related Work

Continuous and anytime data analysis imposes requirements for *adaptability* of learning methods that are simply not addressed by traditional data mining techniques. Conventional methods of unsupervised learning address the issue of statistical fluctuations of the incoming data by means of continual retraining of models that is computationally intractable or inappropriate in time-critical scenarios. Clustering results of batch-oriented methods are available only after their full completion, and must be started again from scratch in order to react on environmental variations.

To address this issue of effective online learning, various approaches have been considered in the literature (refer to online supplementary materials for a complete overview of related work [10]). Decentralized clustering algorithms were proposed to speeds up centralized learning by dividing it onto a set of processors and allowing them to learn concurrently with an additional centralized algorithm of aggregating partial mining results to the global solution [21], [12], [23]. To handle the complexity of the problem, approximate clustering algorithms were developed to search for a feasible solution in incomplete decision space, by applying approximation heuristics that reduce the problem dimension, but lead to worse result quality [7]. Clustering methods for unsupervised learning of streaming data were developed, which support incremental update of the mining result by applying additional repairing methods [1], [3]. Uncertainty of the operating environment is approached by feedback-directed clustering algorithms that apply reinforcement learning techniques to guide the search towards better cluster quality [4], [13]. The distributed-constraint reasoning formalism was proposed to approach optimization and learning problems in a decentralized manner, which is better suited to deal with changes in a localized fashion [5], but can be too expensive in large-scale dynamic environments [16] and restrictive to provide a fast response to environmental variations [22].

As opposed to previous work, we present a different anytime *multi-agent approach* to online unsupervised learning, which is different from conventional methods by being *dynamic*, *incremental* and *distributed*, rather than parallel . We demonstrate that the task of continuous unsupervised learning, when formulated as a *dynamic distributed resource allocation problem*, can be effectively approached by a decentralized market-based method of multi-agent negotiation [11].

14.2 Continuous Online Unsupervised Learning in Complex Uncertain Environments

14.2.1 Market-based Algorithm of Continuous Agglomerative Hierarchical Clustering

As opposed to previous work, we propose a different multi-agent approach to continuous online learning of streaming data by modeling the task of unsupervised clustering as a *dynamic distributed resource allocation problem* [15]. The online learning algorithm implements the concept of *clustering by asynchronous message-passing* [6], whereby the any-time solution to the continual constrained optimization problem of clustering is obtained (inferred) by satisfying a dynamic distributed constraint network of agent interests. Thus, the continual distributed learning process is carried out by means of asynchronous quasi-parallel processes of negotiation between the competitive agents of records and clusters, defined for data elements. Mining agents negotiate (act) with each other in the virtual learning marketplace in order to satisfy their individual goals and maximize their criteria values. Searching for the most profitable allocation variants (semantic links with the highest utility) to enhance their satisfaction levels with minimal costs, the agents of clusters and records dynamically establish and reconsider ontological relationships with other agents, thereby dynamically establishing ontological multilevel virtual market communities.

A distributed computational environment for self-interested agents (a virtual learning marketplace) is formally defined according to the game-theoretic notation of a *marketplace system* [8], [14]. To develop a multi-agent system capable of operating in dynamic distributed environments, we solve the task of online computational mechanism design in distributed computational settings, which assumes that *self-interested agents* can arrive at and depart from the multi-agent system dynamically over time, and there is no a trusted central mechanism to control their behavior. *Computational mechanism design* provides a mathematical framework that defines each agent's decision-making model and specifies the protocols that govern the agent interactions (the market mechanism through which agents interact) [17], [19]. A market-based algorithm of continuous agglomerative hierarchical clustering designs a multi-agent system in which rational self-interested agents with privately known preferences interact in a way that leads to equilibriums with desired system-wide properties (socially desirable outcome). Fig. 14.1 depicts an overview flowchart for the market-based algorithm of continuous agglomerative hierarchical clustering algorithm.

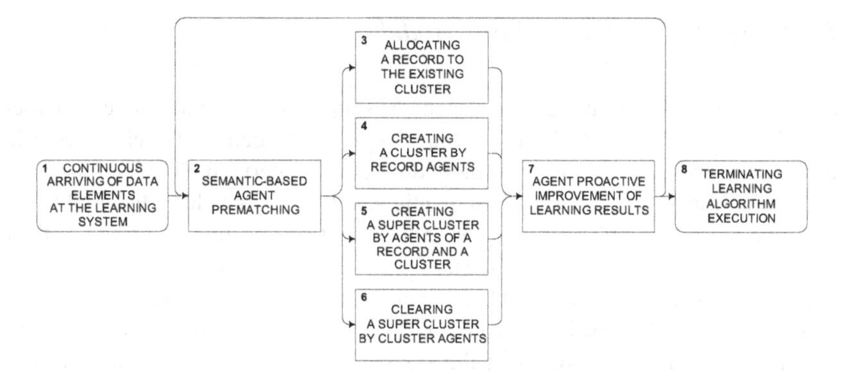

Fig. 14.1: A flowchart for the market-based algorithm of continuous hierarchical clustering

A global decision (macroscopic solution) of the dynamic distributed data clustering is implicitly achieved (performed) by the competitive agents that maintain a dynamic balance among the interest of all participants in the interaction according to the following algorithmic process (cf. Fig. 14.1): continuous arriving of data elements at the system (algorithmic step #1), locating candidate agents for allocation negotiations (algorithmic step #2), satisfying a dynamic *distributed constraint network* of agent interests for each type of agent negotiations (algorithmic steps #3, #4, #5, #6), agent proactive improvement of learning results (algorithmic step #7), and terminating the execution of the learning algorithm (algorithmic step #8). The complete description of each algorithmic step of the method is omitted in the chapter due to space limitations (refer to online supplementary materials).

The implemented auction-based negotiation method of agent negotiations is based on a modified Contract-Net Protocol, where agents dynamically submit bids based on the cost of possible allocation variant [20]. Each participant of the negotiation evaluates new allocation options and sends an approval only when the criteria value of a new agent state with a new established link and broken previous relationships with other agents (if any) is better than the value of the current agent state. If new ontological instance is created as a result of a negotiation process of the synthesis type a new mining agent of a corresponding type is produced in the virtual clustering marketplace and assigned to it.

The proposed computationally efficient multi-agent algorithm for online agglomerative hierarchical clustering is different from conventional unsupervised learning methods by being *distributed*, *dynamic*, and *continuous*. Distributed clustering process provides the ability to perform efficient run-time learning from both centralized and decentralized data sources without an additional centralized algorithm of aggregating partial mining results. Both the input dataset of decentralized sources and decision criteria for learning (e.g. similarity matrices and expert knowledge) are not fixed and can be changed at run-time during execution of the dynamic algorithm. Clustering results of the *adaptive learning algorithm* are available at any time and continuously improved to achieve a global quasi-optimal solution to the optimization problem, trading-off operating time and result quality.

14.2.2 Agent Decision-making Model

Goal-driven behavior of autonomous agents is supported by the developed microeconomic *multi-objective decision-making model*, which defines for each agent in the virtual marketplace ontology its individual goals, criteria, preference functions, and decision-making strategies [9]. Semantic agents of records have the goal to establish the most profitable allocation with the agents of clusters according to their individual agent criteria ("to be allocated"). To accomplish the allocation goal, a record agent can either send a membership application to the existing cluster to join it (algorithmic step #3) or be allocated to a new cluster, which can be created as a result of a negotiation process of the synthesis type with either another record agents or existing cluster agents (algorithmic steps #4 and #5). To support hierarchical clustering, there are two different goals defined for a cluster agent within its decision-making model (bidirectional $\lambda - \pi$ inference). The first goal of a cluster agent ("allocate") is to establish links with the agents of records to create the most profitable ontological cluster of the best quality. The second goal of a cluster agent ("to be allocated") has the same notion as the goal of a record agent, to establish the most profitable allocation with the agents of clusters, and is defined through the task of establishing the most effective relationship of the "part-of" type with another cluster agent (algorithmic step #6). Fig. 14.2 illustrates the agent learning goals and basic types of agent negotiations in the virtual clustering marketplace.

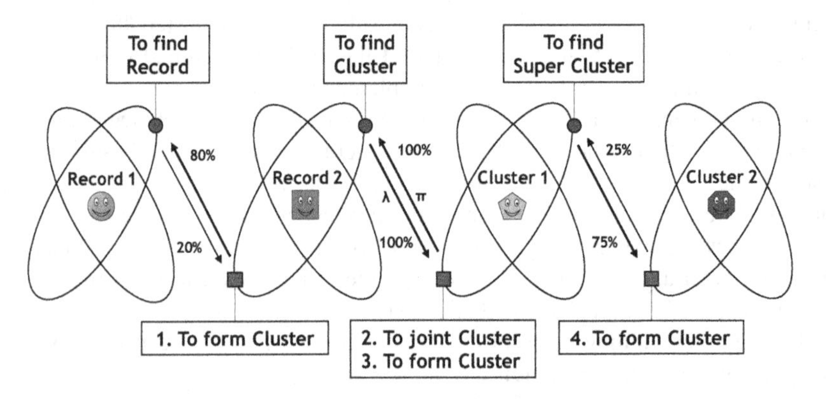

Fig. 14.2: Hierarchical architecture of a virtual clustering marketplace: agent goals and learning tasks

The developed multi-objective decision-making model makes it possible for the learning algorithm to continually operate on the basis of *dynamic learning criteria*, and to be suitable for online exploratory data analysis in complex uncertain environments with incomplete knowledge about parameters of the learning problem (cf. Fig. 14.3). Competitive agents of records and clusters act in the virtual clustering marketplace to satisfy their individual goals and maximize their criteria values according to the chosen decision-making strategy. Currently supported agent decision-making strategies are based on the following agent criteria: the Euclidian distance-based measure of similarity, the Chebychev similarity metric, and the angle metrics defining polarization ("shape") of agent communities (multilevel and multicultural) in decision-space. Agent decision-making strategies can be applied dynamically at run-time to the whole agent society (global level), to a single agent (individual level), or to agent groups in different areas of the virtual clustering marketplace (several polarization vectors).

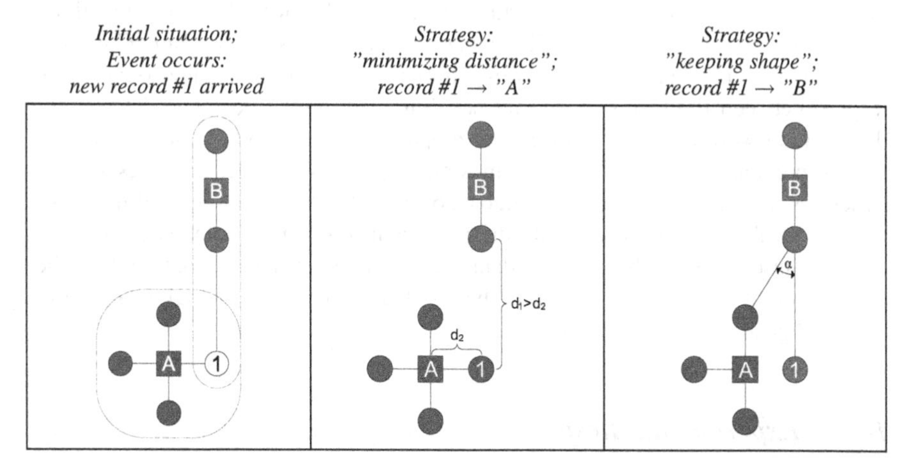

Fig. 14.3: Dynamic support for various agent decision-making strategies and learning criteria

Utilizing the agent criterion, based on the Euclidian distance-based measure of similarity, the learning algorithm relies on the ability to calculate the centroid of each cluster. For the input spaces where it is not possible, we propose to use the Chebychev similarity metric as a basis for a similarity measure between objects. The Chebychev similarity measure is based on the definition of the Chebychev distance metric, where the distance between two vectors is the greatest of their differences along any coordinate dimension. Using the Chebychev similarity metric as a similarity measure has the advantage of being less computational expensive in comparison to the distance-based metric of similarity since there is no need for estimating parameters of all records allocated to the cluster during negotiations. The complete distributed learning model, representing a full set of agent learning criteria for three types of similarity metrics (the Euclidian distance-based similarity, Chebychev similarity metric, and the angle metrics), consists of 24 equations in total and is omitted due to space limitations (refer to online supplementary materials [10]).

In order to be effective in solving time-sensitive data mining problems in complex uncertain environments, the developed multi-agent learning system additionally addresses the following challenges of online learning of streaming data: the processing of large number of input records and online tractability, the continual directed adaptation of the learning system parameters to environmental variations and a fast dynamic response to them in a real-time fashion, and the communication complexity of dynamic large-scale networks of autonomous agents [8].

14.3 Experimental Analysis

14.3.1 Datasets

The proposed multi-agent method of online learning was experimentally evaluated for continuous agglomerative hierarchical clustering of both synthetic datasets and datasets from the following real-world domains: the RoboCup Soccer competition and a gene expression datasets. The experimental datasets for the RoboCup Soccer domain were obtained by analyzing files of the final 2006 game (simulation league) between the teams "Brainstormers" and "WrightEagle". The datasets for data mining were obtained using a data preparation framework, which parses log files of previous RoboCup Soccer games and generates knowledge representation structures of agent action scenes suitable for data mining purposes (459 instances with 10 attributes) [8]. To evaluate the performance of the developed solution to cluster datasets of high-dimensional data, we used a reduced cancer dataset [2]. The acute myeloid leukemia (AML)/acute lymphoblastic leukemia (ALL) dataset contains 192 gene and 73 patient samples.

14.3.2 Experimental Results

The reported experimental results demonstrate the applicability and efficiency of the implemented adaptive multi-agent learning system for continuous online clustering of both synthetic datasets and datasets from the following real-world domains: the RoboCup Soccer competition, and gene expression datasets from a bioinformatics test bed. The major experimental results of the conducted experimental analysis are summarized in Table 14.2 and graphically presented as fourteen charts (cf. Fig. 14.4 – Fig. 14.5 and Fig. 14.8 – Fig. 14.17 listed in Appendix), which demonstrate dynamics of the distributed learning process within and across various dimensions of the performance radar (cf. Fig. 14.5). Table 14.2 consists of four column groups, and reports solution quality for different algorithm parameters and agent decision-making strategies.

Both centralized and distributed (local) performance metrics was used to evaluate solution quality. We use the *Cophenetic (Pearson) coefficient* as a centralized performance metrics to measure quality of hierarchical clustering (computed in the shared memory primarily for comparison purposes) [18]. We also consider the agent decision-making criteria to be distributed performance metrics to evaluate the solution quality. Such "personified" performance indicators allow for identifying quality "bottle-necks" across all clustering hierarchy. The table reveals the dominance of different performance metrics when applying certain agent decision-making strategies, and also demonstrates which parameters of the multi-agent algorithm should be used to increase system performance along specific dimensions of the performance radar (the absolute best and worst parameter values for each performance metrics are emphasized in bold and italic types respectively).

Table 14.2: Performance comparison of different agent strategies

Learning Strategy	Euclidian Similarity	Top Candidates	Reference Point	Level Penalties	Stochastic Prematching	Proactive Roles	Continual Learning	Chebychev Similarity	Angle Metrics
Operating Time									
T, ms	3375	**2781**	3165	3450	*3553*	3484	3297	11937	2859
Cophenetic (Pearson) coefficients									
PMR	**0.9029**	0.8779	0.8866	*0.7873*	0.8939	0.9487	0.9029	0.6395	0.7227
PBC	0.9029	**0.9353**	0.9313	0.9277	*0.8939*	0.9624	0.9029	0.8835	0.7227
PBA	*0.9414*	0.9799	0.9799	**0.9965**	0.9650	0.9813	0.9414	0.9957	0.7790
PM	0.4213	0.4249	0.4261	**0.4270**	*0.4166*	0.5807	0.4213	0.3501	0.2813
Cluster Agent Values "Contains"									
CBA_CS	0.5974	0.5974	0.5974	0.5974	0.5974	0.5974	0.5974	0.9605	0.9930
CBC_CS	0.2273	0.2273	0.2273	0.2273	0.2273	0.3142	0.2273	0.8259	0.7690
CM_CS	**0.1715**	0.1621	0.1665	*0.1596*	0.1630	0.2086	0.1715	0.7086	0.7102
CMR_CS	0.0334	0.0334	0.0334	0.0334	0.0334	0.0334	0.0334	0.2489	0.3880
CWA_CS	0.0184	0.0184	0.0184	0.0184	0.0184	0.0334	0.0184	0.2489	0.0290
Cluster Agent Values "Contained"									
CBA_CD	0.1663	*0.1239*	**0.3046**	**0.3046**	**0.3046**	0.3098	0.1663	0.7383	0.8286
CBC_CD	0.0974	*0.0663*	**0.2273**	0.1149	0.1250	0.1373	0.0974	0.6851	0.6649
CM_CD	0.0618	0.0467	**0.0835**	*-0.4568*	0.0772	0.0654	0.0618	0.5479	0.5321
Record Agent Values "Contained"									
RBA	0.5974	0.5974	0.5974	0.5974	0.5974	0.5974	0.5974	0.5974	0.9930
RM	0.2273	0.2273	0.2273	0.2273	0.2273	0.2644	0.2273	0.2273	0.7690
RWA	0.0184	0.0184	0.0184	0.0184	0.0184	0.0325	0.0184	0.0184	0.0290

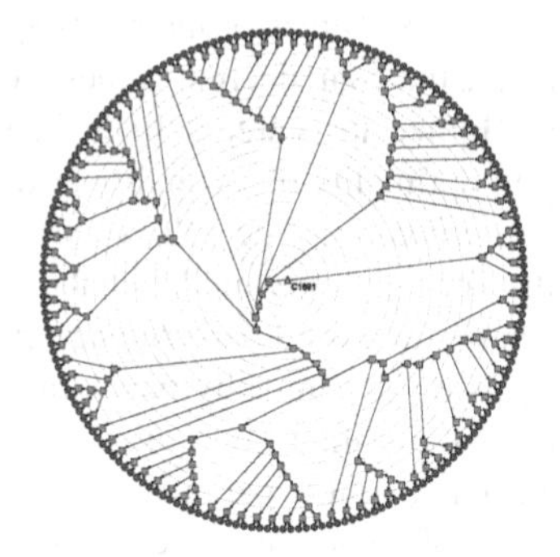

Fig. 14.4: A radial dendrogram of learning results for clustering the gene expression dataset with 192 genes and 73 patients (the equilibrium state)

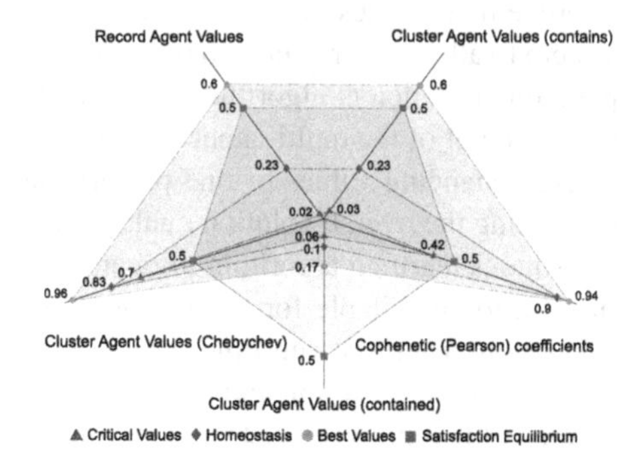

Fig. 14.5: A performance radar of major agent decision-making criteria (the multi-criteria model of quality metrics for the agent society)

The first set of experimental charts presents learning results for clustering the gene expression dataset and RoboCup Soccer dataset, and solution quality of the multi-agent algorithm across different performance metrics, agent decision-making criteria, and different parameters of the multi-agent algorithm. Learning results for clustering the gene expression dataset from the bioinformatics testbed are presented in Fig. 14.4 as a interactive radial hierarchical dendrogram, which at any point in the computation graphically represents a dynamic reconfigurable network of the semantic instances of mining agents in the virtual learning marketplace. Red circles and blue squares of the radial dendrogram represent clustered record and cluster agent respectively, and a green triangle in the center of the dendrogram constitutes the root cluster agent at the top level of the learning hierarchy. Additionally, it can be seen from the dynamic radial dendrogram that the semantic instances of mining agents are not numbered successively in a stable equilibrium state which is explained by the incremental nature of the distributed learning process. The multi-agent system responds to an external event by locally reorganizing only those areas of the global decision space that are affected by the event (*incremental optimization* in the local context). All decisions and established links are not fixed in the system and locally reconsidered when needed during reaction on environmental perturbations.

Fig. 14.5 represents the overall performance radar of five major quality metrics (agent decision-making criteria) for the agent society in the virtual learning marketplace, which define several areas of solution quality: critical and best values, homeostasis and satisfaction equilibrium. Critical values of the performance dimensions define the area of solution quality, where mining agents are not satisfied with established relationships with other agents and actively look for other allocation options to enhance their satisfaction levels with minimal costs. On the contrary, best values define the area of solution quality, where autonomous agents have absolute best satisfaction levels for each performance dimension. Satisfaction equilibrium defines the area of solution quality, where satisfaction levels of the agents are sufficiently high such that they turn to the inactive states and do not exhibit proactivity to improve their values, thereby releasing computational resources of the multi-agent system to allow other agents to be active and enhance their satisfaction values (computationally efficient algorithm implementation). Homeostasis area defines the stationary point of the multi-agent system with Pareto optimality of its quality criteria, where enhancing values of one performance metric could not be achieve without decreasing the overall solution quality. The multi-criteria quality model enables the continuous learning algorithm to operate on the basis of several performance criteria, and to be suitable for exploratory data analysis by dynamically balancing the evaluation criteria at run-time during execution of the online multi-agent system (supporting the selection and regulation of appropriate learning parameters).

Fig. 14.6 depicts the distribution of quality levels of the learning hierarchy (cuts of the agent society) across different performance metrics and over operating time of the multi-agent algorithm. The anytime multi-agent learning algorithm not only extracts an optimal multi-level partitioning of data using various measures of similarity, but also defines optimal segmentations of the input data set into clusters by dynamically selecting the quality levels (cuts) of the learning hierarchy for each performance metric defined in the multi-criteria quality model. Thus, performance comparison of Euclidian and Chebychev similarity metrics as agent learning criteria is presented in Fig. 14.13, which demonstrates that their major performance metrics have the same final configuration of quality levels, and suggests that due to described advantages of the Chebychev similarity metric it should be used where appropriate. Fig. 14.7 demonstrates the result of a conducted experiment on the analysis of the distribution of performance metrics by hierarchical levels of the agent society, and represents a stable equilibrium state of the distributed learning process (static situation). By comparing the learning results of a conventional hierarchical single-linkage clustering algorithm with performance metrics of the developed continual multi-agent algorithm, we can conclude that solution quality obtained by the multi-agent learning system in its final stable configuration is as good as one of the batch algorithm.

Fig. 14.14 illustrates the dynamic characteristics of agent instances during the continual distributed learning process and emphasizes the agent *ripple effect*, which is a decision reconsideration chain that improves the overall clustering results due to *agent proactivity*. The grey shaded square area of the chart, which is formed by the intersection of the vertical (time of the ripple effect during which agents improve their satisfaction) and horizontal (a value gained as a result of the agent proactivity) grey areas, emphasizes the property of the multi-agent algorithm to proactively trade-off solution quality and operating time. A back dotted line of the chart represents the dynamics of the Cophenetic coefficient over operating time of the multi-agent algorithm, and is displayed on its own scale to demonstrate incremental improvement of solution quality as a result of the ripple effect. A red line of the chart represents the number of non-allocated record agents, which is reduced over time and approaches a zero value in the final stable configuration of the multi-agent system. Behavior of the red line inside the shaded square area demonstrates the moment of switching agent proactivity, during which the system transit from one dynamic state of balance into a new economically more effective one by the breaking of previously established ontological relationships between record and cluster agents and establishing new semantic links. Thus, it can be seen from the diagram that the number of non-allocated record agents becomes zero in the beginning of the grey shaded area (all agents are allocated). However, during the proactivity stage the agents, seeking to increase their satisfaction values, break previous ontological relationships (temporary becoming unplanned) and lead the multi-agent system to a new configuration of the dynamic equilibrium with better solution quality (the number of non-allocated record agents becomes zero again at the end of the grey shaded area, but during the period of agent proactivity the value of the Cophenetic coefficient is increased as a result of the agent ripple effect).

Fig. 14.9 demonstrates the results of the conducted experiments on introducing the proactive agent "Record-Cluster" role into the agent decision-making model. The chart reveals a significant increase of the solution quality across all performance metrics (though taking slightly more time to settle down to a new quasi-optimal state) when the record agents not only exhibit reactive behavior by simply responding to system events, but also proactively search for profitable allocation variants. Fig. 14.11 demonstrate that cluster centroids, produced by the developed multi-agent algorithm, can be considered as *significant representative features* ("reference points") for the algorithm since the latter provides representative clustering results with the satisfactory overall quality of partitioning. Fig. 14.10 demonstrates the ability of the multi-agent learning algorithm to avoid local optima and enhance the overall learning results by means of the incremental stochastic agent pre-matching algorithm ("random sampling" of the search space), which restrictively regulates the agent pre-matching radius to incrementally increase a depth of agent vision and stochastically select candidate agents within the agent pre-matching radius. Fig. 14.15 represents performance metrics of cluster agents evolving in the virtual learning marketplace (with quality levels shown as numbers above the lines, and final values of the performance metrics in the stable equilibrium state of the multi-agent system shown on the right side side of the chart).

The second set of experimental charts demonstrates the performance of the developed multi-agent algorithm to conduct continual online learning of stream data. Fig. 14.16 demonstrates the functional dependence of online learning performance on the number of records being continuously clustered. It can be seen from the diagram that in situations in which a new record arrived when previous records have been allocated, the time required to incrementally incorporate a new record into the hierarchical learning structure is not exponential (and approximately linear) due to the implemented agent memory and directed search mechanisms. Each allocation moment, incremental planning a new record takes different amount of time, since mining agents do not consider all allocation options at once, but only those available in their local context, and subsequently increase their field of vision to incrementally improve the initial (previous) solutions. Thus, curve dips (troughs) of the graph demonstrate the situations where new allocation happens with minimum re-learning of previously established relationships (the ripple effect of minimal length), and picks of the graph indicate the moments when planning a new agent leads to considerable reallocation of previously formed agent links. A red line of the chart represents the accumulative number of reallocation of previously planned agents for each stable equilibrium state, which is increased over time.

Fig. 14.17 presents the results of the conducted comparison analysis of the developed continuous multi-agent method and a conventional hierarchical single-linkage clustering algorithm (the Alias "LingPipe" software library). The chart demonstrates that for the incremental approach the time required to react on changes and maintain the clustering hierarchy valid reduces as the number of agents affected by environmental variations decreases (the length of the ripple effect is context dependent), while re-learning time remains approximately constant for the batch algorithm. Thus, the conducted comparison analysis demonstrates the strong advantage of the developed incremental multi-agent approach over the classical batch clustering algorithm in dynamic settings as a result of the developed matching memory mechanism ensures a quick response time to changes by directly adapting of only those areas of the global decision space that are affected by them. Nevertheless, it should be noted that due to its distributed nature the continuous multi-agent algorithm takes more time for batch data mining in static settings than the centralized algorithm (the blue line is higher than the read one in the beginning of the chart). Fig. 14.8 provides a performance comparison chart for clustering in a dataset in batch and continuous learning modes. This chart and Fig. 14.12 demonstrate that the algorithm ensures deterministic learning results for various input sequences of arriving records at the system (the results converge to the same values in the final stable equilibrium state of the distributed learning process).

14.4 Summary, Conclusion and Future Work

14.4.1 Summary

This work considers the problem of continuous data mining of streaming data in complex dynamic environments under conditions of uncertainty. The primary goal of the developed multi-agent approach is continuous learning in distributed environments from decentralized data sources across a heterogeneous data environment with a view to effectively responding to environmental dynamics and performing online data analysis using various dynamic learning criteria and measures of similarity. The main contributions of this research can be summarized as follows.

1. With a view to responding to rapid changes in the environment, we developed a multi-agent method of continuous online learning, by modeling the task of unsupervised clustering as a *dynamic distributed resource allocation problem.* A game-theoretic decentralized market-based method of competitive and implicit multi-agent negotiation, and an asynchronous message-passing algorithm are developed to obtain an implicit global quasi-optimal solution to the distributed constrained learning problem, which requires market-based negotiation between different self-interested agents, defined for data elements, to satisfy a dynamic distributed constraint network of agent interests by maintaining a dynamic balance among the interests of all participants in the interaction.
2. A knowledge-based competitive *multi-agent learning system* is developed to enable the data-driven self-organizing distributed process of dynamic continuous data mining. The implemented multi-agent platform of the system provides a distributed computational environment (virtual learning marketplace) and a run-time support for asynchronous quasi-parallel negotiations between the agents in a virtual marketplace. To consider personal preferences and expert knowledge, the developed virtual marketplace ontology (a semantic knowledge base located in the shared memory) has been set up to contain conceptual knowledge of the problem domain and to support the dynamic regulation of various control parameters of the learning system at run-time during its execution, such as different agent decision-making criteria, active agent negotiation roles, nature of the operating environment and various properties of the learning algorithm.
3. A multi-objective *agent decision-making model* is developed to support goal-driven behavior of autonomous agents in the virtual learning marketplace, which defines for each agent in the virtual marketplace ontology its individual goals, criteria, preference functions, and decision-making strategies. The multi-objective decision-making model enables the learning algorithm to continually operate on the basis of non-standard optimization criteria, and to be suitable for online ex-

ploratory data analysis in complex uncertain environments using various measures of similarity for situations with incomplete knowledge about parameters of the learning problem.

4. The proposed multi-agent method of online learning was experimentally evaluated for continuous agglomerative hierarchical clustering of both synthetic datasets and datasets from the following real-world domains: the RoboCup Soccer competition and a gene expression datasets from a bioinformatics test bed. The conducted comparison analysis demonstrates the superior advantage of the incremental multi-agent learning approach over conventional batch clustering algorithms.

14.4.2 Conclusions and Future Directions

We support our conclusions by conducting the experimental analysis, which demonstrate the applicability and efficiency of the developed continuous multi-agent learning system to respond to environmental dynamics and to perform online data analysis using various dynamic learning criteria and measures of similarity. The reported experimental results demonstrate the strong performance of the developed multi-agent learning system for continuous agglomerative hierarchical clustering of both synthetic datasets and datasets from the real-world domains.

According to the conducted experimental analysis, we conclude that the proposed online multi-agent approach has the advantage over conventional unsupervised learning methods of being *dynamic*, *incremental* and *continuous*. The input dataset, decision-making criteria and various control parameters of the learning system are not fixed and can be changed at run-time during execution of the dynamic algorithm. Clustering results of the adaptive learning algorithm are available at any time, and continuously and incrementally improved to achieve a global quasi-optimal solution to the optimization problem, trading-off operating time and result quality. All decentralized decisions of the method are not fixed and locally reconsidered when needed in response to environmental events, taking advantage of domain semantics.

The implemented multi-objective decision-making model enables the learning algorithm to continually operate on the basis of non-standard optimization criteria and suitable for online exploratory data analysis using various measures of similarity. Additionally, the implemented adaptive agent pre-matching mechanism of regulating a depth of agent vision makes it possible for an agent to restrictively consider only those allocation options that are inside its limited field of vision, thereby preventing enumeration of all allocation possibilities and making learning of a large amount of data computationally tractable.

Nonetheless, the current implementation of the multi-agent learning system has the following limitations. The conducted comparison analysis of the developed multi-agent method and a conventional hierarchical single-linkage clustering algorithm (the Alias "LingPipe" software library) demonstrates the superior performance of the incremental approach in comparison to the batch algorithm. The developed matching memory mechanism on the agent protocol level ensures a quick response time to changes by directly adapting of only those areas of the global decision space that are affected by them. However, keeping such records of agent statuses in the shared memory of the system in order to decrease communication costs introduces the increase in the memory complexity of the algorithm. To reduce memory consumption, a special service role of the World agent periodically validates agent matching memories to delete non-valid history records of instances with dead agents.

The conducted experiments demonstrate that introducing the proactive agent "Record-Cluster" role significantly increases the solution quality across all performance metrics. The effective regulation of agent activities in the distributed environment is crucial to obtain a global quasi-optimal solution to the problem of better quality and to enhance the performance of the distributed learning algorithm. The developed agent activity control framework provides the decentralized mechanism of regulating proactivity for certain types of agent negotiation roles based on the observation of the hormonal level in the multi-agent environment. Nevertheless, balancing proactivity of agent negotiation roles in the virtual learning marketplace is a challenging issue. Thus, some performance charts of the conducted experimental analysis reveal a slight decrease in solution quality during the moments of switching agent proactivity for certain types of agent negotiation roles. Although the strong advantage of the developed incremental multi-agent approach over classical batch clustering algorithms in dynamic settings was demonstrated, the conducted experimental analysis revealed the current limitation of the developed learning algorithm to efficiently perform massive data processing of high-dimensional data (e.g. genome-wide gene expression data) in centralized and batch settings due to communication costs of additional message-passing and decision synchronization algorithmic steps. Nevertheless, to increase the performance of the developed approach for batch data processing in static settings, we suggest conducting combinatorial auctions in the local agent context and propose a hybrid online learning approach that combines distributed constraint optimization techniques and decision theoretic approaches [9].

We consider the developed solution to be an important step in our research towards the development of effective online learning algorithms in dynamic uncertain environments, and plan to explore how to learn non-stationary dynamics. Future work will be directed towards addressing the following challenging problems that have arisen from this research: (1) extending the adaptive learning approach to support online automatic *semi-supervised classification* by continuously deducing semantic-based classification rules from clustering results and performing automatic rule-based classification and subsequent pattern verification at run-time, and (2) developing a *hybrid online learning approach*, which reduces the problem dimension while maintaining the essential characteristics of the original system, and provides a dynamic bidirectional cyclic feedback on using the market-based distributed resource allocation (bottom-up) and Bayesian reinforcement learning of decentralized partially observable Markov decision processes (Dec-POMDPs) (top-down).

Appendix

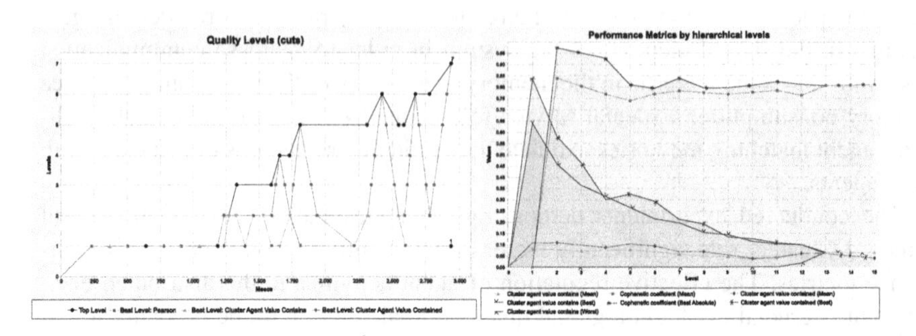

Fig. 14.6: Quality levels (cuts) of the learning hierarchy across different performance metrics

Fig. 14.7: Distribution of performance metrics by hierarchical levels (stable equilibrium state)

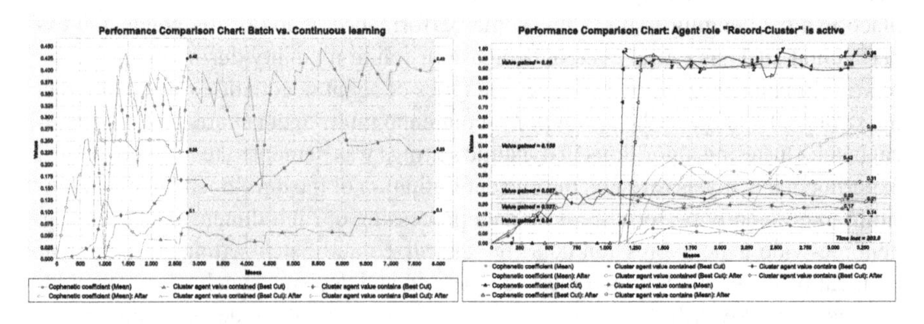

Fig. 14.8: Time required for Batch vs. Continuous learning (performance comparison)

Fig. 14.9: Trading-off operating time & learning quality(agent role "Record-Cluster" is proactive)

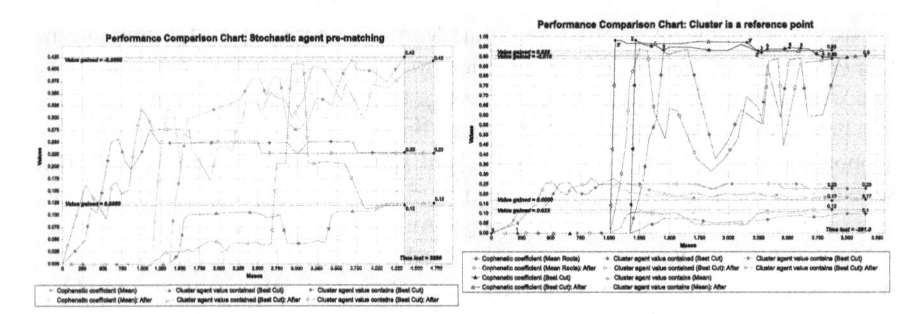

Fig. 14.10: Time required for learning with incremental and stochastic agent prematching

Fig. 14.11: Performance comparison chart: cluster centroids as representative "reference points"

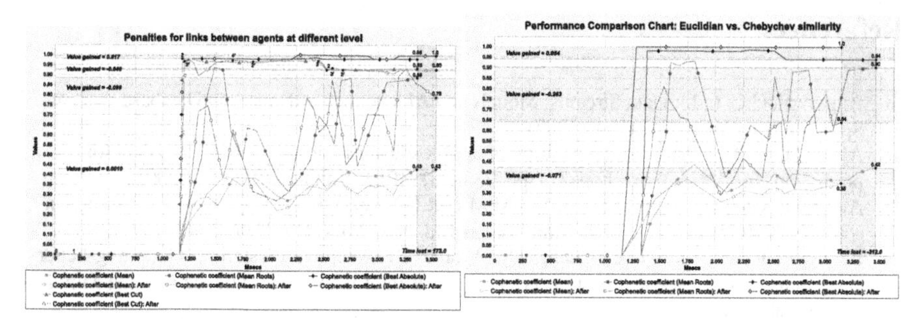

Fig. 14.12: Time required for learning with penalties, established for links between agents at different levels of the learning hierarchy

Fig. 14.13: Performance comparison chart: Euclidian vs. Chebychev similarity metrics as agent learning criteria (with quality levels shown)

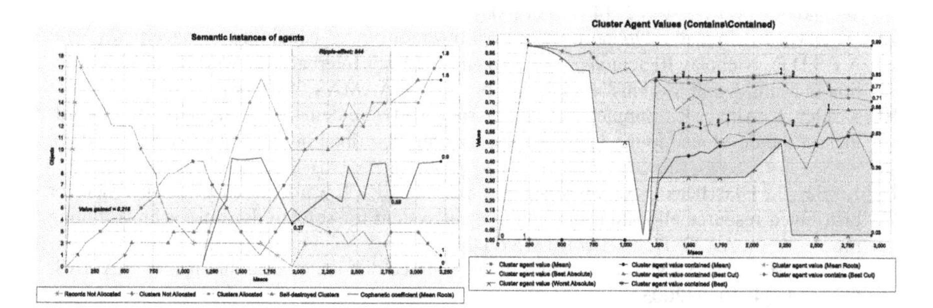

Fig. 14.14: The agent dynamics of the distributed learning process (with incremental improvement of solution quality as a result of the ripple effect)

Fig. 14.15: Performance metrics of cluster agents evolving in the virtual learning marketplace (with quality levels and equilibrium metric values)

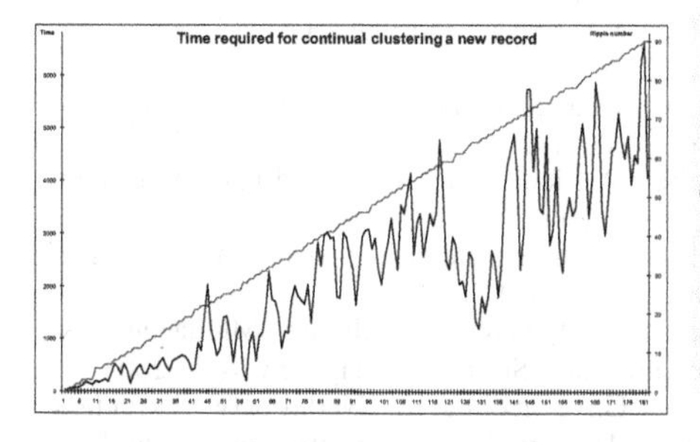

Fig. 14.16: Time required for continual learning a new record when previous records allocated (with the ripple effect for each equilibrium state)

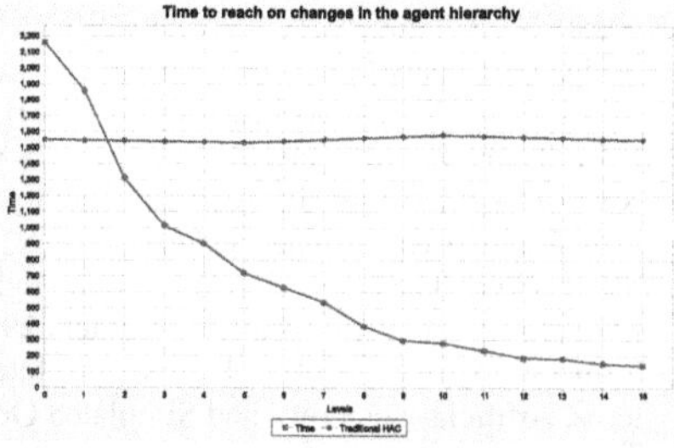

Fig. 14.17: Time required to respond to changes in clustering hierarchy (comparison with the single-linkage clustering Alias "LingPipe" algorithm)

References

1. Aggarwal, C.C. (ed.): Data Streams: Models and Algorithms. Advances in Database Systems. Springer, New York, NY, USA (2007)
2. Al-Shalalfa, M., Alhajj, R.: Application of double clustering to gene expression data for class prediction. In: AINA Workshops (1), pp. 733–738. IEEE Computer Society (2007)
3. Auroop R. Ganguly Joao Gama, O.A.O.M.M.G.R.R.V. (ed.): Knowledge Discovery from Sensor Data. CRC Press Inc - Taylor and Francis Ltd, New York, NY, USA (2008)
4. Bagherjeiran, A., Eick, C.F., Chen, C.S., Vilalta, R.: Adaptive clustering: Obtaining better clusters using feedback and past experience. IEEE International Conference on Data Mining **0**, 565–568 (2005)
5. Davidson, I., Ravi, S.S.: Agglomerative hierarchical clustering with constraints: Theoretical and empirical results. In: PKDD-05, *LNCS*, vol. 3721. Springer (2005)
6. Frey, B.J., Dueck, D.: Clustering by passing messages between data points. Science **315**, 972–976 (2007)
7. Jain, A.K., Murty, M.N., Flynn, P.J.: Data clustering: A review. ACM Comput. Surv. **31**(3), 264–323 (1999)
8. Kiselev, I., Alhajj, R.: An adaptive multi-agent system for continuous learning of streaming data. IEEE/WIC/ACM International Conference on Web Intelligence and Intelligent Agent Technology (WI-IAT'08) **2**, 148–153 (2008)
9. Kiselev, I., Alhajj, R.: Online dynamic optimization under conditions of uncertainty. In: A.P.S.D.R. Nicholas R. Jennings Alex Rogers (ed.) 1th International Workshop "Optimisation in Multi-Agent Systems" (OptMAS), pp. 52–59. AAMAS-08 (2008)
10. Kiselev, I., Alhajj, R.: Supplementary materials, software demonstration, and video recordings of the developed multi-agent learning system, described in this work. Website (2008). `\footnotesize{\ttfamily{http://www.multiagent.org/mining}}`
11. Kiselev, I., Glaschenko, A., Chevelev, A., Skobelev, P.: Towards an adaptive approach for distributed resource allocation in a multi-agent system for solving dynamic vehicle routing problems. In: AAAI-07, pp. 1874–1875 (2007)
12. Klusch, M., Lodi, S., Moro, G.: Issues of agent-based distributed data mining. In: AAMAS-03, pp. 1034–1035 (2003)
13. Likas, A.: A reinforcement learning approach to online clustering. Neural Computation **11**(8), 1915–1932 (1999)
14. MacKie-Mason, J.K., Wellman, M.P.: Handbook of Computational Economics, vol. Volume 2, chap. Chapter 28 Automated Markets and Trading Agents, pp. 1381–1431. Elsevier (2006)
15. Modi, P.J., Jung, H., Tambe, M., Shen, W.M., Kulkarni, S.: Dynamic distributed resource allocation: A distributed constraint satisfaction approach. In: ATAL: Revised Papers from the 8th Intern. Workshop on Intelligent Agents VIII, pp. 264–276. Springer-Verlag, UK (2002)
16. Modi, P.J., Shen, W.M., Tambe, M., Yokoo, M.: Adopt: asynchronous distributed constraint optimization with quality guarantees. Artif. Intell. **161**(1-2), 149–180 (2005)
17. Nisan, N., Roughgarden, T., Tardos, E., Vazirani, V.V. (eds.): Algorithmic Game Theory. Cambridge University Press, New York, NY, USA (2007)
18. Rodrigues, P.P., Gama, J., Pedroso, J.P.: Hierarchical clustering of time-series data streams. IEEE Trans. Knowl. Data Eng. **20**(5), 615–627 (2008)
19. Russell, S.J., Norvig, P.: Artificial Intelligence: A Modern Approach, 2nd international ed. edn. Prentice-Hall, Upper Saddle River, NJ, USA (2003)
20. Smith, R.G.: The contract net protocol: High-level communication and control in a distributed problem solver. IEEE Trans. Computers **29**(12), 1104–1113 (1980)
21. Symeonidis, A.L., Mitkas, P.A.: Agent Intelligence Through Data Mining. Multiagent Systems, Artificial Societies, and Simulated Organizations. Springer-Verlag New York (2005)
22. Theocharopoulou, C., Partsakoulakis, I., Vouros, G.A., Stergiou, K.: Overlay networks for task allocation and coordination in dynamic large-scale networks of cooperative agents. In: E.H. Durfee, M. Yokoo, M.N. Huhns, O. Shehory (eds.) AAMAS, p. 55. IFAAMAS
23. Zhang, S., Zhang, C., Wu, X.: Knowledge Discovery in Multiple Databases. Springer-Verlag (2004)

Chapter 15
Multiagent Systems for Large Data Clustering

T.Ravindra Babu[1], M. Narasimha Murty[2], and S.V.Subrahmanya[3]

Abstract Multiagent system is an applied research area encompassing many disciplines. With increasing computing power and easy availability of storage devices vast volumes of data is available containing enormous amount of hidden information. Generating abstractions from such large data is a challenging data mining task. Efficient large data clustering schemes are important in dealing with such large data. In the current work we provide two different efficient approaches of multiagent based large pattern clustering that would generate abstraction with single database scan, integrating domain knowledge, multiagent systems, data mining and intelligence through agent-mining interaction. We illustrate the approaches based on implementation on practical data.

15.1 Introduction

Computers, for many years, have been conventionally carrying out repetitive tasks as per intended functions ranging from payroll processing to weather predictions and spacecraft orbit computations. Search for intelligent computer has been going on since the advent of computers. With passing time one encountered a number of functions that would require the computer to be able to act on its own, interact with other computing elements in order to achieve its design objectives. Such computing elements lead to **agents**. The number of applications that required such autonomy has also increased over time consisting of multiple agents that can act on their own, react to inputs, cooperate and pro-act to achieve the given objective. The subject has grown into a scientific discipline called **multiagent systems**.

E-Comm Research Lab, Infosys Technologies Limited, Bangalore - 560100, India. e-mail: ravindrababu_t@infosys.com · Department of Computer Science and Automation, Indian Institute of Science, Bangalore - 560012, India. e-mail: mnm@csa.iisc.ernet.in · E-Comm Research Lab, Infosys Technologies Limited, Bangalore - 560100, India. e-mail: subrahmanyasv@infosys.com

L. Cao (ed.), Data Mining and Multi-agent Integration, DOI: 10.1007/978-1-4419-0522-2_15,

Data mining refers to the activity of extracting valid, general and novel abstraction from large amounts of data. Data Mining has a large overlap with domains, such as Machine Learning, Algorithms and Data Structures, Statistics, Artificial Intelligence, Database Management and Data Visualization [1, 63, 46, 23, 7]. A number of related important tasks[30, 32, 40] include the *mining methodology issues*[4, 64, 32, 18] such as data characterization, discrimination, association rule generation, clustering, classification, trend analysis; exploitation of domain or background knowledge; *performance issues* such as efficiency and scalability of data mining algorithms, number of database scans required vis-a-vis amenability to in-memory management etc.

In summary, **Multiagent systems(MAS)** [61, 22] is *a multi-disciplinary area encompassing many disciplines such as artificial intelligence, economics, sociology, management science, philosophy*, including *data mining*[2, 13, 14, 64, 25, 17, 55, 33]. In the following section we discuss about motivation for Agent and Data Mining Interaction (ADMI).

15.1.1 Motivation and Why ADMI

Data Mining deals with large data. A significant part of data mining research focusses on finding scalable algorithms that perform effectively and efficiently in terms of number of scans, speed and space. Consider the issues such as the large data size of the order of peta bytes, data being distributed across different storage devices having different access speeds, data inherently consisting of multiple categories and impracticability of bringing data together. In such scenarios, it is useful to treat such data mining system as multiagent system, with agents driving data mining.

We discuss the need for ADMI. Consider hierarchical clustering of large data containing 'n' patterns with algorithm having computational complexity of O(n^3). We consider 'k' agents to carry out this activity by dividing the number of points such that n=n_1+ n_2+...+n_k. The corresponding algorithm complexity for such multi-agent system is O(n_1^3) +O(n_2^3)+...+O(n_2^3) which is much less than O(n^3). Thus multi-agent driven data mining system is a natural solution to this problem. It offers modularity and flexibility in terms of time and space. *This is a clear case of agents supporting data mining.* We propose two algorithms that make use of such *integration* and illustrate through large data examples about their better accuracy and efficiency.

Agents and Data Mining are both interdisciplinary. A good discussion on agent and data mining integration with future directions can be found in[25]. By integrating agents with data mining the advantages of data mining such as machine learning capability and the ability of multiagent systems like handling social complexity can be made use of.

15.1.2 Current Literature and Proposed Approach

15.1.2.1 Agents Supporting Data Mining

Multiagents for clustering Internet data in real time is carried out by Jung-Eun Park and Kyung-Whan Oh[48]. The intelligent clustering is carried out through a number of steps. Principal components, 3D-Scatter diagram and Kohonen's self organizing map are used to identify optimal clusters and then k-Means[18] is used for clustering. Performance evaluation is based on variance criterion. The scheme was experimented on 4-dimensional data of size 150 and 5-dimensional synthetic data. *For large data, (a)k-Means clustering algorithm is prohibitive because it requires repetitive dataset access, and (b) evaluation of 3D-scatter diagrams for high dimensional data is impractical.* Agogino and Tumer[3] propose a agent-based clustering ensemble method that would work in a distributed data acquisition and failure-prone domains. The scheme generates a clustering ensemble of clusters given a priori.

The scheme is experimented on artificial dataset and a handwritten digit dataset of size 1000 points, each containing 16 features, and a reduced Yahoo dataset using graph based clustering algorithms. F-Trade[13] is another example of successful implementation of agent-mining integration. Tozicka et al [58] propose frame work for agent based distributed machine learning and implement their proposed scheme on a dataset of size 100 of ship surveillance data. Some early multiagent learning systems can be found in [56, 21, 24].

15.1.2.2 Data Mining Supporting Agents

Clustering of agents of similar objectives, based on similarity of characteristics in fully decentralized system is presented in [45, 44]. The method is shown to work well on large generated data points. Garruzzo and Rosaci[26] suggest *clustering of agents* based on similarity value that has lexical, structural and semantic components. The work presents an extensive discussion and survey of the topic. The scheme is experimented on 200 software agents with ability to communicate with each other and using semantic negotiation protocol developed by the same team. Wooldridge and Jennings[62] suggests *clustering of agents* based on similarity of goals. Buccafurri et al[12] present hierarchical clustering method for clustering agents. Rosaci [53] presents an ontology based clustering of customer agents having both similar interests and buying behaviour for e-commerce applications. Piraveenan[49], proposed a predictor for convergence time of cluster formation in a decentralized and dynamic cluster formation in a scale-free multiagent sensor grids. Bekkerman et al [8] proposes multiagent heuristic web searching algorithms for web page clustering agents. Spatio-temporal clustering of two dimensional agents for swap-based negotiation protocols is presented by Golfarelli and Rizzi [27]. Using Potential fields [28] a multiagent algorithm is proposed in [6], where each agent has limited processing power and with communication among agents limited to information through pair-wise communication links between them.

With the objective of providing personalized advertising, multiagent system for web advertising is presented with an elaborate model by Kazienko[36]. The model integrates clustering and data mining.

15.1.2.3 Proposed Approach

An ontology for large data clustering is provided in Fig. 15.1. A useful demon-

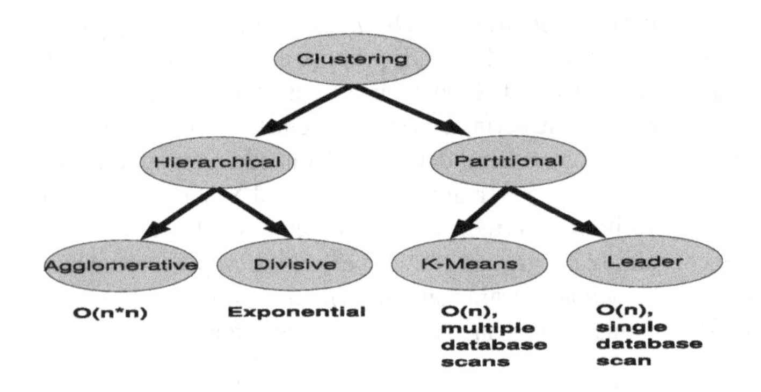

Fig. 15.1: Ontology for Large Data Clustering[32]*(Large data clustering needs to be efficient. Clustering is either hierarchical or partitional. Hierarchical clustering is either agglomerative or divisional. Agglomerative algorithms are of $O(n^2)$ and divisive are of exponential order. Well known partitional algorithms, k-Means-type, require multiple database scans. Leader algorithm is of O(n) and requires single database scan)*

stration of Agent and Data Mining interaction can be found in [13]. A number of research groups are active in these areas[2, 39, 42, 43]. An interesting discussion on distributed artificial intelligence and MAS can be found in [20].

The approach of Agents brings ease to complex systems like Data Mining. Based on the application context, a data mining system can be seen as a multiagent system with all its basic tenets, viz., environment, situated objects, set of agents, defined relations among agents, defined operations among agents. Likewise, data mining can help multiagents by recognizing an agent could be a data mining agent with limited or full functionality. Thus such an agent or agents form solution providers to data mining agents with defined environment. Integration of agents and data mining has been focus in the last few years.

We attempt to solve this problem by **two multiagent systems** as given in Sections 15.2 and 15.3, for large data clustering and classification. They are cases for agents supporting data mining. In the first approach, we define multiple agents to carry out category-wise prototype selection followed by category-wise feature selection. Agents are defined for control and evaluation. In the second approach we integrate results of agents that carry out preliminary analysis and domain knowledge into training to generate a novel decision tree which classifies unseen patterns with a good accuracy. Section 15.4 summarizes the contribution and provides a brief discussion on further work.

15.2 Scheme-1: Multiagent Based Clustering using Divide and Conquer Approach

Large data clustering is a challenging task since ideally the abstraction has to be generated with a single scan of the data; in practice it should limit to a small number of dataset scans. Many clustering algorithms do not up scale well with large data. In the current section, we propose a multiagent system that generates abstraction of large datasets by divide and conquer approach.

The proposed method integrates the following.

- Multiagent System
- Divide and conquer approach
- Prototype Selection
- Feature selection based on frequent item approach[30] and subsequence generation
- k-Nearest Neighbour Classifier

We consider two datasets viz., labeled handwritten data and labeled intrusion detection data to demonstrate the concept. We dwell on motivation for such activity, description of data, discussion on proposed multiagent system and experimental results.

15.2.1 Motivation

Multiagent system is a useful choice for data mining activity requiring abstraction generation. The objective is achieved by defining a number of agents with defined autonomy, inter-agent communication, ability to react to environment and ability to pursue the objective pro-actively. Agents can carry out the tasks of a *pre-facto* data analysis, clustering, feature selection and evaluation. The agents cooperate with each other and communicate in achieving the given objective.

While dealing with high-dimensional, large data, for the sake of abstraction generation and scalability of algorithms, one resorts to either dimensionality reduction or data reduction or both[47]. Approaches to data reduction include clustering, sampling[30, 41] data squashing[19], etc. A number of other approaches like BIRCH[9, 10] generate a summary of original data that is necessary for further use.

Dimensionality reduction is achieved through either feature selection or feature extraction[16].

In the backdrop of these philosophical and historical notes, we examine (a) a large handwritten digit data and (b) intrusion detection data, in terms for feature selection and data reduction.

The compactness achieved by dimensionality reduction is indicated in terms of number of combinations of distinct subsequences [51]. The concepts of Frequent items[29] and Leader cluster algorithms[57] are made use in the work. The k-NNC is the classifier[18].

15.2.2 Description of Handwritten Digit Data and Preliminary Analysis

The handwritten digit data considered for the study consists of 10 classes. The data consists of 10003 labeled patterns, with 6670 training and 3333 test patterns. Each pattern consists of 192 binary features. The number of patterns per class is almost equal. The above training data is further subdivided, for the current study, into training (6000) and validation (670) data. Each pattern can be represented as 16X12 matrix. Table 15.1 contains results of preliminary analysis on number of non-zero features in the training data. A set of typical and atypical patterns and shown in Fig. 15.2.

Table 15.1: Preliminary Analysis on Handwritten data - *Statistics on number of non-zero features*

Class Label	Mean	Standard Deviation	Minimum	Maximum
0	66	11	38	121
1	30	05	17	55
2	64	10	35	102
3	60	10	33	108
4	53	09	24	89
5	61	10	32	101
6	58	09	34	97
7	47	08	28	87
8	67	11	36	114
9	56	08	31	86

15.2.3 Description of Intrusion Detection Data and Preliminary Analysis

Network Intrusion Detection Data[34] was used during KDD-CUP99 contest. Even 10%-dataset can be considered large as it consists of 805049 patterns each of which is characterized by 41 features. We use this dataset in the present study and hereafter

Fig. 15.2: A set of typical and atypical patterns of Handwritten data

we refer to this dataset as 'full dataset' so as to distinguish it from the prototype dataset. We use the dataset for the proposed multiagent system.

The data relates to access of computer network by authorized as well as unauthorized users. The access by unauthorized users is termed as intrusion. Different costs of misclassification are attached in assigning a pattern belonging to a class to any other class. The challenge lies in detecting type of intrusion accurately minimizing the cost of misclassification. Further, whereas the feature values in handwritten data contained binary values, the current data set assumes **floating point values**.

The training data consists of 311029 patterns and the test data consists of 494020 patterns. From the preliminary analysis, it is found that not all features are **frequent**. We make use of this fact during the experiments.

The training data consists of 23 attack types, that form 4-broad classes. The list is provided in Table 15.2. Test data contained more classes than those in the training data. Since the classification of test data depends on learning from training data, the unknown attack types(or classes) in the test data have to be assigned one of *a priori* known classes of training data appropriately as shown in Table 15.3 based on domain knowledge.

In classifying the data, each wrong pattern assignment is assigned a **cost**. The cost matrix is provided in Table 15.4. Observe from the table that cost of assigning a pattern to a wrong class is not uniform. For example, cost of assigning a pattern belonging to class 'u2r' to 'normal' is 3. Its cost is more than that of assigning a pattern from 'u2r' to 'dos', say.

Table 15.2: Attack Types in Training Data

Class	No. of Types	Attack Types
normal	1	Normal
dos	6	back, land, neptune, pod, smurf, teardrop
u2r	4	buffer–overflow, loadmodule, perl, rootkit
r2l	8	ftp-write, guess-password, imap, multihop, phf, spy, warezclient, warezmaster
probe	4	ipsweep, nmap, portsweep, satan

Table 15.3: Assignment of Unknown Attack Types using domain knowledge

Class	Attack Type
dos	processtable, mailbomb, apache2, upstorm
u2r	sqlattack, ps, xterm
r2l	snmpgetattack, snmpguess, named, sendmail, httptunnel, worm, xlock, xsnoop
probe	saint, mscan

Table 15.4: Cost Matrix

Class Type	normal	u2r	dos	r2l	probe
normal	0	2	2	2	1
u2r	3	0	2	2	2
dos	2	2	0	2	1
r2l	4	2	2	0	2
probe	1	2	2	2	0

Further, dissimilarity measure plays an important role. The range of values for any feature within a class or across the classes is large. Also the values assumed by different features within a pattern is also largely variant. This scenario suggests use of Euclidean as well as Mahalanobis distance measures. Both the methods are used in carrying out exercises on random samples drawn from the original data. Based on the study on the random samples, it is observed that Euclidean distance measure provided better classification. Thus, Euclidean measure alone is used further.

With the full data of the given dataset, **NNC provides a classification accuracy of 92.11% with a cost of 0.254086**. This result is made use for further comparisons.

Results reported during KDDCUP'99 are provided in Table 15.5.

Table 15.5: Accuracy of winner and runner up of KDDCUP'99

Class	Winner	Runner-up
Normal	99.5	99.4
Dos	97.1	97.5
R2l	8.4	7.3
U2r	13.2	11.8
Probe	83.3	84.5
Cost	0.2331	0.2356

15.2.4 *Proposed Multiagent System for the Divide and Conquer Method*

We discuss the proposed multiagent system for large data clustering with reference to Fig. 15.3. The system consists of one agent whose functionality is control. It initiates the abstraction activity and has ability to take decision once it receives feedback from Evaluation agent. The control agent communicates with clustering agents. The clustering agents are concurrent. They have autonomy to carry out clustering. The objective is to carry out class-wise clustering and hence of number of clustering agents equals number of classes. After clustering, the prototypes are communicated to feature selection agents. The functionality involves autonomously selecting features by frequent itemset approach[30]. The prototypes with optimal features is communicated to evaluation agent. The functionality of the Evaluation Agent is to compute prediction accuracy of unseen validation dataset using k-Nearest Neighbour Classifier. The result is communicated back to Control Agent. The parameters of the system consists of distance threshold[57] and frequency limit to identify frequent itemsets and thereby optimal features. The control Agent moderates these two parameters based on the feedback from the Evaluator.

The environment consists of Control, Clustering, Feature-selection and Evaluation agents. Each of the agents is capable of autonomous action in this environment to reach its design objectives, viz., **generation of abstraction of a large dataset**. The environment is accessible as at any stage one gets accurate complete information of the state of the environment. The environment is deterministic, episodic and discrete.

Handwritten data has 10 classes or categories of data, viz., 0 to 9. The number of agents in such multiagent system are 22, viz., control agent(1), one clustering agent each per class(10), one feature selecting agent each per class(10) and one evaluation agent(1). The intrusion detection has 5 classes thus consisting of 12 agents.

15.2.4.1 Experimental Results for Handwritten Digit Data

Preliminary analysis on the data provides range of values for distance threshold and feature support. Based on these results, the Control Agent passes appropriate values to clustering and feature selection agents. In both the datasets by virtue of divide and conquer method, the values are defined class-wise. With increasing distance threshold, number of prototypes reduces. Similarly, with increasing support the number of features reduces. For example for a support value of 160, the number of features reduce to about 50%. Final results are provided in Table 15.6.

Table 15.6: Final Results

Support for frequent items(ε)	Distance Threshold(ζ)	No. of Prototypes	Distinct subseq.	CA(kNNC) Valdn Data	CA(kNNC) Test Data
180	3.1	5064	433	93.58%	93.34%

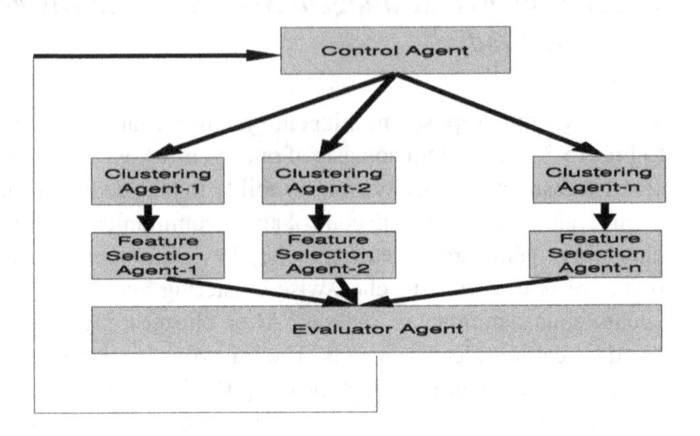

Fig. 15.3: Multiagent System for Divide and Conquer method. *(Control agent provides distance threshold for clustering, support threshold for feature selection. The clustering agents perform category-wise clustering using leader algorithm. Feature selection is performed on leaders. The resultant patterns are made use in classifying validation dataset. The activity continues till classification accuracy constraint is satisfied at the Control Agent. The diagram depicts interagent communication. The system forms a hybrid layered architecture)*

15.2.4.2 Experimental Results for Network Intrusion Detection Data

To begin with we discuss results of preliminary exercises to compute representatives of original data or prototypes. The representatives are obtained based on Leader clustering algorithm. As discussed previously the original data consists of 5 classes of patterns. We find cluster representatives independently for each of the classes and then combine them to form training data consisting of leaders alone. We use the training datasets thus obtained, to classify 411029 test patterns. The experiments are repeated using different distance thresholds.

Table 15.7 contains sizes of data sets for different distance thresholds. Observe that with increasing threshold values, the number of representatives reduces. Table 15.8 contains the results. It is clear from the results that the cost is minimum for class with threshold set of (40,100,2,1,1). Datasets obtained through cases mentioned in Table 15.7 are used in further studies.

The Leader set obtained in Case-3 data of Table 15.7 is considered for the study. Support is applied on the features among the leader set. With support 0% of the patterns, all features are considered. With a support of 10% the number of features reduce to 21 from 38 and with a support of 20% the number of features reduce to 18. The results are tabulated in Table 15.9. It can be seen from the first row of the table, that the costs of assigning a pattern from class "normal" to class "u2r" is 2, "normal" to "dos" is 2 and "normal" to "probe" is 1. Since the cost matrix is not symmetric, cost of assigning "u2r" to "normal" is not same as that of "normal" to "u2r". Also, observe that in spite of reduction in number of features the classification accuracy with NNC has improved with 10% and slightly reduced with 20%. Similarly cost improved with support of 20%. Viewing these results analytically, reduction in number of features while classifying a large number of patterns reduces **storage space** and **computation time**. Further, such a scenario increases classification accuracy and reduces assignment cost.

Table 15.7: Case Study Details

Case	Normal		u2r		dos		r21		probe	
No.	Thr	Pats	Thr	Pats	Thr	Pats	Thr	Pats	Thr	Pats
1	100	3751	100	33	2	7551	2	715	2	1000
2	50	10850	10	48	2	7551	1	895	1	1331
3	40	14891	100	48	2	7551	1	895	1	1331

Table 15.8: Results with prototypes

Case No.	Training Data Size	CA(%)	Cost
1	13050	91.66	0.164046
2	20660	**91.91**	0.159271
3	24702	91.89	**0.158952**

Table 15.9: Results with Frequent Item Support on Case-3(247072 patterns)

Support	No. of features	CA	Cost
0%	38	91.89%	0.1589
10%	21	**91.95%**	**0.1576**
20%	18	91.84%	0.1602

15.2.5 Summary

In the current section, we proposed a Multiagent Data Mining System based on divide and conquer approach. The scheme is applied on two large datasets viz., (a) high dimensional handwritten digit data and (b) large intrusion dataset. Working of the schemes in both the cases is demonstrated. The results obtained using the scheme with handwritten digit has provided a classification accuracy of 93.58% and a compaction of 45%. In case of intrusion detection data, the results obtained are better than those obtained by winner and runner-up during KDDCUP99 contest Table 15.5, viz., **classification accuracy of 91.85%, cost of 0.1586 and compaction of nearly 50%**.

15.3 Scheme-2: Multiagent Based Clustering Using Data Dependent Schemes

Large data clustering which is a data mining activity is carried earlier using multiagent system incorporating divide and conquer approach. In the current section we exploit the domain knowledge that is either available at one's disposal *a priori* or obtained through preliminary data analysis. Based on such knowledge we carry out multi-category classification as multiple binary classifiers. In the current method, we integrate the following.

- Multiagent System
- Domain knowledge
- Leader Clustering
- Adaptive Boosting method or AdaBoost

Like in the previous scheme, large datasets considered for the study are (a) Handwritten digit data and (b) Intrusion detection data. The data description and preliminary data analysis are provided in Section 15.2.

The following sections provide a brief discussion on motivation for such method, background material and proposed method.

15.3.1 Motivation

Multiagent system for data mining activity is a useful integration. The agents are autonomous, reactive to communication and proactive to reach the goal of generating an abstraction. The functionality of individual agents include (a) preliminary analysis, leading to generation of possible grouping of multiple category data, e.g., handwritten digits 3, 5 and 8 share common structure which we integrate such knowledge in designing a classifier; (b) Clustering to generate prototypes from such groups; (c) integrate domain knowledge to design a knowledge-based tree that reduces number of comparisons for a multiple 2-class classifier[52]. Boosting[37, 38] is a general method for improving accuracy of a learning algorithm. It makes use of a base learning algorithm that has an accuracy of at least 50%. Such an algorithm forms a component classifier. Decision rule based on an ensemble of component classifiers provides CA higher than that provided by single use of base learning algorithm. We make use of Adaptive Boosting Algorithm(AdaBoost) [54, 59].

In order to make use of AdaBoost efficiently for large data, we exploit the domain knowledge on the problem under study.

15.3.2 Choice of Prototype Selection Algorithm

In the current work, the base learner is a k-Nearest Neighbour Classifier(kNNC) that employs prototype set generated from training data using Leader clustering algorithm. Prototype selection with the same set of HW digit data using Partition Around Medoids(PAM) [35], CLARA [35] and Leader [57] was earlier reported [50]. It was brought out [50] that Leader algorithm performed better than PAM and CLARA in terms of classification accuracy. Also the computation time for Leader is much less compared to both PAM and CLARA for the same HW data. The complexity of PAM is O(k$(n-k)^2$) and that of CLARA is of O(ks^2+k(n-k)). Leader has linear complexity. Although CLARA can handle larger data than PAM, its efficiency depends on sample size and unbiasedness of the sample.

In the current Section, a comparison of two algorithms, viz., Condensed Nearest Neighbour(CNN) [31, 15] and Leader [57] clustering algorithm is carried out for prototype selection.

In the Leader clustering algorithm [57, 32], the number of leaders depends on the choice of the threshold. As the threshold increases, the number of leaders reduces.

A comparative study is conducted between CNN and Leader, by providing entire 6670 patterns as training data and 3333 as test data for classifying them with the help of kNNC. Table 15.10 provides the results.

Table 15.10: Comparison between CNN and Leader

Distance Threshold	No. of Prototypes	C.A.(%)
CNN		
-	1610	86.77
Leader		
5	6149	91.24
10	5581	91.27
15	4399	90.40
18	3564	90.20
20	3057	88.03
22	2542	87.04
25	1892	84.88
27	1526	81.70

In Table 15.10, C.A. refers to percentage Classification Accuracy. The table, apart from comparing both the methods, demonstrates the effect of threshold on number of prototypes selected, C.A. and processing time. A discrete set of threshold values is chosen to demonstrate the effect of distance threshold. For binary patterns, the Hamming distance provides equivalent information as Euclidean distance metric, while avoiding the need to compute the square root. Hence Hamming distance is chosen as dissimilarity measure, in view of binary patterns. Based on these exercises, leader is chosen to identify prototypes.

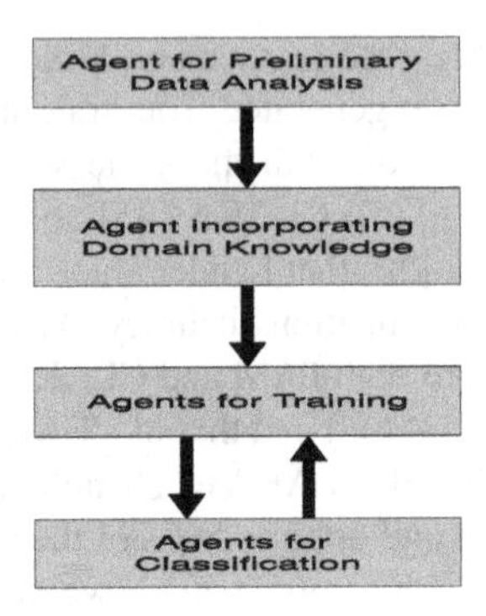

Fig. 15.4: Multiagent System for Data Dependent Scheme. *(The system consists of multiple agents in hybrid layered architecture. The agent for preliminary analysis provides domain knowledge; alternately a priori domain knowledge forms input. Agent for incorporating domain knowledge makes use of this information to form a knowledge-based tree. Agents for training generates leaders as prototypes. The agents for classification use knowledge-based tree to classify the given data in at most 4 comparisons. The interagent communication is indicated in the figure.)*

15.3.3 Proposed Multiagent System for Large Data Clustering based Data Dependent Scheme

The objective of the multiagent system is to generate an abstraction and build prediction ability of unseen patterns with a large data efficiently. The scheme is discussed with reference to Fig. 15.4. The scheme consists of multiple agents. Following are their functionalities. Agent for Preliminary Analysis carries out preliminary analysis on data and suggests possible groupings of view categories. The information is communicated to the Agent that incorporates domain knowledge. The domain knowledge can also be incorporated externally at this stage. Together it forms a knowledge-based decision tree(KB-Tree), as shown in Fig. 15.5, of depth 4. It requires just 4 comparisons to classify 10-category patterns. To elaborate further on KB-Tree, based on domain knowledge on the data, given handwritten data containing classes, (0,1,...,9), is divided into two groups of classes, (0,3,5,6,8) and (1,2,4,7,9) in **stage-a**. In **stage-b**, (0,3,5,6,8) is further subdivided into (0,6) and

(3,5,8); (1,2,4,7,9) into (4,9) and (1,2,7). In **stage-c**, (0,6) is subdivided into (0) and (6); (1,2,7) into (2) and (1,7). In **stage-d**, (3,5) is subdivided into (3) and (5); (1,7) into (1) and (7). Thus, assuming that a pattern with a label '7' is correctly classified, it passes through the path containing, (1,2,4,7,9) at stage-a, (1,2,7) at stage-b, (1,7) at stage-c and (7) at stage-d.

Depending on the groups thus formed, the training agents group the training data and leader based prototypes are generated. The set of leaders is a subset of original large dataset. The classifier agent makes use of KB-Tree, prototypes and kNNC to classify any given pattern using just 4 comparisons. A further discussion on working of KB-tree is provided along with description on experimental results.

It should be noted here that each of the agents are autonomous, communicating, reactive and proactive.

15.3.3.1 Experimental Results with Handwritten Digit Data

A large number of experiments is conducted by varying the values of threshold(η) and value of k in kNNC. At each stage of binary classification into two equivalent class-labels, +1 and -1, only those correctly classified test patterns at that stage are used as input test patterns for subsequent stage. We will explain the classification with reference to the Fig. 15.5. A test pattern labeled '3' when correctly classified passes through the stages of (a) (0,3,5,8,9) vs (1,2,4,7,9), (b) (0,3) vs (5,6,8), (c) 8 vs (3,5) and (d) 3 vs 5. All the experiments are conducted with validation dataset. The parameter set(η and k) that provided best results with validation dataset are used for verification with test dataset. The results are provided in Table 15.11. In the table, set-1 and 2 respectively represent (0,3,5,6,8) and (1,2,4,7,9). It should be noted that the final CA depends on the number of mis-classifications at various stages. In Fig. 15.5, class-wise correctly classified patterns are available at leaf nodes which are denoted in italics. *The 'overall CA' is **94.48%** which is better than the value reported [60] on the same data.* And, NNC of full training data of 6000 patterns against test data provides a CA of **90.73%** and CA with kNNC is **92.26%** for k=5. Thus, *'overall CA' is also better than CA obtained with NNC and kNNC on the complete dataset.* The multi-class classification is carried out with *a decision tree of a **depth of 4***. This should be contrasted with one-against-all and one-vs-one multi-class classification schemes [5].

Further, with increasing η, number of prototypes reduces. For example, in case of Set-1 vs 2, at η=3.2, number of prototypes reduces by **20%**. Secondly, number of training patterns reduces as we approach leaf node. For example, number of training patterns at 0 vs 6 is **1200** as against **6000** patterns at the root of the tree. The number of prototypes for 0 vs 6 at η=4 is 748, which is a reduction by **38%**.

Table 15.11: Results with AdaBoost with kNNC as Component Classifier

Case	Set-1 vs Set-1	(4,9) vs (1,2,7)	4 vs 9	2 vs (1,7)	1 vs 7	(0,6) vs (3,5,8)	0 vs 6	8 vs (3,5)	3 vs 5
k	5	5	8	3	1	5	10	8	3
η	3.2	3.0	3.4	3.0	3.0	4.0	4.0	2.5	3.4
CA	98.3	98.1	96.6	99.7	99.5	99.0	99.4	97.6	96.0

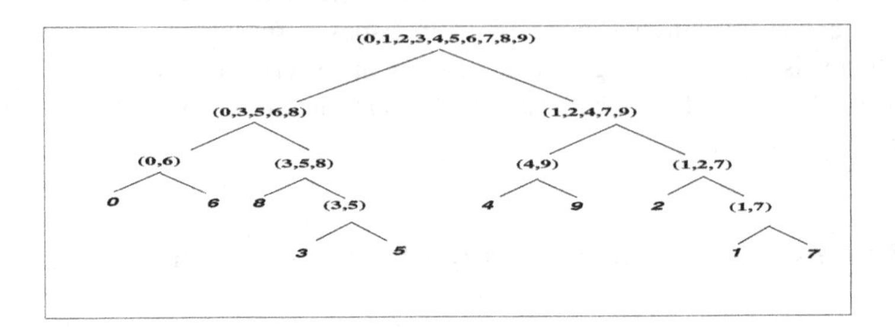

Fig. 15.5: Knowledge-based Decision Tree for Multiclass Classification*(The tree consists of multiple agents to classify a given unseen pattern into different groups of patterns starting from (0,3,5,6,8) vs. (1,2,4,7,9) and further depending on the branching. The communication directions are not explicitly represented here in order for brevity. The diagram corresponds to block of* ***Agents for Classification*** *of Fig. 15.4)*

15.3.3.2 Experimental Results with Network Intrusion Detection Data

AdaBoost is applied on 5-class intrusion data in multiple stages considering 2 categories at a time. This is carried out in 4 stages. The exercise is conducted on 13050 leaders of Case-1 of Table 15.8.

With reference to Fig 15.5, the working of algorithm is explained through the following steps.

1. In the 5-class data, labeled, 0,1,2,3,4, consider 0 vs rest in Stage-a. At this stage, it considers entire data. Classify the data using full test data.
2. From the training data, remove that data corresponding to Class-0 for use as training data at next stage. From test data extract correctly classified data and those corresponding to classes 1-4 for use at next stage as test data.
3. In Stage-b, the two classes are 1 vs (2,3,4). The training data consists of classes 1,2,3 and 4. Test data consists of classes and 1,2,3,4 which were correctly classified at the previous stage.
4. In stage-c, the two classes are 2 vs (3,4). As explained the previous steps, the data corresponds to classes 2,3 and 4.
5. In stage-d, the two classes are 3 vs 4.

Table 15.12 contains results.

Overall Cost cannot be specified for AdaBoost, since under the given cost specification, the cost is defined specifically from one class to another. In the process of 2-class classification, we group multiple classes and hence cost is not computable. However the case demonstrates the applicability of multiagent system to variety of practical data.

Table 15.12: Results with AdaBoost based on Prototypes

Stage	Two-classes	Training Data	Test Data	CA(nnc)
a	0 vs (1,2,3,4)	13050	311029	92.24%
b	1 vs (2,3,4)	9299	229142	99.99%
c	2 vs (3,4)	9266	229133	99.27%
d	3 vs 4	1715	3616	82.40%

15.3.4 Summary

The section contains application of a multiagent system that makes of domain knowledge in providing an efficient means of abstraction of a large handwritten data. The system is described. The datasets considered are handwritten digit data and intrusion detection data. The preliminary analysis carried out in Section 15.2.2 and Section 15.2.3 are valid in the current study too. The systems are a combination of horizontal and vertical layer architectures. We demonstrated working of the proposed system for both types of large datasets.

15.4 Summary and Further work

In the Chapter, we deal with two models for Agents supporting Data Mining activity of large data clustering. The Chapter contains a discussion on background literature of agents supporting data mining and data mining supporting agents. Two schemes for large data clustering that successfully integrate the two areas are provided. The proposed models are described by referring to relevant literature. The schemes are implemented on two real-life large datasets of handwritten digit data and network intrusion detection data. Advantage of treating the problem of large data clustering as agent-mining integration is shown in terms of classification accuracy in case of handwritten data and cost associated with classification in case of intrusion detection data as compared to complete data. A discussion on advantage in computation complexity is provided in Section 15.1.1.

As a further step to the current work, we recognize the choice of clustering algorithm is data dependent[32]. Thus a multiagent system can be designed to carry out clustering using different schemes. The cluster representatives so obtained by different clustering algorithms are shared among the agents to arrive at optimal set of prototypes.

References

1. Abonyi, J. and Feil, B. and Abraham, A.: Computational Intelligence in Data Mining. In:Informatica(Slovenia), **29**, 1, 3–12 (2005)
2. Agent-Mining Interaction and Integration(AMII): http://www.agentmining.org
3. Agogino,A., Tumer,K.: Efficient Agent-Based Clustering Ensembles. AAMAS'06, 1079–1086 (2006)
4. Agrawal, R., Imielinski, T., Swami, A.: Mining association rules between sets of items in large databases. Proc. 1993 ACM-SIGMOD Int. Conf. Management of Data(SIGMOD'93), Washington D.C., 266–271 (1993)
5. Allwein, E.L., Schapire, R.E., Singer, Y.: Reducing Multiclass to Binary - A Unifying Approach for Margin Classifiers. Machine Learning. No.1, 113–141 (2000)
6. Baghshah,M.S., Shouraki,S.B., Lucas,C.: An agent-based clustering algorithm using potential fields. AICCSA, IEEE, 551–558 (2008)
7. Bajaj, C.: Data Visualization Techniques. John Wiley & Sons, John Wiley & Sons, New York (1999)
8. Bekkerman,R., Zilberstein,S., Allan,J.: Web page clustering using heuristic search in the web graph. IJCAI, 2280–2285 (2007)
9. Tian,Z., Ramakrishnan R., Micon, L.: BIRCH:An efficient data clustering method for very large databases. Proceedings of ACM SIGMOD, 103–114 (1996)
10. Bradley, P. and Fayyad, U.M., Reina, C., Scaling clustering algorithms to large databases, Proceedings of 4th Intl. Conf. Knowledge Discovery and Data Mining, AAAI Press, New York, 9–15 (1998)
11. Breban,S., Vissileva,J.: A coalition formation mechanism based on inter-agent trust relationships. In proc. of the 1st Conference on Autonomous Agents and Multi-Agent Systems, Italy, 306–308 (2002)
12. Buccafurri,F., Rosaci,D., Sarne,G.M.L., Ursino,D.: An agent-based hierarchical clustering approach for e-commerce environments. In Proceedings of E-Commerce and Web Technologies, 3rd International Conference (EC-Web 2002), France. Lecture Notes in Computer Science, Vol.2455. Springer, 109–118 (2002)
13. Cao, L., Zhang, C.: F-Trade: An agent-mining symbiont for financial services. AAMAS'07, May 14-18, Hawaii, USA (2007)
14. Cao,L., Yu, P.S., Zhang, C., Zhao, Y., Williams, G.: DDDM2007:Domain Driven Data Mining, SIGKDD Explorations, Vol.9. Issue 2, 84–86 (2007)
15. Cover, T.M.,Hart, P.:Nearest Neighbour pattern classification. IEEE Transactions on Information Theory, Vol 13, 21–27 (1967)
16. Devijver,P.A., J. Kittler, J.: Pattern Recognition: A Statistical Approach, Prentice Hall, Englewood Cliffs (1986)
17. Distributed Data Mining Bibliography, http://www.cs.umbc.edu/~hillol/DDMBIB/
18. Duda, R.O., Hart,P.E., Stork,D.G.: Pattern Classification, John Wiley & Sons, Wiley-interscience (2000)
19. DuMouchel, W., Volinksy, C., Johnson, T., Cortez, C., Pregibon, D.: Squashing Flat Files Flatter. Proc. 5th Int. Conf. on Knowledge Discovery and Data Mining, AAAI Press, San Diego, CA. 6–15, (1999)

20. Durfee, E.H., Rosenschein, J.S.: Distrubuted Problem Solving and Multi-Agent Systems - Comparisons and Examples.ftp://www.eecs.umich.edu/people/durfee/daiw94-dr.ps.Z (1994)
21. Edwards,P., W. Davies, W.: A Heterogeneous multi-agent learning system. In Proc. of the special interest group on cooperating knowledge based systems, 163–184 (1993)
22. Ferber,J.: Multi-Agent Systems. Addison-Wesley, Harlow (1999)
23. Freias,A.A.: Data Mining and Knowledge Discovery with Evolutionary Algorithms. Springer, New York. (2002)
24. Ghosh,J., Strehl,A., Merugu, S.: A concensus framework for integrating distributed clusterings under limited knowledge sharing. In NSF Workshop on Next Generation Data Mining, 99–108 (2002)
25. Integration of Agents and Data Mining http://www-staff.it.uts.edu.au/~lbcao/publication/IntegrationofAgentandDataMining.ppt
26. Garruzzo, S.,Rasaci, D.: Agent Clustering Based on Semantic Negotiatiion. ACM Trans. on Autonomous and Adaptive Systems, Vol.3, No.2, Article 7, 7:1 – 7:40 (2008)
27. Golfarelli, M., Rizzi, S.: Spatio-Temporal Clustering of Tasks for Swap-Based Negotiation Protocols in Multi-Agent Systems.
28. Gomez,J., Dasgupta,D., Nasraoui,O.: A new gravitational clustering algorithm. In Proc. of the Third SIAM International Conference on Data MINING, San Francisco, 83–94 (2003)
29. Han, J., Pei,J., Yin, Y.: Mining frequent patterns without candidate generation, Proceedings of ACM SIGMOD International Conference of Management of Data(SIGMOD 00), Dallas, Texas, USA, 1–12 (2000)
30. Han, J., Kamber, M.: Data Mining: Concepts and Techniques. Morgan Kaufman, San Francisco, CA (2001)
31. Hart, P.E.:The condensed nearest neighbour rule, IEEE Transactions on Information Theory, Vol 14, 515–516 (1968)
32. Jain,A.K., and Murty, M.N. and P.J. Flynn.: Data Clustering: A Review, ACM Computing Review, 264–323 (1999)
33. Jan Tozicka, Michael Rovatsos, Michal Pechoucek: A Framework for Agent-Based Distributed Machine Learning and Data Mining. In Autonomous Agents and Multi-Agent Systems, Article No.96, (AAMAS 2007) ACM Press (2007)
34. KDDCup99 Data, http://kdd.ics.uci.edu/databases/kddcup99/kddcup99.html (1999)
35. Kaufman, L., Rouseeuw,P.: Finding Groups in Data - An Introduction to Cluster Analysis. John Wiley & Sons, New York (1990)
36. Kazienko, P.: Multiagent system for web advertising, www.zsi.pwr.wroc.pl/~kazienko/pub/2005/KazienkoKES05.pdf
37. Kearns,M.,Valiant,L.G.: Leaving Boolean formulae or finite automata is as hard as factoring. Harward University Aiken Computation Laboratory. TR-14-88. (1988)
38. Kearns,M., Valiant,L.G.: Cryptographic limitations on learning Boolean formulae and finite automata. Journal of the Association for Computing Machinery. Vol.41, No.1, 67–95 (1994)
39. Meyer,J., Intelligent Systems Group: http://www.cs.uu.nl/groups/IS/agents/agents.html
40. Mitra,S., Acharya,T.: Data Mining: Multimedia, Soft Computing, and Bioinformatics. John Wiley, New York. (2003)
41. Mitra,P., and Pal, S.K.: Density based Multiscale Data Condensation, IEEE Trans. on Patter Analysis and Machine Intelligence, Vol.24, No.6, 734-747 (2002)
42. Multiagent Research Group:http://www.cs.wustl.edu/$\sim$mas
43. Multiagent Systems: http://www.aaai.org/AITopics/pmwiki/pmwiki.php/AITopics/MultiAgentSystems
44. Ogston,E., Overreinder,R., van Steen,M., Brazier,F.: A method for decentralizing clustering in large multi-agent systems. AAMAS'03, Australia (2003)

45. Ogston,E., Overreinder,R., van Steen,M., Brazier,F.: Group formation among peer-to-peer agents: Learning group characteristics. In 2nd International Workshop, on Agents and Peer-to-peer computing. Lecture Notes in Computer Science, Vol.2872, Springer, 59–70 (2003)
46. Pal,S.K., and Ghosh,A.: Soft computing data mining. In: Information Sciences, Vol.163, No.1–3, pp.1–3 (2004)
47. Pal, S.K.,Mitra, P.: Pattern Recognition Algorithms for Data Mining, Chapman & Hall/CRC (2004)
48. Park,J., Oh,K.: Multi-Agent Systems for Intelligent Clustering. Proc. of World Academy of Science, Engineering and Technology, Vol.11, February 2006, 97–102 (2006)
49. Piraveenan,M., Prokopenko,M., Wang,P. and Zeman,A.: Decentralized multi-agent clustering in scale-free sensor networks. Studies in Computational Intelligence, 115, 485–515 (2008)
50. Ravindra Babu, T., Narasimha Murty, M.: Comparison of Genetic Algorithms Based Prototype Selection Schemes. Pattern Recognition,34(2),523–525 (2001)
51. Ravindra Babu, T., Narasimha Murty, M., Agrawal, V.K.: Hybrid Learning Scheme for Data Mining Applications, Proceedings of the Fourth International Conference on Hybrid Intelligent Systems, IEEE Computer Society, Los Alamitos, California, 266–271 (2004)
52. Ravindra Babu, T., Narasimha Murty, M., Agrawal, V.K.: Adaptive boosting with leader based learners for classification of large handwritten data. Proceedings of the Fourth International Conference on Hybrid Intelligent Systems, IEEE Computer Society, Los Alamitos, California, 266–271 (2004)
53. Rosaci,D.: An ontology-based two-level clustering for supporting e-commerce agents activities. In Proceedings of E-Commerce and Web Technologies, Sixth International Conference (EC-Web 2005). Lecture Notes in Computer Science, Vol.3590. Springer, 31–40 (2005)
54. Schapire,R.E.: Theoretical views of Boosting and Applications. Proceedings of Algorithmic Learning Theory. (1999)
55. Sen, S., Saha, S., Airiau, S., Candale, T., Banerjee, D., Chakraborty, D., Mukherjee,P., and Gursel, A.: Robust Agent Communities. In Autonomous Intelligent Systems: Agents and Data Mining, V. Gorodetsky, C. Zhang, V.A. Skormin, and L. Cao (Editors), pages 28–45, Lecture Notes in Artificial Intelligence, volume 4476, Springer. (2007)
56. Sian,S.: Extending learning to multiple agents: Issues and a model for multi-agent machine learning. In Y. Kodratoff (ed), Machine Learning - EWSL-91, pp 440-456. Springer-Verlag. (1991)
57. Spath,H.: Cluster Analysis - Algorithms for Data Reduction and Classification of Objects, West Sussex, UK, Ellis Horwood Limited. (1980)
58. Tozicka, J., Rovatsos, M., Pechoucek,M.: A Framework for Agent-Based Distributed Machine Learning and Data Mining. AAMAS'07, pp.678–685 (2007), May 14–18. (2007)
59. Viaenne, S., Darrig, R.A., Dedene, G.: A case study of applying boosting Naive Bayes to claim fraud diagnosis. IEEE Transactions on Knowledge and Data Engineering. Vol.16, No.5, 612–620 (2004)
60. Viswanath,P., Narasimha Murty,M. Shalabh Bhatnagar:Overlap Pattern Synthesis with an efficient nearest neighbor classifier. Pattern Recognition. Vol. 38, No.8, 1187–1195 (2005)
61. Weiss,G. (ed). Multiagent Systems - A modern approach to Distributed Artificial Intelligence. The MIT Press (2000)
62. Wooldridge,M., Jennings,N.R.: Towards a theory of cooperative problem solving. In proc. of the Workshop of Distributed Software Agents and Applications, Denmark, 40–53 (1994)
63. Yoshida, K., Pedrycz: Recent developments in hybrid intelligent systems, In:Int. Journal on Hybrid Intelligent Systems,Vol 2, No.4, pp 235–236 (2005) Vol 34, 523–525 (2001)
64. Zhao, Y., Zhang H., Figueiredo, F., Cao, L., Zhang C.: Mining for combined association rules for multiple datasets. Proc. of 2007 International workshop on domain driven data mining. 18–23 (2007)

Chapter 16
A Multiagent, Multiobjective Clustering Algorithm

Daniela S. Santos, Denise de Oliveira, and Ana L. C. Bazzan

Abstract This chapter presents MACC, a multi ant colony and multiobjective clustering algorithm that can handle distributed data, a typical necessity in scenarios involving many agents. This approach is based on independent ant colonies, each one trying to optimize one particular feature objective. The multiobjective clustering process is performed by combining the results of all colonies. Experimental evaluation shows that MACC is able to find better results than the case where colonies optimize a single objective separately.

16.1 Introduction

Clustering is widely used in data mining to separate a data set into groups of similar objects. The importance of clustering is clear in applications related to biology, social sciences, computer science, medicine, and so on. Consequently, many clustering methods have already been developed. One issue with most of these methods is that they rely on central data structures. However, the current use of Internet resources (distribution of data, privacy, etc.) requires new ways of dealing with data clustering. This meets the recent trend around the integration of agent technologies and data mining (see [1] and references therein). In this publication, the authors identify and discuss two main challenges concerning the integration of agents and data mining, namely *data mining driven agent learning* and *agent driven data mining*. In the present chapter we address the latter by proposing the Multi Ant Colony Clustering algorithm (MACC). This is also the line followed in [2] where agent-based data mining was used to integrate knowledge and facilitate the annotation of proteins. Our work thus has addressed some topics listed in [1] such as agent-based distributed data mining, multi-data source mining, and distributed agent-based data gathering and processing. While in [2] we have addressed supervised learning, here we turn to clustering.

Daniela S. Santos, Denise de Oliveira, and Ana L. C. Bazzan
Instituto de Informática, UFRGS
C.P. 15064, 91501-970, P.Alegre, RS, Brazil
e-mail: \{daniela.scherer, denise.oliveira, bazzan}@inf.ufrgs.br

L. Cao (ed.), Data Mining and Multi-agent Integration, DOI: 10.1007/978-1-4419-0522-2_16,

Clustering methods differ not only in many of their most basic proprieties, such as the data type handled, but also in the form of the final partitioning, in the assumptions about the shape of the clusters, and in the parameters that have to be provided. Each clustering algorithm looks for clusters according to a different criterion. This can be a problem because many data sets present different shapes and size of clusters that a single objective clustering is not able to reveal. Moreover, the same data can have more than one relevant structure, each one related to a different cluster definition or to a different refinement level [3].

Recently, several biologically inspired algorithms have been introduced to solve the clustering problem [4, 5, 6, 7, 8]. These algorithms are characterized by the interaction of a large number of simple agents that interact in a multiagent system. These agents can perceive and change their environment locally and they are inspired by ant colonies, flocks of birds, swarms of bees, etc. However, these algorithms have been focussing on single objective clustering.

As an effort to overcome some limitations of single objective clustering algorithms, multiobjective clustering algorithms such as [3, 9] have been proposed. The basic idea of the multiobjective approach is to optimize more than one objective in the same clustering. Using this approach we can find different shapes and sizes of clusters and different types of structures in a data set.

MACC, the algorithm we propose, is inspired by ant colony optimization (ACO) and multiobjective clustering. The central idea of MACC is to simultaneously use several ant colonies, each colony aiming to optimize one objective. In this particular work we focus on two objectives in order to compare our results with previous works on multiobjective clustering. This way, one colony minimizes the compactness, while the other maximizes the connectivity of clusters. The simultaneous optimization of these objectives may lead to better solutions than the results achieved when both objectives are optimized separately.

The remainder of this chapter is organized as follows. Section 16.2 briefly discusses the related work on ant clustering and multiobjective clustering. Section 16.3 introduces and describes the proposed algorithm, MACC. Section 16.4 presents and discusses the results achieved, while Section 16.5 presents the conclusions and outlines future works.

16.2 Related Work

Social insects (e.g. ants, termites, bees, and wasps) distinguish themselves by their self-organization [10, 11]. The use of the social insects metaphor to solve computer problems such as combinatorial optimization, communications networks, or robotics is increasing [12]. Ant colony optimization (ACO) [13] is a class of algorithms based on social insects that has its origins in the ant foraging behavior. This behavior consists in ants depositing a chemical pheromone as they move from a food source to their nest, and foragers following such pheromone trails [12].

As mentioned, clustering approaches inspired by social insects (e.g. [4, 5, 6, 7, 8]) all optimize only a single objective. The algorithm presented in [6] relies mainly on pheromone trails to guide ants to select a cluster for each data object, while a local search is required to randomly improve the best solution before updating pheromone trails. In this algorithm, ants visit data objects one by one in a sequence and select clusters for data objects by considering pheromone information. Pheromone deposition depends on an objective function value and on an evaporation rate.

In [4], the authors propose the Ant Colony Optimization for Clustering (ACOC) which extends the algorithm proposed in [6] by introducing the concept of dynamic cluster centers in the ant clustering process, and by considering pheromone trails and heuristic information together at each solution construction step. This heuristic information is the Euclidean distance between data objects and clusters centers of ants. It serves to guide artificial ants to group data objects into proper clusters.

The algorithm presented by Yang and Kamel in [7] employs three ant colonies to solve an important subset of the clustering problem known as the cluster ensemble problem. In this problem one needs to combine multiple clustering, formed from different aspects of the same data set, into a single unified clustering [14]. This approach is based on a hypergraph to combine clustering produced by three colonies. In each colony ants move with different speeds: constant, random, and randomly decreasing. Each ant colony projects data objects randomly onto a plane and the clustering process is done by ants picking up or dropping down objects with different probabilities. Authors used a different ant clustering model inspired by the cemetery organization behavior proposed by Deneubourg [15].

In [8] Yang and Kamel have presented an extended version of [7]. They added a *centralized* element to compute the clustering: a queen ant agent. This agent receives the results produced by all colonies, calculates a new similarity matrix and broadcasts to each ant colony of the model. Each colony re-clusters the data using the new information received.

Besides these works based on ACO for single objective clustering, we discuss next some multiobjective clustering approaches that are based on evolutionary algorithms. In [9] the authors present a multiobjective evolutionary algorithm called MOCK. This algorithm is able to simultaneously optimize two complementary objectives based on cluster connectedness and compactness. MOCK returns a set of different trade-off partitioning over a range of different numbers of clusters.

Another work is proposed by Faceli et al. [3]. This approach, called multiobjective clustering ensemble, MOCLE, is based on ideas from cluster ensembles and multiobjective clustering. Notice that the cluster ensemble is different from clustering. It is a subset of the clustering problem, which combines multiple clusterings, formed from different aspects of the same data set, into a single unified clustering. The goal is to create a single clustering that best characterizes a set of clustering, without using the original data points already used to generate the clusterings. MOCLE uses the results of several different clustering algorithms and returns a set of solutions. According to the authors these solutions are diverse (revealing the different structures of the data) and concise (may be analyzed by domain experts). Finally, we remark that a multiagent ensemble approach was also proposed [16] but it has a different purpose.

Table 16.1: Comparison between MACC and related ant and multiobjective clustering approaches

	ACOC	Shelokar et at.	Yang and Kamel	**MACC**	MOCLE	MOCK
Data selecting process	stochastic	sequential	stochastic	stochastic	-	-
Multiple colonies	no	no	yes	yes	-	-
Paradigm	ACO	ACO	Cemetery organization	ACO	evolutionary algorithms	evolutionary algorithms
Pheromone update	by elitist ants	by elitist ants	-	by all ants	-	-
Memory of visited data	yes	no	no	no	-	-
Clustering type	single objective	single objective	ensemble	multiobjective	ensemble and multiobjective	multiobjective

Table 16.1 summarizes some differences and similarities between the MACC algorithm, ACO based clustering approaches, and approaches for multiobjective clustering. The main difference between MACC and ACO approaches is that MACC performs the clustering with two colonies optimizing two objectives. MOCK and MACC optimize multiobjective functions in a different way. While MOCK uses evolutionary algorithms, MACC is inspired by social insects.

16.3 MACC – Multi Ant Colony Clustering Algorithm

Given a data set $\mathcal{D}$ containing $|\mathcal{D}|$ data objects with $|\mathcal{X}|$ attributes and a predetermined number of clusters $|\mathcal{Z}|$, the proposed algorithm has to find a clustering configuration combining the results of two objective functions. Other approaches inspired by ACO [4, 6] also use the number of clusters as a parameter for *a priori*.

MACC uses two colonies, each one with K agents working to group a data set according to two different objectives. Each colony C works in parallel over the same data set. However, each one optimizes a different objective function. In this work, we use two different objective functions: compactness (*dev*) and connectedness (*con*), using one colony for each objective. Colony C^1 optimizes the objective based on compactness of clusters. Compactness measures the overall summed distances between objects and their corresponding cluster center. It is defined by Equation 16.1, where $\mathcal{Z}$ is the set of all clusters, o is a data object, c_j is the center (or *centroid*) of cluster Z_j and $d(o,c_j)$ is the Euclidean distance (Equation 16.2) between o and c_j.

$$dev(\mathcal{Z}) = \sum_{Z_j \in \mathcal{Z}} \sum_{o \in Z_j} d(o,c_j) \tag{16.1}$$

$$d(o,c_j) = \sqrt{\sum_{i=1}^{|\mathcal{X}|} (o_i - c_{ji})^2} \tag{16.2}$$

The other colony, C^2, optimizates the connectedness. It is inspired on the Shared Nearest Neighbor clustering algorithm [17], which defines the similarity between pairs of objects in terms of how many nearest neighbors two objects share. In MACC we use this approach to group a set of data in the following way: Each object of the data set has a list $\mathcal{N}$ of neighboring data (size $|\mathcal{N}|$). An agent groups an object o according to the similarity between o and its neighbors that are in the cluster. However these neighbors are considered only if $o \in \mathcal{N}$. The connectedness reflects how frequently neighboring objects have been placed in the same group. It is computed according to Equation 16.3, where nn_{iq} is the $q\text{-}th$ nearest neighbor of o. This definition of connectedness was also employed in [3, 9].

$$con(\mathcal{Z}) = \sum_{i=1}^{|\mathcal{D}|}\sum_{q=1}^{|\mathcal{N}|} a(o_i, nn_{iq}), \text{ where } a(o_i, nn_{iq}) = \begin{cases} \frac{1}{q}, & \text{if } \nexists Z_j \text{such as } o_i, nn_{iq} \in Z_j \\ 0 & \text{otherwise} \end{cases} \tag{16.3}$$

Table 16.2 summarizes the main parameters used by MACC. Colonies C^1 and C^2 use global pheromone matrixes G^1 and G^2 of size $|\mathcal{D}| \times |\mathcal{Z}|$ to store the pheromone values. Each agent carries a local pheromone matrix L of size $|\mathcal{D}| \times |\mathcal{Z}|$ with values of pheromone which are used to group the objects. The whole information about the global pheromone matrix G is not known by the agents. They only know their own local pheromone matrix L, which has a partial view of the global pheromone matrix. Let us denote k_i^n the $i\text{-}th$ agent of colony C^n. Each k_i^1 of colony C^1 carries a list $\mathcal{M}$ of centers of size $|\mathcal{Z}|$ to store their own cluster centers and update them each T steps. Each agent k_i^2 of colony C^2 carries an object $g_i \in \mathcal{D}$ which has a list $\mathcal{N}$ of neighbors. In the clustering process each agent k_i^2 needs to update the neighbors list related to its object g_i so that this list contains data that is similar to the object g_i.

MACC is described in Algorithm 6. It starts with the initialization of the colonies and their agents. The elements of the global pheromone matrix G^1 and the local pheromone matrix L^1 are initialized with zero, indicating no pheromone. The list of centers $\mathcal{M}$ of each k_i^1 is initialized with randomly chosen data. The elements of G^2 and L^2, which calculate the connectivity, are initialized in a different way because Procedure 3, needs non-zero values for G^2 and L^2. Thus G^2 and L^2 are initialized by agents of colony C^2 according to Procedure 1. Each k_i^2 receives a randomly chosen

Table 16.2: Main parameters of the MACC algorithm

Parameter	Description	Parameter	Description
$\mathcal{D}$	data set (size $\|\mathcal{D}\|$)	$\mathcal{M}$	list of centers (size $\|\mathcal{Z}\|$)
$\mathcal{X}$	set of attributes (size $\|\mathcal{X}\|$)	ρ	persistence of the pheromone trail
$\mathcal{Z}$	set of clusters (size $\|\mathcal{Z}\|$)	T	interval size (steps) to redefine $\mathcal{M}$
C	colony	z	number of elements in cluster Z
G^n	global pheromone matrix of colony C^n (size $\|\mathcal{D}\| \times \|\mathcal{Z}\|$)	L^n	local pheromone matrix of colony C^n (size $\|\mathcal{D}\| \times \|\mathcal{Z}\|$)
$\mathcal{N}$	neighbors list (size $\|\mathcal{N}\|$)	K	number of agents in colony C^n

object g_i to carry. The list $\mathcal{N}$ of neighbors of each datum o is also initialized with $|\mathcal{N}|$ randomly chosen elements of the data set $\mathcal{D}$.

After the initialization process, agents of both colonies start to cluster the data. Colonies work separately, each one with an objective. At the beginning, each agent k randomly selects an object o to group. After that, agents in C^1 and C^2 work according to Procedure 2 and 3 respectively.

We now explain the behavior of colony C^1. Each k_i^1 needs to redefine its list $\mathcal{M}$ of centers in a distributed way. The parameter T regulates the execution of this process. Each attribute x_v of the new centroid $c_j \in \mathcal{M}$ is calculated according to Equation 16.4, where z_j is the total number of data in cluster Z_j.

$$x_v = \frac{\sum_{i=1}^{z_j} x_{vi}}{z_j}, \quad v = 1 \ldots |\mathcal{X}|, \quad j = 1 \ldots |\mathcal{Z}|, \tag{16.4}$$

All necessary information to redefine the list of centers is searched from the local pheromone matrix L^1 of the agent k_i^1. Ant k_i^1 calculates the similarity between the object o and each centroid c_j of its list of centers ($\mathcal{M}$) according to Equation 16.2. Each agent adds τ_{oj}, a pheromone concentration of object o associated to the cluster Z_j, in the global pheromone matrix G^1 . This matrix is updated according to Equation 16.5, where ρ is a constant indicating the persistence of the trail, $(0 \leq \rho \leq 1)$; $(1-\rho)$ is the evaporation rate; t is the iteration number; and τ_{oj} is calculated according to Equation 16.6 for C^1.

$$\tau_{oj}(t) = (1-\rho)\tau_{oj}(t-1) + \sum_{k=1}^{K} \tau_{oj}^k \tag{16.5}$$

$$\tau_{oj} = 1 - d(o, c_j) \tag{16.6}$$

After updating the global pheromone matrix G^1, agents of colony C^1 update their local pheromone matrix L^1. They copy the pheromone concentration τ_{oj} relative to object o and each cluster Z_j from the global pheromone matrix G^1 to the local pheromone matrix L^1.

Next we explain the behavior of colony C^2. Agents working in colony C^2 group the objects in a different way. To calculate the pheromone τ_{oj} they consider the neighborhood of the object. They first update the neighbors list of their object g_i they carry. This process is performed by each agent k_i^2 when it selects an object o to group. The agent calculates the similarity between the selected object o and the data of the list of neighbors of its object g_i according to Equation 16.2. If the object o is more similar to the object g_i than to every datum in the list $\mathcal{N}$ of neighbors, then the object o will be added to $\mathcal{N}$ replacing the less similar object.

After updating $\mathcal{N}$ each k_i^2 starts the clustering process. To group an object o in a cluster Z_j it considers the list $\mathcal{N}$ of o. Then, when k_i^2 randomly selects an object o to group, it checks in its local pheromone matrix L^2 how many neighbors of o are in each cluster Z_j and updates G^2 according to Equation 16.5. It calculates the pheromone value τ_{oj} for the object o and cluster Z_j according to Equation 16.7, where o_i indicates the object that needs to be compared with o ($o_i \in Z_j$), P is the number of objects that are neighbors of o and are in the cluster Z_j, and V is the number of objects that are not neighbors of o and are in the cluster Z_j. $\mathcal{N}_o$ and $\mathcal{N}_{o_i}$

are the list of neighbors of object o and o_i respectively, and $d(o,o_i)$ is the Euclidean distance between the object o and o_i.

$$\tau_{oj} = \begin{cases} \frac{\sum_{i=1}^{P} \delta(o)}{V} & \text{if } P < V \\ 1 & \text{otherwise,} \end{cases} \tag{16.7}$$

$$\text{where } \delta(o) = \begin{cases} d(o,o_i) & \text{if } o \in \mathcal{N}_{o_i} \quad \text{and} \quad o_i \in \mathcal{N}_o \\ 0 & \text{otherwise,} \end{cases}$$

After updating the global pheromone matrix G^2, agents of colony C^2 also update their local pheromone matrix L^2. This process is carried out in the same way as for colony C^1 by copying G^2 relative to the object o and to each cluster Z_j to L^2. At the end of the process, each object will have different values of pheromone for each group. The object will belong to the cluster with the highest amount of pheromone.

Algorithm 6: MACC algorithm

1 initialize C^1 and C^2 with K agents;
2 initialize $\mathcal{N}$ with random objects;
3 initialize G^1 with pheromone value equal to 0;
4 initialize L^1 of each agent k_i^1 with pheromone value equal to 0;
5 initialize G^2 and L^2 according Procedure 1;
6 initialize $\mathcal{M}$ of each agent k_i^1 with random objects;
7 initialize parameters $T, \rho, |\mathcal{Z}|$;
8 **repeat** for each *step*
9   **foreach** *colony* C^n **do**
10   **foreach** *agent i* **do**
11   randomly choose an object o to group;
12   **if** *agent i is working from colony* C^1 **then**
13   cluster o using Procedure 2;
14   **else if** *agent i is working from colony* C^2 **then**
15   cluster o using Procedure 3;
16 **until** *numSteps*;
17 combine both global pheromone matrices G^1 and G^2: $G = G^1 + G^2$;

Procedure 1 `initialize` G^2 and L^2

1 **foreach** *agent* k_i^2 **do**
2   randomly choose an object o to group;
3   randomly choose a cluster Z_j to put o;
4   agent initializes G^2 for the object o and cluster Z_j with value 0.1;
5 $L^2 \leftarrow G^2$;

Procedure 2 `compactness`

```
1 if step % T == 0 then
2     foreach cluster Z_j, j ∈ {0,...,|Z|} do
3         foreach attribute x ∈ {0,...,|X|} do
4             calculate x according Equation 16.4;
5 foreach cluster Z_j, j ∈ {0,...,|Z|} do
6     calculate the similarity between o and c_j according to Equation 16.2;
7     update G^1 for the object o and cluster Z_j according to Equation 16.5;
8     update local pheromone matrix L^1 ;
```

Procedure 3 `connectedness`

```
1 foreach object i ∈ {0,...,P} do
2     if d(o,g) < d(g,i) then
3         N_g ← o;
4 foreach cluster Z_j, j ∈ {0,...,|Z|} do
5     update G^2 according to Equation 16.7;
6     update local pheromone matrix L^2 ;
```

16.4 Experiments and Results

We have performed experiments to investigate the quality of the proposed multi-objective ant clustering algorithm using public domain data set. The Iris data set from the Machine Learning Repository [18] was used (as in other multiobjective algorithms). The data set contains 3 classes referring to a type of Iris plant (Setosa, Versicolour and Virginica), with 4 attributes and 50 instances each.

For comparison we use the MOCK and the Yang & Kamel algorithms. Experiments were repeated 50 times (comparison with MOCK) or 10 times (with Yang & Kamel). The MACC algorithm parameter values used are: $T = 40$, $\rho = 0.3$, $|\mathcal{N}| = 20\%$ of $|\mathcal{D}|$, and $K = 3 \times |\mathcal{D}|$, and the number of iterations (*numSteps*) was 2000. These values have chosen after several tests with different values for each parameter.

To assess the performance of the clustering produced according to both objectives (compactness and connectedness) we have performed the experiments in two phases. In the first, we have investigated the quality of results of the two colonies C^1 and C^2 working independently. Each colony performed the clustering using a pheromone matrix (G^1 and G^2 respectively). In the second phase we have studied the results of both global pheromone matrix combined. This combination is performed by adding the pheromone values of both matrices. The main idea of the second phase is to optimize both objectives to find better results than when using the colonies optimizing a single objective separately.

We use the F-measure evaluation function expressed in Equation 16.8; it can take values in the interval [0,1] and should be maximized. This function is used in

[3, 7, 8] as well. The overall F-measure for the clustering is computed by Equation 16.9, where $p(i,j) = \frac{n_{ij}}{n_j}$, $r(i,j) = \frac{n_{ij}}{n_i}$, n_{ij} is the number of objects of class i within cluster j, n_j is the total number of objects within cluster j, and n_i is the number of objects of class i.

$$F(i,j) = \frac{2 \times p(i,j) \times r(i,j)}{p(i,j) + r(i,j)}, \tag{16.8}$$

$$F = \sum \frac{n_i}{D} \times max\{F(i,j)\} \tag{16.9}$$

Table 16.3 shows both phases of the clustering obtained by MACC. As we can see, the objective of MACC was achieved: the multiobjective clustering process (*combined*) works better than each colony working separately (*dev* and *con*). Table 16.4 compares MACC and MOCK while 16.5 Table shows the performance of MACC in comparison to Yang & Kamel algorithm. Entries in Table 16.4 show the sample median and interquartile range (IQR) for the F-measure while Table 16.5 depicts the average and standard deviation for the F-measure. These different measures are used to make both comparisons possible. Table 16.4 shows that MACC outperforms MOCK both when colonies work separately (the first phase) and after the combination of both pheromone matrices (the second phase).

Table 16.5 shows that our approach performs similarly to the Yang & Kamel algorithm, when the combined matrices are used. Moreover, it must be also noticed that the Yang & Kamel algorithm uses a centralized queen ant agent that receives the results produced by all colonies, computes a new similarity matrix and broadcasts to each ant colony of the model, while the agents of MACC do the clustering in a partially distributed way without knowledge about the whole environment.

Table 16.3: F-measure for MACC for the Iris data set (50 repetitions).

Measure	MACC (*dev*)	MACC (*con*)	MACC (*combined*)
Average	0.8857	0.8997	0.9215
Std. deviation	0.0017	0.1033	0.0686

Table 16.4: F-measure for MOCK and MACC for the Iris data set (50 repetitions).

Measure	MACC (*dev*)	MACC (*con*)	MACC (*combined*)	MOCK
Median	0.8853	0.9286	0.9397	0.8346
IQR	0	-0.0609	-0.0676	0.0076

Table 16.5: F-measure for Yang & Kamel and MACC for Iris data set (10 repetitions).

Measure	MACC (*dev*)	MACC (*con*)	MACC (*combined*)	Yang & Kamel
Average	0.8910	0.9219	0.9326	0.9494
Std. Deviation	0.0059	0.0407	0.0407	0.0038

16.5 Conclusion and Future Work

This chapter presented MACC, a new clustering algorithm based on ideas of multi ant colony and multiobjective clustering. The central idea of MACC is to simultaneously use several ant colonies, each optimizing one objective. Here we have performed experiments where one colony minimizes the compactness, while another maximizes the connectivity of clusters. The simultaneous optimization of these objectives leads to better solutions than those achieved when both objectives are optimized separately.

MACC was compared with a multiobjective clustering algorithm, MOCK, and with a multi ant colony, the Yang & Kamel algorithm. Results were extremely encouraging and therefore we plan to use it in other datasets, including some that are related to experiments on microarrays, where the objective is to find correlations among several genes that are expressed.

Besides, we plan to try another way to perform the multiobjective clustering: each colony collaborates with the results of each other by adding pheromone values in the corresponding global pheromone matrix. We also plan to test the optimization of conflicting objectives.

Acknowledgements This research is partially supported by the Air Force Office of Scientific Research (AFORS) (grant number FA9550-06-1-0517) and by the Brazilian National Council for Scientific and Technological Development (CNPq).

References

1. Cao, L., Luo, C., Zhang, C.: Agent-mining interaction: An emerging area. In: AIS-ADM, Springer (2007)
2. Santos, C.T., Bazzan, A.L.C.: Integrating knowledge through cooperative negotiation – a case study in bioinformatics. In Gorodetsky, V., Liu, J., Skormin, V.A., eds.: Proceedings of the International Workshop on Autonomous Intelligent Systems: Agents and Data Mining. Number 3505 in Lecture Notes in Artificial Intelligence, Springer-Verlag (2005) 277–288
3. Faceli, K., Carvalho, A.C.P.L.F., Souto, M.C.P.: Multi-objective clustering ensemble. In: Proceedings of the Sixth International Conference on Hybrid Intelligent Systems (HIS 06), Washington, DC, USA, IEEE Computer Society (2006) 51
4. Kao, Y., Cheng, K.: An ACO-based clustering algorithm. In: Proceedings of the Fifth International Workshop on Ant Colony Optimization and Swarm Intelligence - ANTS 2006. Volume 4150 of Lecture Notes in Computer Science., Brussels, Belgium, Springer (2006) 340–347
5. Lumer, E.D., Faieta, B.: Diversity and adaptation in populations of clustering ants. In: Proceedings of the third international conference on Simulation of adaptive behavior: from ani-

mals to animats, Cambridge, MA, USA, MIT Press (1994) 501–508
6. Shelokar, P.S., Jayaraman, V.K., Kulkarni, B.D.: An ant colony approach for clustering. Analytica Chimica Acta **509** (2004) 187–195
7. Yang, Y., Kamel, M.: Clustering ensemble using swarm intelligence. In: Proceedings of the Swarm Intelligence Symposium (SIS 03), Indianapolis, USA (2003) 65–71
8. Yang, Y., Kamel, M.: An aggregated clustering approach using multi-ant colonies algorithms. Pattern Recongnition **39** (2006) 1278–1289
9. Handl, J., Konwles, J.: Exploiting the trade-off - the benefits of multiple objectives in data clustering. In: Proceedings of the Third International Conference on Evolutionary Multi-Criterion Optimization (EMO 2005), Springer Verlag (2005) 547–560
10. Camazine, S., Deneubourg, J.D., Franks, N.R., Sneyd, J., Theraulaz, G., Bonabeau, E.: Self-Organization in Biological Systems. Princeton University Press, Princeton, N.J. (2003)
11. Gordon, D.: The organization of work in social insect colonies. Nature **380** (1996) 121–124
12. Bonabeau, E., Theraulaz, G., Dorigo, M.: Swarm Intelligence: From Natural to Artificial Systems. Oxford University Press, New York, USA (1999)
13. Dorigo, M., Maniezzo, V., Colorni, A.: Ant system: Optimization by a colony of cooperating agents. IEEE Transactions on Systems, Man and Cybernetics Part B: Cybernetics **26** (1996) 29–41
14. Strehl, A., Ghosh, J.: Cluster ensembles: a knowledge reuse framework for combining partitionings. In: Proceedings of the Eighteenth National Conference Intelligence, Menlo Park, CA, USA, American Association for Artificial Intelligence (2002) 93–98
15. Deneubourg, J.L., Goss, S., Franks, N., Sendova-Franks, A., Detrain, C., Chrétien, L.: The dynamics of collective sorting: Robot-like ant and ant-like robot. In: Proceedings of the First Conference on Simulation of Adaptive Behavior: From Animals to Animats, Canbridge, MA, USA, MIT Press (1991) 356–363
16. Agogino, A., Tumer, K.: Efficient agent-based cluster ensembles. In Stone, P., Weiss, G., eds.: Proceedings of the fifth international joint conference on Autonomous agents and multiagent systems, AAMAS '06, New York, NY, USA, ACM (2006) 1079–1086
17. Ertöz, L., Steinbach, M., Kumar, V.: A new shared nearest neighbor clustering algorithm and its applications. In: Proceedings of the International Conference on Data Mining (2nd SIAM). Volume 4., VA, IEEE Press (2002) 2642–2647
18. Asuncion, A., Newman, D.J.: UCI machine learning repository. University of California, Irvine, School of Information and Computer Sciences (2007) http://www.ics.uci.edu/~mlearn/MLRepository.html.

Chapter 17
Integration of Agents and Data Mining in Interactive Web Environment for Psychometric Diagnostics

Velibor Ilić

Abstract Information technologies are intensively used in modern psychometric. Interactive environment for psychometrics diagnostics enables evaluation of cognitive capabilities using several multimedial tests, collecting information about users, organizing this information in user's personal profiles, visualization, interpretation and analysis of tests results, control over procedure of testing and making conclusions on collected data. Agents supervise user's actions in the interactive environment and they are trying to adjust questionnaires, diagnostic tests, training programs and other integrated tools to user's personal needs making this environment easier for use. Interactive environment contains agents for helping users in process registration, agents for guiding users trough process of diagnostics and training, and agents for helping psychologists in their activities on this system. Internet environment that contains diagnostic tests and questionnaires generates large volumes of data that should be processed. Data mining is integrated in interactive environment for diagnostic of cognitive functions and it's used for searching of potentially interesting information that this data contains. Agents use data mining system to make their decisions more precise.

17.1 Introduction

Agents are computer programs that can assist the user with computer applications to accomplish their tasks. Agents should be able to sense and act autonomously in their environment. Agents present software that, in interaction with environment, is capable to react flexible and autonomous by following assigned goals. Interaction with environment means that agent is capable of responding on input values from sensors, which it reads from environment, and it is capable to take a course of actions in order to change agent's environment [9]. Agent's environment can be real world or virtual (software environment implemented on computer or on Internet). Agents must be able to process data, and for that purpose may have several processing strategies. They should be designed to use simple strategies (algorithms), or they could use complex reasoning and learning strategies to achieve their tasks. There are several

Velibor Ilić, University of Novi Sad, Serbia, e-mail: ilicv@EUnet.rs

L. Cao (ed.), Data Mining and Multi-agent Integration, DOI: 10.1007/978-1-4419-0522-2_17,

techniques that agents can be trained for improving agent performance and better understanding their environment using computational intelligence techniques, such as using evolutionary computing systems, neural networks, adaptive fuzzy logic, expert systems and data mining, etc [19, 26]. The concept of interface agents that collaborate with the user in the same work environment, helping him/her to perform various computer-related tasks, was introduced by in the 90's with a reasonable success. This help may range from hiding the complexity of difficult tasks, training the user or making suggestions about how to achieve specific activities, to directly execute actions on the user's behalf [5]. In order to provide personalized assistance, agents rely on user profiles modeling user information preferences, interests and habits. Inserted in communities of people with similar interests, personal agents can improve their assistance by gathering knowledge extracted from the observed common behaviors of single users. Agents help users find relevant information based on detailed models of their interests contained in user profiles [7].

Systems for intelligent data analysis (data mining or knowledge discovery in databases) represent software tools capable to analyze content of large databases and to find relations between data [21]. Increasing database volumes leads to increasing demands for development efficient software tools for data analysis. Collecting large amount of various data in modern information systems creates need for software that can efficiently retrieve and select information when they are needed [8]. Structure, format and meaning of this data is various and usually it can not be modeled mathematically, that's why overall analysis is complex or sometimes impossible using standard methods [14].

Data mining can be defined as efficient discovering human knowledge and interesting rules from large databases. This technology is motivated by the need for new techniques to help analyze, understand and visualize the huge amount of stored data gathered from scientific and business applications. Data mining involves the semiautomatic discovery of interesting knowledge, such as patterns, associations, changes, anomalies and significant structures from large amounts of data stored in databases and other information repositories. Data mining differs from traditional statistics in several ways. Statistical inference is assumption-driven, in the sense that a hypothesis is formed and validated against the data. By contrast, data mining is discovery driven; patterns and hypotheses are automatically extracted from large databases. Second, the goal of data mining is to extract qualitative models that can easily be translated into business patterns, associations or logical rules. The major data mining functions that have been developed for the commercial and research communities include generalization, summarization, classification, association, prediction-based similarity search, and clustering [15]. Using a combination of machine learning, statistical analysis, modeling techniques and database technology, data mining finds patterns and subtle relationships in data and infers rules that allow the prediction of future results. Combining agent and data mining these two innovative technologies together can improve their performance. Integrating agents into data mining systems, flexibility of data mining systems can be greatly improved. Equipping agents with data mining capabilities, the agents are much smarter and more adaptable [13, 30].

Modern psychometric uses information technologies intensively [6, 20]. Psychologist can use several software tools that can enable fast and precise diagnostics [16, 24], and there are several tools that can be used for therapy [2, 28]. This software tools can be used alone or as addition to standard diagnostic and therapeutic methods. Advantage of implementation diagnostic tests on Internet environment is that it enables diagnostic of large number of patients. Computer version of the tests directly stores results in database and automatically process data, which eliminates costs and errors that can appear during analyzing results of paper version of tests. Applying computerized adaptive testing (CAT) can provide interaction with users in a real time and generates optimal tests for individual user [17]. Implementation of psychological tests on computers provides evaluation and creation of reports immediately after testing. This enables users to get faster feedback information about achieved results [25]. The overall time testing decreases, safety of testing procedure increases and testing can be performed more often than in classic form [4]. Framework for Integrated Testing (FIT) represents online environment that contains diagnostic tests or tests for knowledge evaluation. FIT environment enables collaboration of different categories of users: patients, psychologists and software developers. Diagnostic tests on Internet environment generate large amount of information that need to be processed [3, 29]. Data mining can be efficiently used for processing this kind of data. Agents integrated in web environment can use data mining for extracting knowledge from collected data to enhance their performance [18, 22].

17.2 Interactive Environment for Psychometrics Diagnostics

Interactive environment for psychometrics diagnostics enables evaluation of cognitive capabilities using several multimedial tests, collecting information about users, organizing this information in user's personal profiles, training programs (therapy), visualizations, interpretation and analysis of test results, control over procedure of testing (individual users and group of users) and making conclusions on collected data [11]. Interactive environment for psychometrics diagnostics is based on combination of two technologies: content management system (CMS) and computer supported cooperative work (CSCW) [27]. Environment for psychometric diagnostic and psychotherapy represents a place for gathering different categories of users, which could use long distance diagnostic of cognitive abilities. Beside tools for diagnostics and therapy, environment enables methods for communication and interaction of persons interested for evaluation of their cognitive capabilities (patients) and field experts (psychologists/therapists). Open architecture of this environment allows adding easily new tools and services [23].

Agents guide the users trough available tools on the interactive environment for psychometric diagnostic. They monitor the user's activities on environment, provide help and assistance when it's required, trying to find the best way how to present the next problem or instructional sequence; diagnose problems and provide corrective feedback. Based on users' profiles, agents trying to anticipate what users should do next in the diagnostic and training process and they respond immediately to take corrective action. Agent is trying to replicate relation one-to-one between psychologist and patient on environment.

Overview of all subsystems of interactive environment for psychometrics diagnostics can be seen on Fig. 17.1. Main subsystems of interactive web environment for psychometrics diagnostics are: subsystem for collecting information about users, multimedial psychometrics tests, subsystem for evaluation and preprocessing results, subsystem for interpretation and visualization results, training programs, support for web seminars, forum, web calendar and wiki.

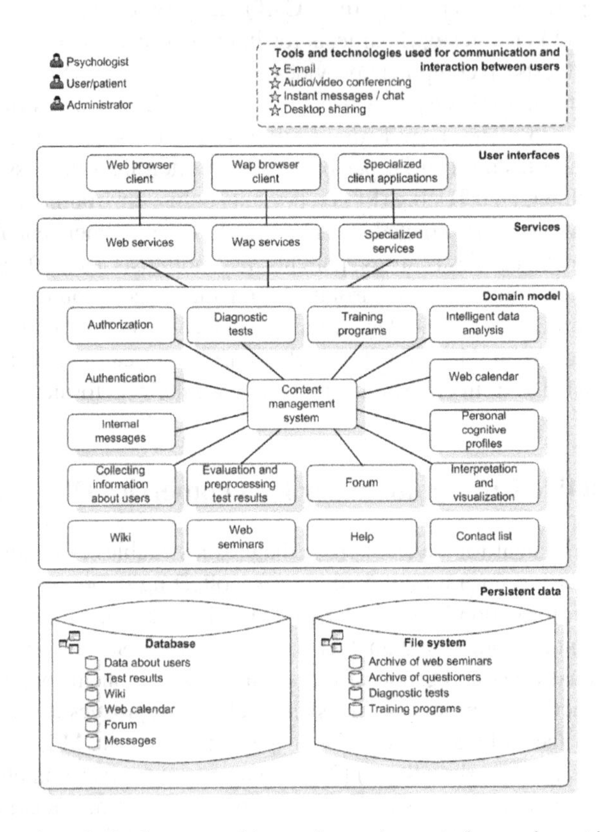

Fig. 17.1: Overview of all subsystems of interactive environment for psychometrics diagnostics

17.3 Collecting Information about Users

Subsystem for collecting information about users is activated when user visits environment for psychological diagnostics for the first time. Subsystem contains several questionnaires for collecting demographics, medical and psychological data about users. Agent A1 assists in creating the initial profile. Setting the initial profile will influence the further updates and usage of it, therefore, special care should be taken at this stage to assist the user in completing the questionnaire. At this stage, the agent will use stored knowledge about typical users characteristics to set values in the users profile.

Agent A1 supervises users during their first visit the environment for psychometric diagnostic and helps them to fill questionnaires (Fig. 17.2). Answers from questionnaires are evaluated in subsystem for reviewing and preprocessing before recording in database. Agent A1 checks over the answers and depending on answer in some question, it can add some additional questions. Agent A1 provides dynamical questionnaires and it has a goal to create more precise user's profile. This method provides mechanism for avoiding other users to waste their time on irrelevant questions. Agent A1 can provide help and assistance with text, animation or video if it is necessary. A first-time user, who is registering in the environment for psychometric

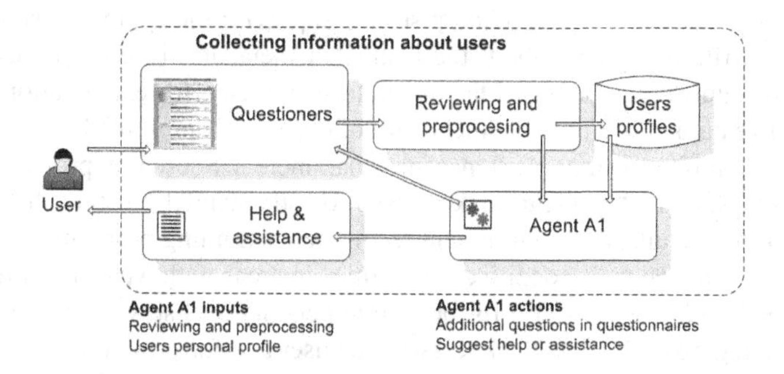

Fig. 17.2: Assisting users to fill questionnaires during their first visit the environment

diagnostic, might not want to give all the information during the first visit. Forcing users to complete a really long questionnaire could cause that they give up at start, before they get access to tests and training programs. The information that the user supplies can go in the user's profile directly, but many parameter values will remain empty because of the lack of direct user's input. To fill the missing information in the user's profile, the agent can apply a set of rules based on stereotypes. Agent A1 will help users to understand the implications of the presented questions and to make correct choices. When user fills set of base questions, agent A1 can allow user to stop filling questionnaires, even he/she hasn't answered on all questions. User can do some tests even registration is not finished and user's profile is not completed. But, some tests require user to fill questions that haven't been filled during the registration. In that moment, agent A1 activates questionnaires again asking users to fill missing questions.

17.4 Guiding Users Trough Diagnostic Tests and Training Programs

Multimedial diagnostic tests are used for evaluation cognitive user's capabilities such as attention and working memory. The tests are similar to simple computer games. Users have to solve given task, while reactions, exactness of procedures, speed and other parameters are monitored in the background. Using such tests, it can be possible to perform patients testing on distance [10]. For example, the users have to detect meaningful patterns in a noisy background. The success in such a task depends crucially on the capability to suppress noisy stimuli by increasing inhibition levels.

Diagnostic tests generate detail data, based on that data complete process of testing for each user can be reconstructed. Results that are generated by diagnostic tests (raw data) aren't suitable for comparing and analyzing and it is necessary to process them before storing in database. Subsystem for reviewing and preprocessing evaluates these tests results before storing in database. Metadata are added to perform easier managing results of the testing. This metadata layer contains additional information of tests results. Psychologist may supervise the testing personally and he/she can write complains about the testing in metadata if he/she notices some irregularities during the testing. Diagnostic tests are used for examination of various cognitive parameters and the final result of each test is presented as single value (score) that combines values of all measured parameters. Score is expressed as value in interval from 0 to 100 and this makes easier to understand the test results. According to achieved results, the environment recommends training programs for users to practice and enhance their cognitive capabilities in areas with weaker test results.

User modeling can be useful for providing personalized services to a particular user, providing proactive feedback to assist the user and for presenting the information in a way suitable to the user's needs [1]. Environment contains large number of different diagnostic tests and training program. This can confuse new users and cause lack of their motivation. Agent A2 helps new users during process of testing to avoid this situation.

User interfaces play an important role in achieving user acceptance. Environment guides testing and user's actions by the test selection page. Agent A2 controls content of this page (Fig. 17.3). When user starts to use environment, agent A2 doesn't have much information about certain user and in this case usually suggests one of the predefined list of tests that best fit to that user according to information from user's profile.

As environment collects more information about some users, agent A2 has more possibilities to make more precise suggestions for user's next steps. Agent A2 monitors events on environment such as: results of latest test, previous results of the same test (if user used that test previously), results of other diagnostic tests, information from user's profile and information from data mining. Based on this information, agent A2 tries to make decisions about further optimal steps for that user. Agent A2 can suggests users: to go on next test, to repeat same test, to do additional test for detecting same cognitive capabilities, training programs, to provide help and additional information (text, animation or video), or to suggest live communication between user and psychologist. Agent A2 is trying to provide optimal difficulties of diagnostic tests and training programs, it gradually increases difficulties as user accomplish given tasks. Agent A2 uses data mining subsystem to make more efficient decisions for suggestions about optimal actions of particular user based on available information. Beside that, agent A2 uses data mining to analyze its own previous decisions from the similar situation and uses that for self improvement in next decision process. Interaction between users and environment are changing in

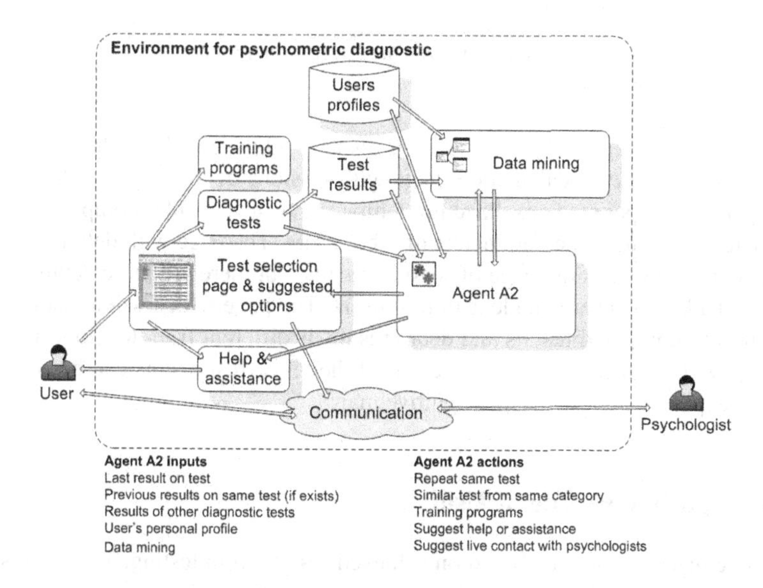

Fig. 17.3: Guiding users trough process of diagnostics and training

time. Agents guide new users until they finish basic set of tests necessary for completing user's psychological profile. After completing user's psychological profile, agents are reducing their involvement by giving suggestions to users and allowing them to select tests and training programs.

17.5 Interpretation and Visualization of Results

Subsystem for interpretation and visualization of results generates the reports based on information collected in database (table about users and table with test results). Using this subsystem, users can visually compare their own results with results of other users. Subsystem for interpretation and visualization generates the reports of test results with multiple detail levels. First level of reports shows changing test results per sessions and compares user's results of the last tests with results of other users. If user is interested for more detail results, he can look on the next pages: session results, trend sessions, results comparing, and more detail results. The Fig. 17.4 shows the organization structure of results pages. The page with brief results is the same for every tests and it shows only the final test result (score). On this page, user can compare his/her latest achieved result with own results from previous sessions, and also he/she can compare his/she results with other user's results (displayed on

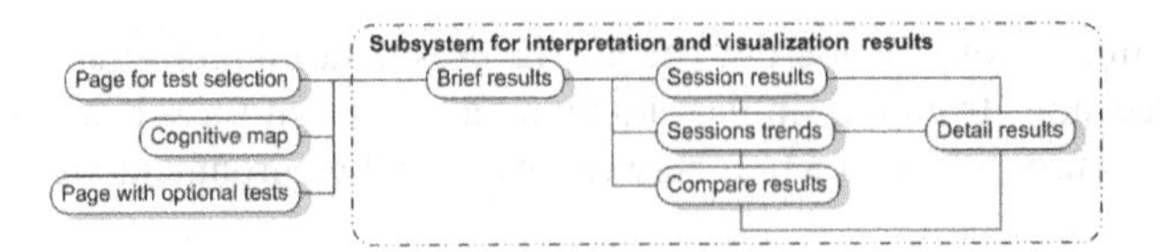

Fig. 17.4: Organization of pages with results

two graphs). The page with session results displays all the detail significant tests parameters used for score calculation. Page with sessions trends displays chaining of those parameters per sessions. Page for comparing results enables comparing user's parameters with the same parameters of other users. The page with detail results is the same by content as the page of session results, but there are more details and it contains table with every single action of users. The page with session results, sessions trends, compared results and detail results is different from test to test, while the page with brief result is the same for all the tests. Test results can be accessed trough the page with tests list or cognitive map.

17.6 Cognitive Profile of User

Cognitive map is generated based on achieved results from testing. Test results are displayed visually and this makes their understanding easier. Personal cognitive profile contains summary of all achieved user's results from diagnostic testing. According on user's profile, user can get comparative summary of his/her advantages and weaknesses. The report is generated from results achieved during the testing which are stored in database. Report contains only final result (score) of single tests that are displayed numerically and graphically with progress bar. At the bottom of the page, there is a graph in flower shape and its petals represents test results, blue color represents user's results and red color represents average results of other users.

This report allows users to have a complete overview of achieved results at the same place. Users, interested in additional information about a single test and explanation of how final result is evaluated, can take a look at the pages with more detailed reports. Page with cognitive map is linked with previously described subsystem for interpretation and visualization of test results (brief results, session results, session trends, page for comparing results, detail results).

17.7 Analysis of Psychometric Data

Data mining represents process of extracting potentially interesting information from data. Data mining, as subsystem of interactive environment for psychometric diagnostic of cognitive functions, can be used by psychologist for finding the correlation between user's data (demographic, psychological and medical data) and results gained from diagnostic testing [12].

To make explanation of procedure of data analysis easier, information from database is represented as matrix (Fig. 17.5). Matrix is gained by combining information from table about users and table with test results. This software is capable to choose one or several fields from matrix (databases), to perform data segmentation based on selected field and to search correlations with other data from matrix's fields. One of the main goals of analysis is to find correlations between results achieved in diagnostic tests and demographic, psychological and medical data. Database can be analyzed through table segmentation by one or several fields. By

	Demographic data				Medical data				Psychological data				Test results			
	D_1	D_2	...	D_j	M_1	M_2	...	M_j	P_1	P_2	...	P_j	T_1	T_2	...	T_j
Record 1	d_{11}	d_{12}	...	d_{1j}	m_{11}	m_{12}	...	m_{1j}	p_{11}	p_{12}	...	p_{1j}	t_{11}	t_{12}	...	t_{1j}
Record 2	d_{21}	d_{22}	...	d_{2j}	m_{21}	m_{22}	...	m_{2j}	p_{21}	p_{22}	...	p_{2j}	t_{21}	t_{22}	...	t_{2j}
.	.	.	...	.	.	.	...	.	.	.	...	.	.	.	...	.
.	.	.	...	.	.	.	...	.	.	.	...	.	.	.	...	.
.	.	.	...	.	.	.	...	.	.	.	...	.	.	.	...	.
Record i	d_{i1}	d_{i2}	...	d_{ij}	m_{i1}	m_{i2}	...	m_{ij}	p_{i1}	p_{i2}	...	p_{ij}	t_{i1}	t_{i2}	...	t_{ij}
.	.	.	...	.	.	.	...	.	.	.	...	.	.	.	...	.
.	.	.	...	.	.	.	...	.	.	.	...	.	.	.	...	.
.	.	.	...	.	.	.	...	.	.	.	...	.	.	.	...	.
Record n	d_{n1}	d_{n2}	...	d_{nj}	m_{n1}	m_{n2}	...	m_{nj}	p_{n1}	p_{n2}	...	p_{nj}	t_{n1}	t_{n2}	...	t_{nj}

Fig. 17.5: Database represent as matrix

segmenting, the records are grouped according fields values, while symbol * marks free fields. Each group of records is analyzed by evaluation of correlation between free fields and the field used for grouping records. For example, in table that contains four fields (X_1, X_2, X_3 and X_4), term (X_1, *, *, *) means that the records from database are segmented based on function on field X_1. In following step, correlation between selected field (X_1) and free fields, in this case (X_2, X_3, X_4), are examined. Fig. 17.6 shows tree with all possible combinations of fields used for segmentation and free fields. Database table shown on Fig. 17.7 contains demographic, psycho-

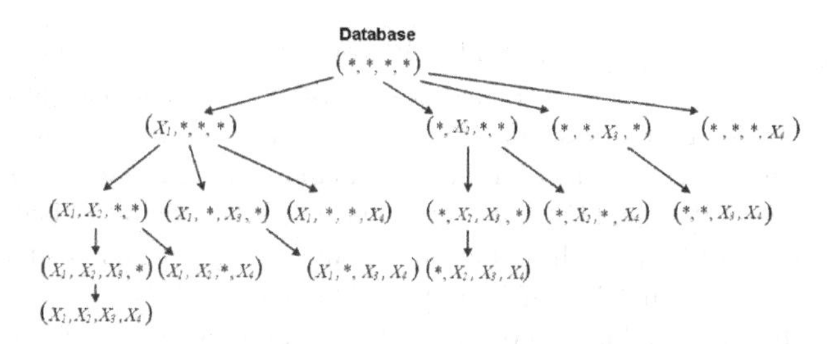

Fig. 17.6: Tree with all possible combinations

logical, medical data and test results. This table can be segmented, for example, by content of field D_1 and according syntax from Fig. 17.6 it can be written as (D_1, *, *, ... ,*). Field D_1 shown on Fig. 17.7, contains three groups (*G1*, *G2*, *G3*) and based on this classification, three subcategories can be created and further analyzed.

This procedure enables examination of degree of correlation between selected field and other (free) database fields. In this example, database is segmented by content of field D_1, but segmentation can be performed by any other field or combination of fields. Free fields in subcategories are marked as X_1, X_2, ... X_n. Boolean data sets can be segmented on two categories (true/false). Data sets that contain fields where users select one of several items from list can be segmented by those items. Other data sets can be segmented on more then one way (i.e. numeric fields or date fields). Numeric fields can be segmented by defined intervals or using fuzzy functions. Date fields can be segmented by years, months, weeks, seasons or some other defined periods. If we want to analyze database automatically, environment should "know" all types of segmentation related to each database fields. Field X_1 in subcategories S1

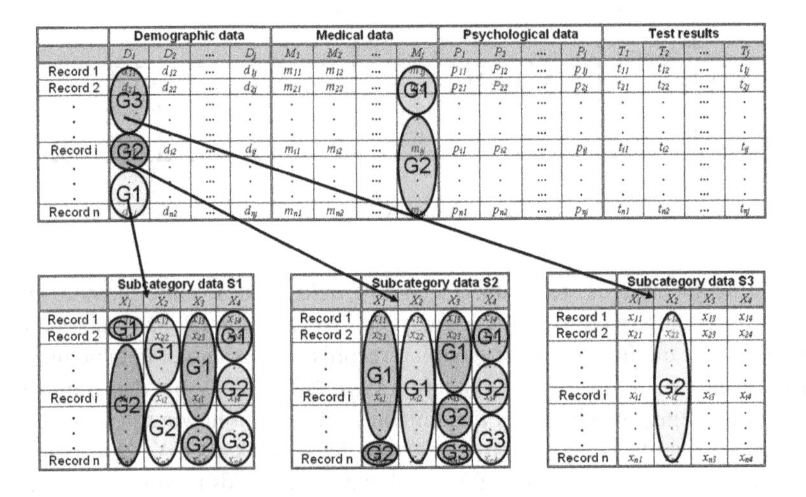

Fig. 17.7: Example of data segmentation

and S2 have differences in number of elements of group *G1* and *G2*. Subcategory S1 contains two groups of elements *G1* and *G2* in field X_2, while subcategories S2 and S3 contain only one group of data (*G1* or *G2*). It can be noticed that subcategories S2 contains new group of data *G3* in field X_3 that hasn't appeared in subcategories S1. Fig. 17.7 shows that field D1 doesn't influence significantly on X_4 because fields from subcategories S1 and S2 have the same number of groups and elements. Both groups contain approximately equal number of elements.

Fig. 17.8 shows procedure of collecting information about users by using questionnaires and diagnostic test until performing conclusions based on collected data. Data are stored in database after users fill questionnaires and perform diagnostic tests. This data are processed in appropriate form before storing in databases. In process of analyzing databases (data mining), software is trying to find relations between data about users and their achieved test results.

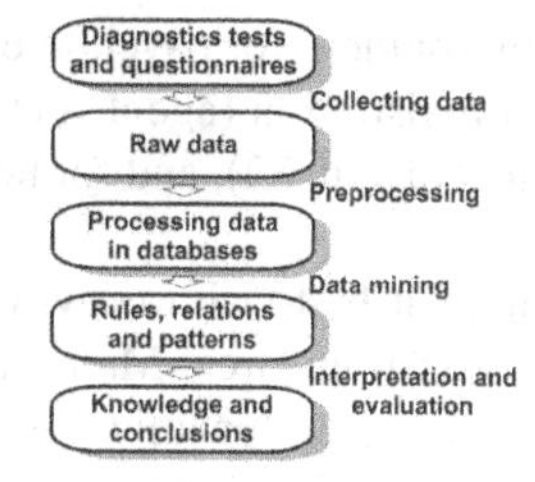

Fig. 17.8: Procedure from collecting information to knowledge extracting

17.8 Helping Psychologist in Their Activities

Environment for psychometric diagnostic should be considered as a tool that helps psychologists in their work but not as solution that completely avoids them. Psychologists evaluate test results and user's profiles, and they can communicate with users when that is necessary. Agent A3 (Fig. 17.9) helps psychologists with their activities in the environment. Agent A3 monitors information from user's profile and results of diagnostic tests that could be interesting for psychologist as reports or it can suggests live communication between psychologist and user. Agent A3 selects information that is generated by data mining and reports psychologist about it. Agent A3 uses data mining to improve its own decision process.

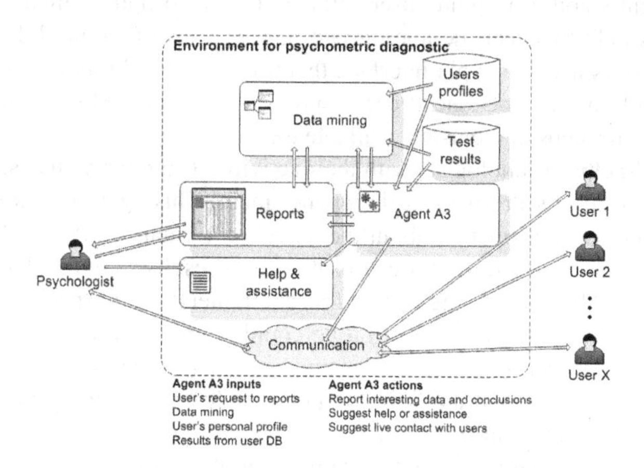

Fig. 17.9: Helping psychologists in their activities

17.9 Agent's Role in Environment for Psychometric Diagnostic

The role of the agents is to provide task-related feedback and assistance to the users and to guide the users through diagnostic and training process and help them to reach his/her goals. Agents, in the environment for psychometric diagnostic, are used for helping users during process of registration (agent A1), guiding users trough diagnostic tests and training programs (agent A2), and for helping psychologists in their work on system (agent A3).

Agent A1 helps users during their first visit to environment for psychometric diagnostic and helps them to fill questionnaires. Main goal of agent A1 is to create more precise user's profile by providing dynamical questionnaires, checking over the answers and adding some additional questions. Agent A1 isn't connected with data mining subsystem and it doesn't have possibility to enhance its own performance, this agent just follows predefined set of rules.

Agent A2 guides testing and user's actions by controlling the test selection page based on real-time events. This agent uses user's profile to make personalized test selection page. Agent doesn't force users to follow advices and suggestions strictly, these can be ignored. Agent A2 makes decisions based on predefined rules, but it also uses data mining subsystem to make more efficient suggestions. Agent A2 relies on users' profiles and their achieved results to provide optimal difficulties of diagnostic tests and training programs, also it increases difficulties gradually as user accomplish given tasks. Agent tries to learn from users' performance. Based on users' achieved results, agent receives positive or negative marks. Environment tracks users' activities (order of using diagnostic test and training programs) and their progress (achieved results) and records it in web logs. Analysis of web log files has several goals to evaluate efficiency of diagnostic tests and training programs, to improve agent's ability to guide users efficiently and to find optimal list of tests. Agent A2 is able to select the appropriate interface for different ability levels of users. New users need a simple interface that provides the ability to get help when required, while more experienced users may need minimal help and may want to skip some of the steps in the using available tools.

Agent A3 helps psychologists with their activities (monitoring users, evaluating test results and user's profiles) in the environment. This agent offers to the psychologist overview information about the users' performances. Agent A3 searches for interesting information from user's profiles, test results stored in databases, and results generated by data mining and reports about them. Agent specifies the importance of each advice. Importance of rating could be modified by the frequency of appearance and by user reactions. The agent uses this information to degrade the potential output to a less important level. Importance of some suggestions could have a varying level of importance over time.

The basic idea is that agents monitor given set of information and process them trying to find suitable suggestions for users. Initially, the agents start with general information about users. As a user interacts with the diagnostic tests, questionnaires and other tools on environment, he/she provides more information about itself. All this information about the user constitutes the user profile. Once the agent gets certain degree of competence, it can deliver pieces of advice that are expected to be relevant for the actual user's context. An important aspect of the user-agent interaction is that the agent usually does its work in an autonomous way, although the user can still require the agent's guidance explicitly or even bypass the agent. Another important source of knowledge is the user's feedback. This means that the user can express his/her degree of satisfaction with respect to a recommendation, or he/she can even want to inspect alternative solutions. Agents can collect this feedback that will lead to improving the accuracy of future suggestions. Equipping agents with data mining capabilities, the agents are much efficient and more adaptable. In this way, the performance of whole environment can be improved.

17.10 Conclusion

Technology integrated into web environment can improve the efficiency of psychometric diagnostics, psychotherapy, collaboration and communication between users. Environment can be used on several ways considering relation patient-environment-psychologists (therapist), patient can use system alone (or with minimal interaction with psychologist), patient and psychologist can use available tools and use environment for communication (synchronous or asynchronous) and psychologists can use tools from environment as additional tool in classic diagnostics and therapy.

Agents supervise user's actions in the interactive environment and they are trying to adjust questionnaires, diagnostic tests, training programs and other integrated tools to personal user's needs making this environment easier for use. Agents use data mining to make more efficiently decisions for suggestions about optimal actions of particular user based on available information. Agents use data mining to analyze their own decisions from the similar situation and use that for self improvement in next decision process.

Integration of data mining in environment for psychometric diagnostic simplifies procedure for analyzing of collected data and reduces time for processing results. Data mining allows processing of the large amount of data and reports potentially interesting data without user's intervention.

For psychologists, diagnostic tests from interactive environment represent easy available diagnostic tool for identification of persons with disorder of cognitive functions. Result analysis of diagnostic tests can identify persons with this kind disorder. Information from cognitive profile of patients can be used for individual and problem oriented therapy prescription.

For patients interested in quick evaluation of their cognitive capabilities, environment for psychometric diagnostic enables comparing achieved results with results of other users and they can identify their cognitive advantages and weaknesses on the cognitive map.

References

1. Andreevskaia, A., Abi-Aad, R., Radhakrishnan, T.,: "Agent-Mediated Knowledge Acquisition for User Profiling", Intelligent Agents for Data Mining and Information Retrieval, Idea Group Inc (2004)
2. Berger, T.: "Computer-Based Technological Applications in Psychotherapy Training", Journal of Clinical Psychology 60(3), pp. 301-315. (2004)

3. Buchanan, T., Smith, J.L,: "Using The Internet For Psychological Research: Personality Testing on the World Wide Web", British Journal of Psychology, 90 125-144. (1999)
4. Bugbee Jr., A.C,: "The equivalence of paper-and-pencil and computer-based testing", Journal of Research on Computing in Education, 28(3) 282-290. (1996)
5. Díaz Pace J. A., Berdún L. S., Amandi A. A., Campo M. R.,: "Towards Advising Design Patterns Application through Interface Agents", Proc. of Argentine Symposium on Software Engineering, pp. 1-18 (2005)
6. Emmelkamp, P. M.G.,: "Technological Innovations in Clinical Assessment and Psychotherapy", Psychother Psychosom 2005;74, pp 336-343 (2005)
7. Godoy D., Amandi A.,: "An Agent-Based Recommender System to Support Collaborative Web Search Based on Shared User Interests", J.M. Haake, S.F. Ochoa, and A. Cechich (Eds.): CRIWG 2007, LNCS 4715, pp. 303-318 (2007)
8. Hand, P., Mannila H., Padhraic S.,: "Principles of Data Mining", The MIT Press, ISBN: 026208290x (2001)
9. Ilić, V.,: "Evolutionary Neuro Autonomous Agents", Neurel 6th - IEEE Conference of Neural Network in Electrical Engineering, pp 37-40, (2002)
10. Ilić, V., "Integration adaptive psychometric tests in interactive web environment", ICCC2008, IEEE 6th International Conference on Computational Cybernetics, pp. 277-281, Stará Lesná, Slovakia, (2008)
11. Ilić, V.,: "Interactive environment based on Internet technologies for psychometric diagnostics, psychotherapy and collaboration", WI-IAT 2008 Workshops proceedings, 2008 IEEE/WIC/ACM International Conference on Web Intelligence and Intelligent Agent Technology, IEEE Computer Society Press, Sydney, (2008)
12. Ilić, V.,: "Model of data analysis on interactive web environment for psychometric diagnostics of cognitive functions", ICCC2008, IEEE 6th International Conference on Computational Cybernetics, pp. 277-281, Stará Lesná, Slovakia, (2002)
13. Lee I., (2005) "Data Mining Coupled Conceptual Spaces for Intelligent Agents in Data-Rich Environments", R. Khosla et al. (Eds.): KES 2005, LNAI 3684, pp. 42-48, Springer-Verlag Berlin Heidelberg
14. Kantardzic, M.,: "Data Mining: Concepts, Models, Methods, and Algorithms", John Wiley and Sons Massachusetts, Massachusetts Institute of Technology (2003)
15. Kim, J. S.,: "Customized Recommendation Mechanism Based on Web Data Mining and Case-Based Reasoning", Intelligent Agents for Data Mining and Information Retrieval, Idea Group Inc (2004)
16. McGhee, D. E., Lowell, N.,: "Psychometric Properties of Student Ratings of Instruction in Online and On-Campus Courses", New Directions for Teaching and Learning, no 96, pp 39-48 (2003)
17. Meijer, R.R., Nering, M.L,: "Computerized Adaptive Testing: Overview and Introduction", Applied Psychological Measurement, 23 (3) 187-194. (1999)
18. Mitkas P. A., Kehagias D., Symeonidis A. L., Athanasiadis I. N.,: "A Framework for Constructing Multi-Agent Applications and Training Intelligent Agents", Fourth International Workshop on Agent-Oriented Software Engineering AOSE 2003, Melbourne, Australia, 2003.
19. Mohammadian, M., Jentzsch, R.,: "Computational Intelligence Techniques Driven Intelligent Agents for Web Data Mining and Information Retrieval", Intelligent Agents for Data Mining and Information Retrieval, Idea Group Inc (2004)
20. Newman, M. G.,: "Technology in psychotherapy: An introduction", Journal of Clinical Psychology, no 2, vol 60, pp 141-145 (2004)
21. Owrang, M. O., "Discovering Implicit Knowledge from Data Warehouses", Encyclopedia of Communities of Practice in Information And Knowledge Management, pp 131-137, Idea Group Inc (2006)
22. Patel M., Duke M.,: "Knowledge Discovery in an Agents Environment", J. Davies et al. (Eds.): ESWS 2004, LNCS 3053, pp. 121-136, Springer-Verlag Berlin Heidelberg, (2004)
23. Pauleen, D.,: "Virtual Teams: Projects, Protocols, and Processes", Idea Group Publishing, ISBN:1591402255 (2004)

24. Percevic, R., Lambert, M. J., Kordy, H.,: "Computer-supported monitoring of patient treatment response", Journal of Clinical Psychology, no 3, vol 60, pp 285-299 (2004)
25. Schmitz, N., Hartkamp, N., Brinschwitz, C., Michalek, S., Tress, W.,: "Comparison of the standard and the computerized versions of the Symptom Check List (SCL-90-R): a randomized trial", Acta Psychiatrica Scandinavica. 102, 147-152. (2000)
26. Sklar, E., Blair, A. D., Funes, P., Pollack, J.,: "Training Intelligent Agents Using Human Internet Data", Proceedings of the First Asia-Pacific Conference on Intelligent Agent Technology, World Scientific, 354-363. (1999)
27. Suh, P., Addey, D., Thiemecke, D., Ellis, J.,: "Content Management Systems", Glasshaus, ISBN:190415106X (2003)
28. Tate, D. F., Zabinski, M. F.,: "Computer and Internet applications for psychological treatment: Update for clinicians", Journal of Clinical Psychology, no 2, vol 60, pp 209-220 (2004)
29. Wright, R. P.,: "Mapping cognitions to better understand attitudinal and behavioral responses in appraisal research", Journal of Organizational Behavior, no 3, vol 25, pp 339-374 (2004)
30. Zhang C., Zhang Z., Cao L.,: "Agents and Data Mining: Mutual Enhancement by Integration", Proceedings of the International Workshop on Autonomous Intelligent Systems: Agents and Data Mining, LNAI, Springer, (2005)

Chapter 18
A Multi-Agent Framework for Anomalies Detection on Distributed Firewalls Using Data Mining Techniques

Kamel Karoui,Fakher Ben Ftima and Henda Ben Ghezala

Abstract The Agents and Data Mining integration has emerged as a promising area for disributed problems solving. Applying this integration on distributed firewalls will facilitate the anomalies detection process. In this chapter, we present a set of algorithms and mining techniques to analyse, manage and detect anomalies on distributed firewalls' policy rules using the multi-agent approach; first, for each firewall, a static agent will execute a set of data mining techniques to generate a new set of efficient firewall policy rules. Then, a mobile agent will exploit these sets of optimized rules to detect eventual anomalies on a specific firewall (intra-firewalls anomalies) or between firewalls (inter-firewalls anomalies). An experimental case study will be presented to demonstrate the usefulness of our approach.

18.1 Introduction

Firewalls are the cornerstones of network security; they provide secure access to and from the network. Its function is to examine each incoming and outgoing packet and decide whether to accept or to discard it [2]. This decision is made according to a sequence of ordering rules. These rules are explicitly written to filter out any unwanted traffic coming into or going from the secure network [4]. However, the management of firewall rules has been proven to be complex, error-prone, costly and inefficient for many large-networked organizations. The effectiveness of firewall security may be limited or compromised by a poor management of firewall policy

Kamel Karoui
ENSI, University of Manouba,Tunisia e-mail: kamel.karoui@insat.rnu.tn

Fakher Ben Ftima
ENSI, University of Manouba,Tunisia e-mail: fakher.benftima@infcom.rnu.tn

Henda Ben Ghezala
ENSI, University of Manouba,Tunisia e-mail: henda.hg@cck.rnu.tn

L. Cao (ed.), Data Mining and Multi-agent Integration, DOI: 10.1007/978-1-4419-0522-2_18,

rules. Therefore, these rules are in a constant need of updating, tuning and validating to optimize firewall security.

A set of data mining techniques and algorithms will be applied to manage firewalls rules for anomalies detection; the data mining process, over network traffic log, may reveal any violation against the current firewall polices [12]. This leads us to remove, aggregate or reorder these rules to optimize the firewall policies.

To attain this goal, we will integrate the multi-agent paradigm to facilitate the mining tasks execution between firewalls. A multi-agents system is a set of software agents or software programs that perform tasks on behalf of users with some degree of autonomy, intelligence and mobility.

On a specific firewall, our approach consists on the following steps; Firstly,a static agent analyzes traffic and log of packets and organize them. Secondly, it applies algorithms and techniques to generate a new set of efficient firewall policy rules. Finally,a mobile agent correlates these optimized rules to detect eventual anomalies between them.

For anomalies detection on distributed firewalls, our approach is the following; the mining and detection process presented above will be generalized for a set of firewalls that is; the steps mentioned above will be executed between firewalls integrating a multi-agent system to guarantee reliable inter-communication and make the inter-firewalls anomalies detection easier.

18.2 Background

There has been a great amount of research and work in the area of firewall and policy based security management . Several models have been proposed for the anomaly detection using its firewall policy rules [4] [5] [7] [9].In our contribution, we use data mining techniques to generate structured and optimized firewall policy rules and detect eventual anomalies on distributed firewalls. Three main areas are related to our work; (1) firewall policy rule anomaly detection, (2) data mining techniques, and (3) multi agent paradigm. We refer to [1] [3] [13] [21] for more details on firewalls, the formalization of their rules, inter-firewalls anomalies and intra-firewalls anomalies detection. As far as, we refer to [11] [14] [16] [18] for multi-agent paradigm details and to [6] [10] [17]for data mining approach.

18.3 Data Mining Techniques and Their Application on Firewalls

In this section, we will present the overall process of mining log file to generate policy rules and its architecture. The process consists of the following iterative components in sequence [6]:

A.Analysis of Firewall Policy Rules

Initially, the administrator specifies the set of attributes used on the firewall policy rules for the network's security organization. The first attribute is the packet "protocol" either TCP or UDP. The next attributes are two pairs of "IP address" and its "port" for source and destination of packet. Finally, the last attribute is the "action" packet being accepted or dropped by firewall. An attribute may not be present in a firewall rule to imply all or any. An attribute may be a range such as the range of TCP port from 1024 to 65535 or the range of IP address in a local area network by mask.

B.Association Rule Mining(ARM)

This step consists in collection and extraction of attributes presented in step (A) from the collected log data file. The extracted attributes consist of protocol (TCP or UDP), source IP, source port destination IP, destination port and action (accept or deny). Furthermore, these attributes are defined as nominal to avoid any functional significance for its values. A sample line of Linux firewall log is shown in Fig 18.1. Also, the protocol is limited to be either TCP or UDP for our study.

```
Dec 01 10:44:51 utd52075 kernel: ACCEPT_LOGIN= OUT=eth0 SRC=129.110.96.80
DST=129.110.31.7 LEN=62 TOS=0x00 PREC=0x00 TTL=64 ID=10849 DF PROTO=UDP
SPT=32789 DPT=53 LEN=42
```

Fig. 18.1: An Example of Linux Firewall Log

C.Mining firewall Log using Frequency(MLF)

MLF is an algorithm that reads each line of firewall log file, extracts the attributes for each log record, counts its occurrence and outputs the count for each unique combination of these attribute-values. The frequency of each rule discovered is kept and summed up for these rules are being aggregated and used for its probability and statistical processing. Thus, each log record of a packet in firewall log file is then processed to be a primitive rule. Here primitive rule is a specific firewall policy rule where all of its attributes are instantiated or specified with its value being observed in firewall log file. Thus, the initial process of extracting the attributes of a packet of each log record is to discover and generate its corresponding primitive firewall rule, with the instantiated attributes in step (B) which are; (1) protocol (2) source IP, (3) source port, (4) destination IP, (5) destination port, and (6) the action to accept or deny a packet satisfying these attributes. MLF algorithm is shown in Fig 18.2.

D.Filtering-Rule Generalization (FRG)

Filtering-Rule Generalization (FRG) is an aggregation algorithm to generate a minimum number of firewall policy rules for efficiency and Anomaly detection purpose. FRG generates a decision tree where each level or branching represents one of the attributes of a primitive rule from log record of a packet which is an instance of a rule to be processed. Each of the rules extracted from firewall log file on step (C) is then examined and analyzed to be branched . Then, these rules are grouped to be aggregated by combining the common fields to get the superset rules that match with a group of unique rules. The Filtering Rule Generalization (FRG) algorithm in pseudo code is shown in Fig 18.3.

E. Correlation of rules

Correlation analysis has been a cornerstone of regression models and data mining analysis in general. It reveals and quantifies an effect of linear dependency between two variables. In our case, the technique of correlation consists in checking the co-

```
Input: Firewall Log file
Output: Unique Rules and their Frequency
1. Packet# ← 0
2. FOR EACH Line in Firewall Log file
3.          FirewallRule[Packet#] ← Protocol, SRC-IP, SRC-Port, DST-IP, DST-Port, Action
4.          Increment Packet#
5. END FOR
6. FOR EACH i WHERE 0 ≤ i < Packet#
7.          Frequency ← 0
8.          FOR EACH j WHERE i < j < Packet#
9.               IF FirewallRule[i] = FirewallRule[j]
10.                   Increment Frequency
11.              END IF
12.         END FOR
13.         IF FirewallRule[i] NOT discovered previously
14.                   Write FirewallRule[i] and Frequency
15.         ELSE
16.                   Continue
17.         END IF
18. END FOR
```

Fig. 18.2: Mining firewall Log using Frequency (MLF) Algorithm

herence between rules resulted from step (D) to detect eventual anomalies between them by examining and comparing the different fields of rules semantically.

18.4 Integration of Agents and Data Mining

Data mining systems are typical complex systems and difficult to construct due to many techniques and iterative steps involved in the process [7]. Agents offer a new and often more appropriate route to the development of complex systems, especially in open and dynamic environment. Employing agent techniques to improve distributed data mining systems can bring the following benefits [15]:

- Remaining the autonomy of data sources; a data mining agent can be considered as a modular extension of a data management system to deliberatively handle the access to the underlying data source in accordance with given constraints on the required autonomy of the system, data and model. This is in full compliance with the paradigm of cooperative information systems.
- Facilitating interactive distributed data mining; pro-actively assisting agents can limit the amount of data that a user has to supervise and interfere with the running mining process.
- Improving dynamic selection of sources and data gathering; one challenge for data mining systems used in open distributed data environments is to discover and select relevant sources. In such settings data mining agents can be applied to adaptively select data sources according to given criteria such as the expected amount, type and quality of data at the considered source and actual network and data mining server load.
- Having high scalability to massive distributed data; one option to reduce network and data mining application server load is to let data mining agents migrate to each of the local data sites in a distributed data mining system which they can

```
Input : Rules
Output : The Tree
1.FOR EACH attribute {Action, Protocol, DST Port, SRC Port, SRC IP and DST IP}
2.  IF attribute doesn't exist in the tree AND attribute is NOT DST Port
3.    Create new branch with attribute value
4.  ELSE
5.    IF DST Port exists in the table
6.      IF DST Port doesn't exist in Tree
7.        Create new branch with DST Port Value
8.      ELSE
9.      Follow existing branch with DST Port Value
10.   ELSE IF DST Port does not exist in Table
11.         IF DST Port doesn't exist in Tree
12.          Create new branch with DST Port Value
13.         ELSE
14.          Determine Range of DST Port
15.          Update existing branch with DST P ort Range Value
16.         END IF
17.     END IF
18.     IF SRC Port exists in the table
19.       IF SRC Port doesn't exist in Tree
20.         Create new branch with SRC Port Value
21.       ELSE
22.        Follow existing branch with SRC Port Value
23.       ELSE IF SRC Port does not exist in Table
24.         IF SRC Port doesn't exist in Tree
25.           Create new branch with SRC Port Value
26.         ELSE
27.           Determine Range of SRC Port
28.           Update existing branch with SRC Port Range Value
29.         END IF
30.      END IF
31.      IF SRC IP doesn't exist in Tree
32.        Create new branch with SRC IP Value
33.      ELSE
34.        Determine Superset of SRC IP
35.        Update existing branch with SRC IP Superset Value
36.      END IF
37.      IF DST IP doesn't exist in Tree
38.        Create new branch with DST IP Value
39.      ELSE
40.        Determine Superset of  DST IP
41.        Update existing branch with DST IP Superset Value
42.      END IF
43.   END IF
44. END LOOP
45. PRINT the generated Rule (Each path of the tree will produce a new generated Rule).
```

Fig. 18.3: Filtering-Rule Generalization (FRG) Algorithm

perform mining tasks locally, and then either return with or send relevant preselected data to their originating server for processing.

- Stimulating multi-strategy distributed data mining; for some complex application settings, appropriate combination of multiple data mining techniques can be more beneficial than applying just on a particular one. Data mining agents can learn in due course of their deliberative actions to choose depending on the type of data retrieved from different sites and tasks to be pursued.
- Enabling collaborative data mining; data mining agents can operate independently on data gathered at local sites, and then combine their respective models. They can agree to share potential knowledge as it is discovered, in order to benefit from additional opinions of other agents.

In the following sections, we will exploit advantages oh these two complementary technologies to detect anomalies on distributed firewalls.

18.5 The Principal Contribution

The development of distributed firewalls and the emergence of data mining techniques lead us to propose a novel system based on the multi-agents approach to perform anomalies detection on firewalls. There are two kind of firewalls anomalies: intra-firewalls anomalies and inter-firewalls anomalies.An intra-firewall policy anomaly is defined by [9] :

- The existence of two or more filtering rules that may match the same packet.
- The existence of a rule that can never match any packet on the network paths that cross the firewall.

There are five possible intra-firewall policy anomalies: Shadowing anomaly, Correlation anomaly, Generalization anomaly, Redundancy anomaly and Irrelevance anomaly.

An inter-firewall anomaly may exist if any two firewalls on a network path take different filtering actions on the same traffic [7].For any traffic flowing from sub-domain D_x to sub-domain D_y, an anomaly exists if one of the following conditions holds:

- The most-downstream firewall accepts a traffic that is blocked by any of the upstream firewalls.
- The most-upstream firewall permits a traffic that is blocked by any of the downstream firewalls.
- A downstream firewall denies a traffic that is already blocked by the most-upstream firewall.

There are four possible inter-firewall policy anomalies: Shadowing anomaly, Spuriousness Anomaly, Redundancy anomaly and Correlation anomaly.

Hereafter, we present the architecture of our system composed of two kinds of agents:

- A static agent, encapsulting in its code data mining algorithms, organize and optimize traffic and log rules.
- The returned results by the static agent, will be encapsulated by a mobile agent to detects intra-firewalls anomalies or inter-firewalls anomalies by migration between distributed firewalls .

18.5.1 *Intra-Firewalls Anomalies Detection Model*

In the case of intra-firewalls anomalies detection, the administrator activates a software agent SA_i (Software Agent i) in a specific firewall Fw_i (firewall i). SA_i analyses the firewall log, deduces firewall policy rules by using (ARM) and (MLF) techniques and reduces the number of policy rules by using the (FRG) algorithm. Regularly, the administrator sends a mobile agent called CA (Correlator Agent) with anomalies description to the firewall Fw_i. This latter correlates the set of organized and optimized rules prepared by SA_i to detect eventual anomalies and returns back to the administrator with results.

18.5.2 Inter-Firewalls Anomalies Detection Model

In the case of inter-firewalls anomalies detection, the system architecture presented above will be applied for distributed firewalls, that is; the administrator activates the software agent SA_i in every firewall Fw_i on the system (1). Then, SA_i analyses the firewall log, deduces firewall policy rules by using (ARM) and (MLF) techniques and reduces the number of policy rules by using the (FRG) algorithm (2).Theses tasks are executed simultaneously on all firewalls hosts. Regularly, the administrator sends a mobile agent (CA) to the first firewall Fw_1 which encapsulates the set of organized and optimized rules prepared by SA_1 and migrates to the next firewall (3). It correlates the list of rules with those on Fw_2 (prepared by SA_2) to detect eventual anomalies (4) and passes to the next firewall. It repeats the same processes (steps (3) and (4)) until finishing a complete tour of the system and returns back to the administrator with results (see Fig 18.4).

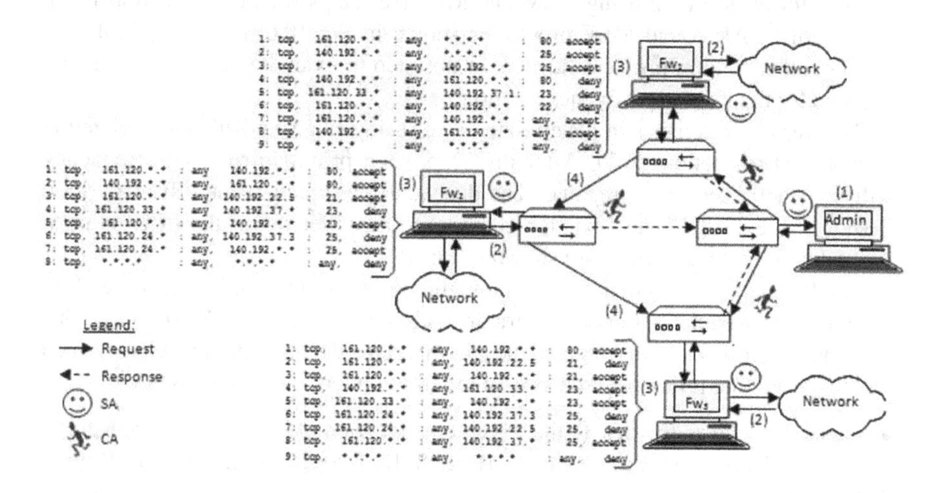

Fig. 18.4: Firewalls' anomalies detection system using data mining techniques

18.6 Case Study

Based on the approaches presented in previous sections , we will present an experimental case study; we have implemented a ring network composed of an administrator machine and three firewalls Fw_1, Fw_2 and $Fw_3$5(see Fig 18.4). These machines are equipped with Core 2 Duo processor with 1,6 MHZ frequency and 1GB of RAM. We used the platform BeeGent [21] to implement the multi-agents approach and the firewalls IPTABLES [20] for firewall rules description. We implemented our system under Linux FEDORA6 operating system. Hereafter, we present the mining steps and experimental results.

18.6.1 Intra-Firewalls Anomalies Detection Results

To detect anomalies on a specific firewall, the administrator activates a SA_i on Fw_i to execute tasks. We have chosen Fw_2 as firewall. SA_2 analyses the corresponding firewall log, filter the firewall policy rules by using (ARM) and (MLF) techniques and reduces the number of policy rules by using the (FRG) algorithm(see section 18.5.1). More precisely, SA_2 follows these steps:

The first step consists in the generation of firewall traffic log file for Fw_2 (see step A in section 18.3). The firewall log data contains 20 log records. To reduce the number of states for our research and prototype, we will consider only the six major fields in the firewall policy rules: protocol, source IP address, source port, destination IP address, destination port and action. The absence of any of the above attributes in the policy rule indicates that such a rule is not impacted by that attribute. A listing of Fw_2 data log file which is used for our study is shown in Fig 18.5. The second step consists in the processing of the firewall log dataset to extract packet features for data mining using ARM and MLF (see steps B and C in section 18.3). The result of ARM and MLF process is shown in Fig 18.6. It's also possible to use any other field in protocol headers in this step to extend its preprocessing and analysis for the next step of generalization.

The third step consists in generalizing rules using Filtering-Rule Generalization (see step D in section 18.3). After this step, each primitive rule with frequency will be generalized or aggregated. The decision tree in action for FRG algorithm is shown in Fig 18.7. The next step consists in the collection of the FRG outputs; the resulting optimized rules are shown in Fig 18.8.

The last step consists in detection anomalies between rules on Fw_2 (see step E in section 18.3). Regularly, the administrator sends a mobile agent (CA) to execute this task and returns back with results to the administrator. The detection results returned by the CA are the following (see system execution interfaces in Fig 18.9(a)):

- Generalization anomalies: (Rule 7 is a generalization of Rule 6), (Rule 8, Rule1), (Rule 8, Rule 2), (Rule 8, Rule 3), (Rule 8, Rule 5), (Rule 8, Rule 7), (Rule 5, Rule 4)
- Redundancy anomalies :(Rule 1 is redundant to Rule3)

18.6.2 Inter-Firewalls Anomalies Detection Results

In the case of inter-firewalls anomalies detection, the mining steps presented above (see section 18.6.1) will be executed simultaneously on all firewall hosts. Regularly,

```
COMMIT
#completed on Mon Dec 01 10:45 :14 2008
#generated by iptables-save v 1.2.11 on Mon Dec 01
+mangle
:PREROUTING ACCEPT [265144:123826498]
:INPUT ACCEPT [259110: 123137467]
:FORWARD ACCEPT [0:0]
:OUTPUT ACCEPT [185663:69820616]
:POSTROUTING ACCEPT [182657:69005504]
COMMIT
#completed on Mon Dec 01 10:45:14 2008
#generated by iptables- savev 1.2.11 on Mon Dec 01
+filter
:INPUT ACCEPT [169897:115595620]
:FORWARD ACCEPT [0:0]
:OUTPUT ACCEPT [180367::68855166]
:RH-Firewall-1-INPUT -[0:0]
-A -s 161.120.66.123 -p tcp -m tcp --sport 235 --dport 23 -d 161.120.37.139 -j ACCEPT
-A -s 161.120.5.20 -p tcp -m tcp --sport 23 --dport 60 -d 163.75.26.123 -j DROP
-A -s 161.120.5.20 -p tcp -m tcp --sport 80 --dport 1211 -d 155.220.130.10 -j DROP
-A -s 161.120.24.25 -p tcp -m tcp --sport 1160 --dport 25 -d 140.192.37.3 -j DROP
-A -s 161.120.24.36 -p tcp -m tcp --sport 1163 --dport 25 -d 140.192.37.3 -j DROP
-A -s 161.120.30.15 -p tcp -m tcp --sport 25 --dport 80 -d 140.192.55.17 -j ACCEPT
-A -s 161.120.24.235 -p tcp -m tcp --sport 25 --dport 25 -d 161.120.27.165 -j ACCEPT
-A -s 123.122.28.136 -p tcp -m tcp --sport 25 --dport 1110 -d 163.75.26.123 -j DROP
-A -s 161.120.49.81 -p tcp -m tcp --sport 111 --dport 23 -d 161.120.37.61 -j ACCEPT
-A -s 161.120.38.12 -p tcp -m tcp --sport 23 --dport 21 -d 140.192.22.5 -j ACCEPT
-A -p tcp -m tcp --sport 23 -d 140.192.22.5 -j ACCEPT
-A -s 161.120.22.135 -p tcp -m tcp --sport 67 --dport 21 -d 140.192.22.5 -j ACCEPT
-A -s 161.120.26.127 -p tcp -m tcp --sport 53 --dport 80 -d 140.192.13.200 -j ACCEPT
-A -s 161.120.33.12 -p tcp -m tcp --sport 1170 --dport 23 -d 140.192.37.203 -j DROP
-A -p tcp -m tcp --sport 1170 --dport 23 -d 140.192.37.203 -j DROP
-A -s 161.120.33.3 -p tcp -m tcp --sport 1160 --dport 23 -d 140.192.37.100 -j DROP
-A -s 140.192.26.107 -p tcp -m tcp --sport 80 --dport 80 -d 161.120.36.11 -j ACCEPT
-A -s 140.192.130.20 -p tcp -m tcp --sport 80 --dport 80 -d 161.120.24.127 -j ACCEPT
-A -s 123.122.28.136 -p tcp -m tcp --sport 23 --dport 70 -d 155.220.130.10 -j DROP
-A -s 161.120.24.11 -p tcp -m tcp --sport 23 --dport 25 -d 140.120.34.22 -j ACCEPT
COMMIT
# completed on Mon Dec 1 10:45:14 2008
```

Fig. 18.5: A firewall data log file

the administrator sends a mobile agent (CA) to the first firewall Fw_1 which encapsulates the set of organized and optimized rules prepared by SA_1 and migrates to Fw_2. It correlates the list of rules with those on Fw_2 (prepared by SA_2) to detect eventual anomalies and passes to Fw_3. It repeats the same process and returns back to the administrator with results (see section 18.5.2 and Fig 18.4). The detection results returned by CA are the following (see system execution interfaces in Fig 18.9(b)):

- Shadowing anomalies :(Rule2 on Fw_3 is shadowed by Rule3 on Fw_2), (8/Fw_2, 4/Fw_3), (7/Fw_2,7/Fw_1), (5/Fw_2, 5/Fw_1)
- Spuriousness anomalies: (Rule2 on Fw_2 allows spurious traffic to Rule4 on Fw_1), (2/Fw_2, 9/Fw_3), (5/Fw_3, 4/Fw_2), (3/Fw_3, 3/Fw_2), (5/Fw_1, 4/Fw_2)
- Redundancy anomalies: (Rule 6 on Fw_3 is redundant to Rule 6 on Fw_2), (9/Fw_3, 6/Fw_1)

18.7 Conclusion

In this paper, we exploited the advantages of the multi-agent approach and data mining integration to ameliorate and make the anomalies' detection in distributed

Nb of count	Prtcl	S_ip	S_port	D_ip	D_port	Action
1	tcp	161.120.33.3	1160	140.192.37.100	23	deny
1	tcp	161.120.33.12	1170	140.192.37.203	23	deny
1	tcp	161.120.24.25	1160	140.192.37.3	25	deny
1	tcp	160.120.24.36	1163	140.192.37.3	25	deny
1	tcp	161.120.5.20	23	163.75.26.123	60	deny
1	tcp	123.122.28.136	23	155.220.130.10	70	deny
1	tcp	161.120.5.20	80	155.220.130.10	1211	deny
1	tcp	123.122.28.136	25	163.75.26.123	1110	deny
1	tcp	161.120.24.11	23	161.120.34.22	25	accept
1	tcp	161.120.24.235	25	161.120.27.165	25	accept
1	tcp	161.120.49.81	111	161.120.37.61	23	accept
1	tcp	161.120.66.123	235	161.120.37.139	23	accept
1	tcp	161.120.38.12	23	140.192.22.5	21	accept
1	tcp	161.120.22.135	67	140.192.22.5	21	accept
1	tcp	161.120.30.15	25	140.192.55.17	80	accept
1	tcp	161.120.26.127	53	140.192.13.200	80	accept
1	tcp	140.192.26.107	80	161.120.36.11	80	accept
1	tcp	140.192.130.20	80	161.120.24.127	80	accept

Fig. 18.6: The MLF technique

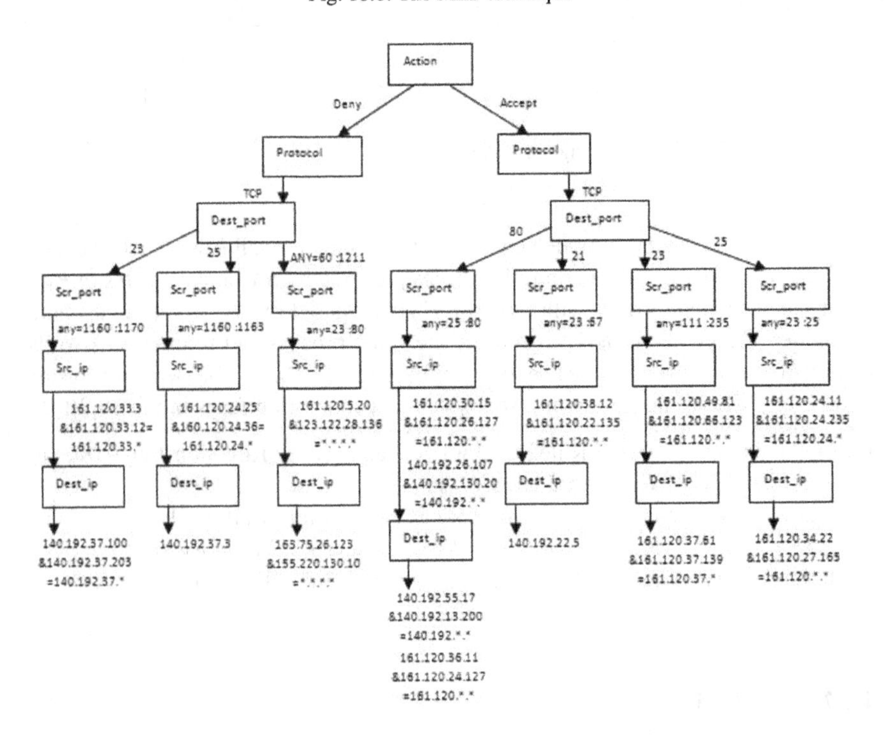

Fig. 18.7: The FRG decision tree

```
1: tcp, 161.120.*.* : any 140.192.*.* : 80, accept
2: tcp, 140.192.*.* : any, 161.120.*.* : 80, accept
3: tcp, 161.120.*.* : any, 140.192.22.5 : 21, accept
4: tcp, 161.120.33.* : any 140.192.37.* : 23, deny
5: tcp, 161.120.*.* : any, 140.192.*.* : 23, accept
6: tcp, 161.120.24.* : any, 140.192.37.3 : 25, deny
7: tcp, 161.120.24.* : any, 140.192.*.* : 25, accept
8: tcp  *.*.*.*       : any,  *.*.*.*     : any, deny
```

Fig. 18.8: The firewall optimized rules

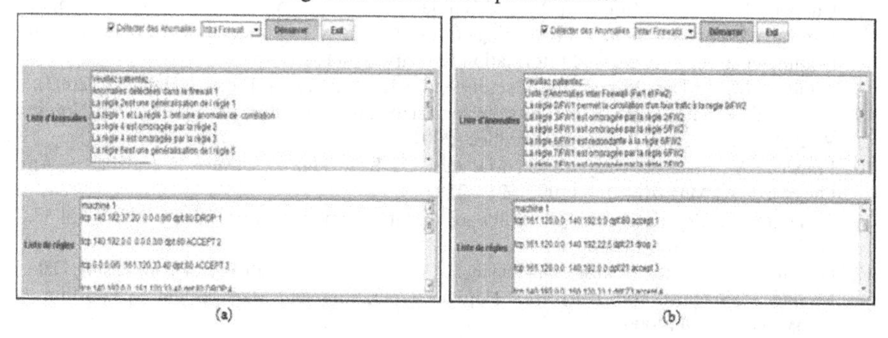

Fig. 18.9: Iter-anomalies and intra-firewalls anomalies detection results

firewalls easier. We have presented a prototype implementing algorithms and techniques described above; the obtained results confirm our approach. Integration of agent and data mining techniques allowed us to:

- To detect anomalies in distributed firewalls that it was not possible using only data mining techniques.
- To detect new types of anomalies that it was difficult to discern with classical detection method.

Our case study was accomplished in a specific network with the environment constraints (laboratory). We expect to validate the efficiency of our approach on more complex network architectures. Obviously, more efforts are required to explore this new trend of research;for example, we can use benefits of this integration to detect anomalies on heteregenous distributed systems composed of various security components such as anomalies detection system and intrusion detection system.

References

1. Al-Shaer, E., Hamed, H.: Discovery of policy anomalies in distributed firewalls. Sch. of Comput. Sci., Telecommun. and Inf. Syst.(2004) DePaul Univ, USA.

2. Bellovin, M., Chewsick, R.:Network firewalls. IEEE Communications Magazine, (1994), pages 50-57.
3. Bellovin, M.: Distributed Firewalls.Special Issue on Security, (1999) , ISSN 1044-6397.
4. Benelbahri, A., Bouhoula, A.:Tuple Based Approach for Anomalies Detection within Firewall Filtering Rules. ISCC (2007). 12th Volume, 1-4 Page(s):63 - 70.
5. Ben Ftima, F. Karoui, K. , Ben Ghezala, H.: A Secure Mobile Agents Approach for Anomalies Detection on Firewalls, Proceedings of the The 10th International Conference on Information Integration and Web-based Applications and Services, ACM, (iiWAS2008), November 24 - 26, (2008), Linz, Austria.
6. Chen, M.S., Han, J., Yu, P. S.: Data Mining: an over view from a database perspective, IEEE Trans .On knowledge And data Engineering, Vol 8, (1996), 866-883.
7. Cuppens, F., Cuppens-Boulahia, N., and Garcia-Alfaro, J. (2005). Detection and Removal of Firewall Misconfiguration. In Proceedings of the 2005 IASTED International Conference on Communication, Network and Information Security, 154-162.
8. Dunham, M. H.: Data Mining-Introductory and Advanced Topics, Prentice Hall, NJ, (2003).
9. Eronen, P., Zitting, J.: An Expert System for Analyzing Firewall Rules. Procs of 6th Nordic Workshop on Secure IT-Systems-NordSec (2001).
10. Frawley, W., Piatetsky-Shapiro, G., Matheus, C.: Knowldgz Discovery in Databases: An Overwiew, AI Magazine, fall (1992), 213-228.
11. Genesereth, M.R., Ketchpel, S.P.: Software Agents, Communications of the ACM, vol.37, no.7, pp. 48-53, July (1994).
12. Golnabi, K., Min, R., Khan L., Al-Shaer, E. : Analysis of Firewall Policy Rule Using Data Mining Techniques, 10th IEEE/IFIP Network Operations and Management Symposium - (NOMS06), April (2006)
13. Ioannidis, S., Keromytis, A., Bellovin, S., Smith, J.: Implementing a Distributed Firewall. Proceedings of CCS'00, November (2000).
14. Karoui, K.: MA Overview, published in encyclopaedia of Multimedia Technology and Networking, (2005), Idea Group.
15. Klusch, M. ,Lodi, S., Moro, G.: The Role of Agents in Distributed Data Mining: issues and benefits, proceedings of the IEEE/WIC international conference on Intelligent Agents Technology, IEEE CS Press, (2003), 211-217.
16. Lange, D.B., Oshima, M.: Seven Good Reasons for Mobile Agents, Communications of the ACM, vol. 42, no. 3,pp. 88-89, March (1999).
17. Lee, W. A Data Mining Framework for Constructing Features and Models for Intrusion Detection, Ph.D. Dissertation, Columbia Univeristy, (1999).
18. Murch, R. ,Johnson, T.: Intelligent Software Agents, Prentice-Hall, (1999).
19. Russell, R.: Linux iptables HOWTO, v0.0.2, (1999).
20. Toshiba Corporation. (2001). Beegent Multi-Agent Framework.
21. Wack, J., Cutler, K. , Pole, J.: Guidelines on Firewalls and Firewall Policy. NIST Recommendations, (2002)

Chapter 19
Competitive-Cooperative Automated Reasoning from Distributed and Multiple Source of Data

Amin Milani Fard[1]

Abstract Knowledge extraction from distributed database systems, have been investigated during past decade in order to analyze billions of information records. In this work a competitive deduction approach in a heterogeneous data grid environment is proposed using classic data mining and statistical methods. By applying a game theory concept in a multi-agent model, we tried to design a policy for hierarchical knowledge discovery and inference fusion. To show the system run, a sample multi-expert system has also been developed.

19.1 Introduction

Reasoning is perhaps the most powerful and useful method for information systems which is applied to foundational issues in distributed AI. Considering the fast growth of data contents size and variety, finding useful information from collections of scattered data in a network have been extensively investigated in past decades. Practical databases are now becoming very huge containing billions of records and therefore knowledge discovery from databases (KDD) techniques were introduced as the process of nontrivial extraction of implicit, previously unknown, and potentially useful information from data [1].

In this paper a multi-expert competitive mechanism for knowledge discovery process in heterogeneous distributed information systems based on our previous work [2] is investigated. The structure of the paper is as follows. Section 19.2 describes related works on expert systems and distributed information retrieval in brief. The proposed multi-agent systems architecture and agents behaviors are declared in section 19.3; and a sample run result is dedicated to section 19.4. We finalize our work with a conclusion and future work part in section 19.5.

School of Computing Science, Simon Fraser University, BC, Canada
milanifard@cs.sfu.ca

L. Cao (ed.), Data Mining and Multi-agent Integration, DOI: 10.1007/978-1-4419-0522-2_19,

19.2 Related Work

Expert systems (ES) are software tools to help experts and specialists for partial knowledge substitution and decision making. ES work based on a knowledge base composed of Facts and Rules, and an inference engine. *MYCIN* [3] was actually the first successful ES designed in 1970 at Stanford University with the purpose of assisting physician in diagnosis of infectious blood diseases and antibiotics. It was never actually used in practice not because of any weakness in its performance but much because of ethical and legal issues related to the use of computers in medicine, in case its diagnosis is wrong.

There are two main methods of reasoning when using inference rules, forward chaining and backward chaining. Forward chaining [4] starts with the available data and uses inference rules to extract more data until reaching a goal. Because the data determines which rules are selected and used, this method is called data-driven. Backward chaining [4], on the other hand, starts with a list of goals (or a hypothesis) and works backwards from the consequent to the antecedent to see if there is data available that will support any of these consequents. Because the list of goals determines which rules are selected and used, this method is called goal-driven. Both of the methods explained are often employed by expert systems. With the growth of distributed processing approaches, methods for combining multiple expert systems knowledge, known as multi-expert systems (MES), have been widely studied in the last decade [5].

Distributed information retrieval (DIR) aims at finding information in scattered sources located on different servers on a network. This problem, also known as federated search, involves building resource descriptions for each database, choosing which databases to search for particular information, and merging retrieved results into a single result list [6], [7]. Some applications of information retrieval from multiple sources include meta-search engines, distributed genomic search, newsletter gathering and etc.

19.3 The Proposed Approach

In order to reach an organized deduction, a hierarchy is performed. At the first phase, association rules are generated from each database. In the second phase rules and facts compose a knowledge base and makes local deduction. The third phase combines these results into a single list ordered by relevancy.

Our proposed multi-agent system architecture is based on Java Agent DEelopment (JADE) framework [10]. JADE is a software development framework aimed at developing multi-agent systems and applications in which agents communicate using FIPA[1] Agent Communication Language (ACL) messages and live in containers

[1] Foundation for Intelligent Physical Agents (http://www.fipa.org)

which may be distributed to several different machines. JADE uses RMI[2] method for communication. One of the most important characteristics of this tool is that programmer is not required to handle variables and functions concurrency as it is done automatically by the system. JADE is capable of linking Web services and agents together to enable semantic web applications. The Web Services Integration Gateway (WSIG) [11] uses a Gateway agent to control the gateway from within a JADE container. Interaction among agents on different platforms is achieved through the Agent Communication Channel. Whenever a JADE agent sends a message and the receiver lives on a different agent platform, a Message Transport Protocol (MTP) is used to implement lower level message delivery procedures [12]. Currently there are two main MTPs to support this inter-platform agent communication - CORBA IIOP-based and HTTP-based MTP.

Since we aim to design a large-scale knowledge mining system for heterogeneous separated networks, agent communications has to be handled behind firewalls and Network Address Translators (NATs). Although the current JADE MTP does not allow agent communication through firewalls and NATs, fortunately the problem can be solved by using the current JXTA implementation for agent communication [13]. JXTA is a set of open protocols for P2P networking. These protocols enable developers to build and deploy P2P applications through a unified medium [14]. Consequently JADE agent communication within different networks can be facilitated by incorporating JXTA technology into JADE [13]. A multi-agent system for intelligent information retrieval in heterogeneous networks have been proposed in a previous work [15] and upon that architecture a microorganism DNA pattern search trough web-based genomic engine [16], and a web-based criminal face recognition system [17] was proposed.

In this work, we also use a same infrastructure to solve agent communication problem in a heterogeneous network such as data Grids. In our proposed architecture, we use five different types of agents, each having its own characteristics as the followings:

a) Manager Agent (MA): MA has the responsibility of managing the whole system including other agents creation. The creation node determination is influenced by different criterion such as CPU power, available processor load, total memory amount, used memory amount, traffic around node, and etc.

b) Broker Agent (BA): These agents will deliver the query from user to Inference Agents. The query is in the form of facts to be included in IA knowledge base.

c) Association Rules Miner Agent (ARMA): ARMAs are used to discover useful association rules and convert them to First Order Logic (FOL) to be included in the local IA knowledge base. The local IA is the one which is responsible for its local LAN ARM agents.

[2] Remote Method Invocation

d) Inference Agent (IA): Inference Agents use FOL based rules gathered by ARMAs, FOL based facts from BA, and apply the inference mechanism (here forward chaining) and return their inference result to the response agent.

e) Response Agent (RA): This agent is responsible of showing the result of retrieved information. To do so, RA collects IAs results and combine them using Dempster-Shafer method, and then writes them on the screen ordered by relation percentage.

The proposed multi-agent architecture is shown in Fig. 19.1. As the most innovative design parts were done on ARMA, IA, and RA, we focus on their details in the following sections

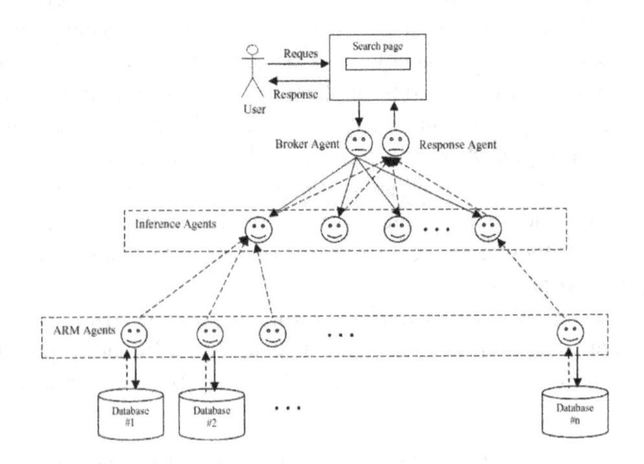

Fig. 19.1: The proposed multi-agent knowledge mining architecture

19.3.1 ARM Agent Behavior

Having created each ARMA, they would start mining according to their predefined behavior. The well-known *Apriori* Association rule mining algorithm [18] is used for this matter. Two main parameters in this algorithm are *MinSup* and *MinCon*, which denote minimum acceptable support and confidence respectively. Although mostly these parameters are set to 50% , in our thesis we used a game theory-based mechanism to define almost optimal *MinSup* and *MinCon* values for them.

19.3.1.1 Apriori Algorithm

The *Apriori* algorithm [18] computes the frequent itemsets in the database through several iterations. Iterations i computes all frequent i-itemsets (itemsets with i elements). Each iteration has two steps: *candidate generation* and *candidate counting and selection*. In the first phase of the first iteration, the generated set of candidate itemsets contains all i-itemsets. In the counting phase, the algorithm counts their support searching again through the whole database. Finally, only i-itemsets (items) with s above required threshold will be selected as frequent. Thus, after the first iteration, all frequent i-itemsets will be known. Basically, all pairs of items are candidates. Based on knowledge about infrequent itemsets obtained from previous iterations, the *Apriori* algorithm reduces the set of candidate itemsets by pruning apriori those candidate itemsets that cannot be frequent. The pruning is based on the observation that if an itemset is frequent all its subsets could be frequent as well. Therefore, before entering the candidate-counting step, the algorithm discards every candidate itemset that has an infrequent subset [19].

```
1) L_1={Large 1-itemsets};
2) for (k=2; L_{k-1}≠∅; k++) do begin
3)   C_k=apriori-gen(L_{k-1}); // new candidates
4)   forall transactions t∈D do begin
5)     C_t=subset(C_k,t);   // candidates contained in t
6)     forall candidates c∈C_t do
7)       c.count++;
8)   end
9)   L_k={c∈C_k|c.count ≥ minsup}
10) end
11) Answer = ∪_k L_k;
```

Fig. 19.2: Apriori Algorithm

19.3.1.2 Game theory approach

Game theory [20] provides us with the mathematical tools to understand the possible strategies that utility-maximizing agents might use when making a choice. The simplest type of game considered in game theory is the *single-shot simultaneous-move* game. In this game, all agents must take one action simultaneously. Each agent receives a utility that is a function of the combined set of actions. This is a good model for the types of situations often faced by agents in a multi-agent system where the encounters mostly require coordination [21].

In the one-shot simultaneous-move game we say that each agent i chooses a strategy $s_i \in S_i$, where S_i is the set of all strategies for agent i. These strategies represent the actions the agent can take. When we say that i chooses strategy s_i we mean that it chooses to take action s_i. The *Nash equilibrium* in an n-player game is a set of strategies, $\sigma = \{\sigma^1, ..., \sigma^n\}$, such that, given that for all i, player i plays σ^i, no player j can get a higher payoff by playing a strategy other then σ^j. [22] It has been shown that every game has at least one Nash equilibrium, as long as mixed strategies are allowed. If the system is in equilibrium then no agent will be tempted to take a different action. A common classic example for equilibrium point is *two prisoners dilemma* (PD) in which there are two prisoners kept in separated cells. If both confess they would be sentenced three years in jail. If one confess and the other one dont (testify the one who confess is guilty) the first one goes 4 years in jail and another one would release. And if none of them confess they would be sentenced for a less important crime and will go one year in the jail each. The matrix in Fig. 19.3 shows that the best action a player can do is not to confess and the Nash equilibrium is (dont confess, dont confess).

	confess	Do not confess
Confess	3 - 3	0 – 4
Do not confess	4 - 0	1 – 1

Fig. 19.3: The two prisoners dilemma

Our proposed game model is to find best *MinCon* and *MinSup* for ARMAs so that the total system performance improves. To model the game lets assume in the real world a computer multi-conference is going to be held and our players aim is to publish at least a paper in the conference anyway possible. We also assume players are multi expert so they can publish paper in different fields of computer science and engineering. Also we know that in each field only a limited number of papers would be accepted. As a result each player tries to submit any paper in any field he can and with any quality. So if another player does not present a paper in a particular filed that the other one did, there would be a 100% chance of acceptance in contrast with the one who did not proposed. However, if both players submit papers since we do not have any information about the quality, there would be an equal 50% chance for each player. Intuitively the game equilibrium would be the case in which each player submits any paper and with any quality he can. (We assume different paper submission does not have negative effects such as time wasting and etc.) The scenario in our work is somehow the same for ARM agents which are in charge of gathering first level knowledge. They will not get any payoff if can not present any amount of knowledge. This is considered by choosing a suitable value for *MinCon* and *MinSup* in the run-time. Here we propose a lemma explaining the competitive knowledge presentation.

Lemma 19.1. *Competitive Knowledge Presentation: There exist equilibrium for knowledge presentation game in which players would select their knowledge with the highest possible confidence and support even if is less than common suggested 50% MinCon and MinSup.*

Proof. Let A be the payoff for mutual knowledge presentation, B the payoff for the presenter and D for not-presenter in case one presents and the other does not, and C the payoff when both do not present. In case A, both players receive a reward of 50% which is chance of presenting their work. In case B, the one who presented the knowledge would get 100% and the other player receives the payoff 0%; and in case C both players receive payoff 0%. Obviously if you think the other player will not present then you should present to give a payoff of 100% and in case he other guy also presents, you must present to have a 50% chance to win the game. No matter what option your opponent chooses, you should present your own work. According to equilibrium condition in PD, certain following conditions have to hold: $B \geq A \geq C \geq D$. Also an even chance of being exploited or doing the exploiting is not as good an outcome as both players mutually presenting. Therefore, the reward for A should be greater than the average of the payoff for the B and the D. That is, the following must hold $A \geq (D+B)/2$. And we see that our model yields these equations, therefore it has equilibrium, the same as PD. □

This mechanism assures that all ARMAs do their best to propose something as a first level knowledge. The ARMA game matrix in Fig. 4 shows agents payoff in winning percentage regarding their knowledge presentation.

	Present knowledge	Do not present knowledge
Present knowledge	50% - 50%	100% - 0%
Do not present knowledge	0% - 100%	0% - 0%

Fig. 19.4: proposed ARMA game matrix

One of the best examples that this approach would be successful is Diagnosis Expert Systems in which diagnostic test among different experts are different and maybe symptoms represent different disease with respect to conditions such as country, race, age and etc.

19.3.2 Inference Agent Behavior

The extracted association rules from ARMAs will be sent to IAs with their corresponding support and confidence. These *rules* plus the *fact* query sent by the BA will construct the IA knowledge base. By this time agents are able to inference results. Expert systems mostly use backward chaining; however, we decided to use forward chaining due to our system model which is goal-driven. The forward chaining starts by adding new fact P to the knowledge base and finds all conditional combinations having P in assumption.

19.3.3 Response Agent Behavior

The response agent finally obtains all results from IAs and then combines them using *Dempster-Shafer theory*. These results will be then proposed to the user in an ordered list according to their relevancy.

Dempster-Shafer theory of evidence: The Dempster-Shafer (DS) theory was first introduced by Dempster (1968) [23] and then extended by Shafer (1976) [24], but the kind of reasoning the theory uses can be found as far back as the seventeenth century. This theory is actually an extension to classic probabilistic uncertainty modeling. Whereas the Bayesian theory requires probabilities for each question of interest, belief functions allow us to base degrees of belief for one question on probabilities for a related question. These degrees of belief may or may not have the mathematical properties of probabilities; how much they differ from probabilities will depend on how closely the two questions are related. This theory has been used in information retrieval in [25], [26], [27].

This theory is a generalization to the Bayesian theory of probability. The most significant differences between DS theory and probability theory are the explicit representation of uncertainty and evidence combination mechanism in which made it more effective in document processing fields [25]. In information retrieval the uncertainty occurs in three cases:

- Existence of different evidences regarding relation of a document to a query
- Unknown number of evidences regarding relation of a document to a query
- Existence of incorrect evidences regarding relation of a document to a query

In the DS theory of evidence, *Belief* is a value to express certainty of a proposition. This belief is calculated with respect to a density function $m : \rho(U) \rightarrow [0,1]$, called *Basic Probability Assignment* (BPA), where m(A) represents *Partial Belief* amount of A. Note that $m(\phi)=0$ and $\sum_{A \in U} m(A) = 1$.

To measure the Total *Belief* amount of $A \subset U$ the *belief function* is defined as:

$$Bel(A) = \sum_{\forall B \subset A} m(B) \tag{19.1}$$

Shafer defined *doubt* amount in A as the belief in A', and the *plausibilty function* as the total belief amount in A

$$Dou(A)=Bel(A') \tag{19.2}$$

$$Pl(A)=1\text{-}Dou(A) = \sum_{B \subseteq U} m(B) - \sum_{B \subseteq A'} m(B) = \sum_{B \cap A = \phi} m(B) \tag{19.3}$$

Pl(A) is actually the high boundary of belief in A so that the correct belief in *A* is in the interval of [*Bel(A)*,*Pl(A)*]. Dempster's rule of combination is a generalization of Bayes' rule. This rule strongly emphasizes the agreement between multiple sources and ignores all the conflicting evidence through a normalization factor. Let m1 and m2 are the BPAs in a frame of discernment. The combination BPA is calculated in the following manner:

$$m(A) = m_1 \otimes m_2 = \frac{\sum\limits_{B \cap C = A} m_1(B)m_2(C)}{\sum\limits_{B \cap C \neq \phi} m_1(B)m_2(C)} \tag{19.4}$$

19.4 System Run Sample

In order to show the knowledge mining process in our system, we used a self-generated patient database, according to a disease symptoms website[3] . Since using diseases with completely separated symptoms are not desirable as a test case, we chosen those with common signs which are: *Cold, Flu, Bronchitis, Allergy, Asthma, Sinusitis, Strepthroat,* and *Gastroenteritis*. Some of the extracted association rules done by ARMAs are as bellow:

$\{runnynose\} \longrightarrow \{cold\} sup : 0.58, conf : 1.0$
$\{sorethroat\} \longrightarrow \{cold\} sup : 0.55, conf : 1.0$
$\{headache\} \longrightarrow \{cold\} sup : 0.62, conf : 1.0$
$\{cough\} \longrightarrow \{cold\} sup : 0.72, conf : 1.0$
$\{sorethroat fever\} \longrightarrow \{cold\} sup : 0.34, conf : 1.0$
...
$\{chills\} \longrightarrow \{flu\} sup : 0.62, conf : 1.0$
$\{runnynose\} \longrightarrow \{flu\} sup : 0.58, conf : 1.0$
$\{fever\} \longrightarrow \{flu\} sup : 0.51, conf : 1.0$
$\{chillsmuscleache\} \longrightarrow \{flu\} sup : 0.24, conf : 1.0$
$\{muscleachecough\} \longrightarrow \{flu\} sup : 0.34, conf : 1.0$
...
$\{itchy\} \longrightarrow \{allergy\} sup : 0.62, conf : 1.0$
$\{sneeze\} \longrightarrow \{allergy\} sup : 0.79, conf : 1.0$
$\{itchyrunnynose\} \longrightarrow \{allergy\} sup : 0.24, conf : 1.0$
$\{sneezeitchy\} \longrightarrow \{allergy\} sup : 0.44, conf : 1.0$
...
$\{wheezing\} \longrightarrow \{bronchitis\} sup : 0.72, conf : 1.0$
$\{breathshortness\} \longrightarrow \{bronchitis\} sup : 0.62, conf : 1.0$
$\{feverwheezing\} \longrightarrow \{bronchitis\} sup : 0.41, conf : 1.0$
...
$\{eyepain\} \longrightarrow \{sinusitis\} sup : 0.51, conf : 1.0$
$\{eyepain\} \longrightarrow \{sinusitis\} sup : 0.51, conf : 1.0$
$\{cough\} \longrightarrow \{sinusitis\} sup : 0.62, conf : 1.0$
$\{coughheadache\} \longrightarrow \{sinusitis\} sup : 0.44, conf : 1.0$
...

[3] http://familydoctor.org/

$\{headache\} \longrightarrow \{strepthroat\} sup: 0.79, conf: 1.0$
$\{sorethroat\} \longrightarrow \{strepthroat\} sup: 0.58, conf: 1.0$
$\{fever\} \longrightarrow \{strepthroat\} sup: 0.72, conf: 1.0$
$\{feverheadache\} \longrightarrow \{strepthroat\} sup: 0.55, conf: 1.0$
...
$\{breathshortness\} \longrightarrow \{asthma\} sup: 0.58, conf: 1.0$
$\{wheeze\} \longrightarrow \{asthma\} sup: 0.79, conf: 1.0$
$\{cough\} \longrightarrow \{asthma\} sup: 0.62, conf: 1.0$
$\{wheezecough\} \longrightarrow \{asthma\} sup: 0.44, conf: 1.0$
$\{wheezebreathshortness\} \longrightarrow \{asthma\} sup: 0.44, conf: 1.0$
...

The produced knowledge base by the IA is then:
IF has X itchy THEN has X allergy(S:0.62,C:1.0)
IF has X sneeze THEN has X allergy(S:0.79,C:1.0)
IF has X itchy has X sneeze THEN has X allergy(S:0.44,C:0.72)
IF has X wheezing THEN has X bronchitis(S:0.72,C:1.0)
IF has X breathshortness THEN has X bronchitis(S:0.62,C:1.0)
...

Now the query containing facts of *"headache"* and *"fever"* is added to the KB:
has patient headache & has patient fever

In this sample the two IAs use forward chaining and send the two results to the RA shown below:

IA#1:
has patient gastroenteritis(s:0.51,c:1.0)
has patient strepthroat(s:0.72,c:1.0)
has patient strepthroat(s:0.55,c:1.0)
has patient cold(s:0.20,c:1.0)

IA#2:
has patient bronchitis(S:0.51,C:1.0)
has patient flu(S:0.51,C:1.0)
has patient sinusitis(S:0.79,C:1.0)
has patient gastroenteritis(s:0.62,c:1.0)
has patient gastroenteritis(s:0.20,c:1.0)
has patient strepthroat(s:0.79,c:1.0)
has patient cold(s:0.62,c:1.0)
has patient cold(s:0.51,c:1.0)

Finally and in the last phase, RA implies D-S combination method, shown in Table 19.1, and returns the final result to the user. After computing the D-S combination table, the final probable amount would be: m(Gastroenteritis)=0.001318070061 , m(Cold)= 0.001318070061, m(Strepthroat)= 0.001314439014, m(Bronchitis)= 0, m(Flu)= 0, and m(Sinusitis)= 0. Therefore the most probable diagnosis through this process would be *Gastroenteritis*, *Cold*, *Strepthroat,Bronchitis*, *Flu*, and *Sinusitis* respectively.

Table 19.1: D-S combination table

	Bronchitis: 0.0181	Flu: 0.0181	Sinusitis: 0.0181	Gastroenterit: 0.0363	Strepthroat: 0.0181	Cold: 0.0363
Bronchitis: 0	0	0	0	0	0	0
Flu: 0	0	0	0	0	0	0
Sinusitis: 0	0	0	0	0	0	0
Gastroenterit: 0.0357	0.00064617	0.00064617	0.00064617	0.00129591	0.00064617	0.00129591
Strepthroat: 0.0714	0.00129234	0.00129234	0.00129234	0.00259182	0.00129234	0.00259182
Cold: 0.0357	0.00064617	0.00064617	0.00064617	0.00129591	0.00064617	0.00129591

19.5 Conclusion and Future Work

In this work, the KDD process and data mining methods was investigated and then a new multi-agent architecture for grid environment have been proposed. The most innovative techniques in our design of the system include game-theory modeling for competitive knowledge extraction, hierarchical knowledge mining, and Dempster-Shafer result combination. A distributed diseases diagnosis expert system was designed and implemented upon the proposed method in which results shows its performance in knowledge gathering and inferencing.

Regarding project novelty and open research areas in the field, there are surely good potentials to complete hierarchical knowledge mining process including usage of approximation algorithms in knowledge extraction, modeling the process as a mixed strategic games and finding Nash equilibrium, and developing an infrastructure for a high performance grid-based search engine

Acknowledgements Author would like to thank contribution of Prof. Mahmoud Naghibzadeh, Hossein Kamyar, and Saied Moghaddaszadeh from Ferdowsi University of Mashhad, Iran, for their modest comments and help in part of system implementation.

References

1. Frawley, W., Piatetsky-Shapiro G., Matheus C.: Knowledge Discovery in Databases: An Overview, pp: 213-228. AI Magazine (1992)
2. Milani Fard A., Kamyar H., Naghibzadeh M. :Multi-expert Disease Diagnosis System over Symptom Data Grids on the Internet: World Applied Sciences Journal, ISSN 1818-4952, Volume 3 Number 2, (2008)
3. Shortliffe, E. H.: MYCIN: A rule-based computer program for advising physicians regarding antimicrobial therapy selection, Ph.D. dissertation, Stanford University (1974)
4. Russell, S.J., Norvig, P.: Artificial Intelligence: A modern Approach, Prentice Hall (1995)

5. Waterhouse, S.R.: Classification and regression using mixtures of experts, Ph.D. dissertation, Department of Engineering, Cambridge University (1997)
6. Si, L., Callan, J.: A Semisupervised Learning Method to Merge Search Engine Results. ACM Transactions on Information Systems, Vol. 21, No. 4, 457-491, (2003)
7. Callan, J.: Distributed information retrieval. In: Advances in Information Retrieval, W. B. Croft, Ed., pp. 127150, Kluwer Academic Publishers (2000)
8. Pasi, G., Yager, R. R.: Document retrieval from multiple sources of information. In: Uncertainty in Intelligent and Information Systems, Bouchon-Meunier, B., Yager, R. R., Zadeh, L. A. (eds.), pp. 250-261, World Scientific, Singapore (2000)
9. Baeza-Yates, R., Ribeiro-Neto B.: Modern Information Retrieval. Addison-Wesley, (1999)
10. Bellifemine, F., Caire, G., Trucco, T., Rimassa, G.: JADE Programmers Guide. (2006)
11. JADE Board: JADE WSIG Add-On Guide JADE Web Services Integration Gateway (WSIG) Guide. 03 March, (2005)
12. Cortese, E., Quarta, F., Vitaglione, G., Vrba, P.: Scalability and Performance of the JADE Message Transport System. (2002)
13. Liu, S., Kngas, P., Matskin, M.: Agent-Based Web Service Composition with JADE and JXTA. In: SWWS 2006, Las Vegas, USA, June 26-29, (2006)
14. Gradecki, J. D.: Mastering JXTA: Building Java Peer-to-Peer Applications JohnWiley (2002)
15. Milani Fard, A., Kahani, M., Ghaemi, R., Tabatabaee, H.: Multi-agent Data Fusion Architecture for Intelligent Web Information Retrieval. In: International Journal of Intelligent Technology Volume 2 Number 3 (2007)
16. Mohebbi, M., Akbarzadeh T., M. R. , Milnai Fard, A.: Microorganism DNA PatternSearch in a Multi-agent Genomic Engine Framework. In World Applied Science Journal, Volume 2 Number 5, Sep-Oct (2007)
17. Tabatabaee, H., Milani Fard, A., Akbarzadeh T., M. R.: Cooperative Criminal Face Recognition in Distributed Web Environment. In: The 6th ACS/IEEE International Conference on Computer Systems and Applications (AICCSA-08), Doha, Qatar, March 31 - April 4 (2008)
18. Agrawal, R., Imielinski, T. , Swami, A. N.: Mining Association Rules between Sets of Items in Large Databases. In: SIGMOD, 22(2):207-16, June (1993)
19. Kantardzic, M.: Data Mining: Concepts, Methods, and Algorithms. JohnWiley (2003)
20. Von Neumann, J., Morgenstern, O.: Theory of Games and Economic Behavior. Princeton University Press (1944)
21. Vidal, J. M.: Learning in Multiagent Systems: An Introduction from a Game-Theoretic Perspective. In: Alonso, E. (eds.) Adaptive Agents. LNAI 2636. Springer Verlag (2003)
22. Mor, Y., Goldman, C. V., Rosenschein, J. S.: Learn your Opponent's strategy (in Polynomial Time)!. In: Adaptation and Learning in Multi-Agent Systems, IJCAI95 Workshop, Montreal, Canada, August 1995, In: Proceedings. Lecture Notes in Artificial Intelligence Vol. 1042, G. Weiss and S. Sen (eds.) Springer Verlag, (1996)
23. Dempster, A.: A generalization ofba yesian inference. In: Journal of the Royal Statistical Society, 30:205-247 (1968)
24. Schafer, G.: A mathematical theory of evidence., Princetown University Press (1976)
25. Verikas, A., Lipnickas, A. ,Malmqvist, K., Bacauskiene M., Gelzinis, A.: Soft combination of neural classifiers: A comparative study In: Pattern Recognition Letters, 20:429–444 (1999)
26. Krogh, A., Vedelsby, J.: Neural network ensembles, cross validation and active learning. In: G. Tesauro, D.S. Touretzky, T. K., Leen, (eds.) Advances in Neural Information Processing Systems, volume 7, pp. 231-238. MIT Press, Cambridge, MA (1995)
27. Kuncheva, L. I.: Combining Pattern Classifiers: Methods and Algorithms. John Wiley (2004)

Chapter 20
Normative Multi-Agent Enriched Data Mining to Support E-Citizens

Stanislaw A. B. Stane and Mariusz Zytniewsk

Abstract The current trends in the development of multi-agent systems indicate the possibility to apply the concept of multi-agent systems employing ontologies for encoding the systems domain knowledge and procedural knowledge. Within such structures, the use of knowledge discovery models may represent an enhancement to the systems functionality in the context of discovering relations that can support the users activities. Distributed data mining (DDM) concepts ([19, 11, 18]) demonstrate that multi-agent systems are capable of using knowledge discovery processes in a variety of ways in the context of the process being supported by the agent as well as in order to expand its knowledge. If this is the case, the knowledge discovery becomes an intrinsic component of the agents learning process. The application of norms to support the work of agents, associated with the idea of normative multi-agent systems ([30, 2, 5, 4]), may significantly boost the performance of such systems by directing the agents actions and determining the desirable states of the agent itself as well as of the group it is part of. The chapter aims to discuss the key aspects of the development of multi-agent systems and knowledge discovery systems, and to present a proposal for an architecture of multi-agent systems supported by knowledge discovery systems.

20.1 Introduction

It could be argued that we currently see the emergence of the third generation of systems oriented on data mining processes. The first phase in their evolution was concentrated on the development of algorithms to support the data mining process, and the mainstream of research aimed to identify the possible types of analysis to be implemented under different systems. To a large extent, efforts were directed on the creation and analysis of data mining algorithms. The second stage of the evolution

Karol Adamiecki University of Economics in Katowice, POLAND
e-mail: {stanek, zyto}@ae.katowice.pl

L. Cao (ed.), Data Mining and Multi-agent Integration, DOI: 10.1007/978-1-4419-0522-2_20,

focused on defining development methodologies for knowledge discovery processes that would extend the data mining process itself (such that involves the use of a kind of algorithm) to incorporate steps involving the preparation and evaluation of the data and the outcomes. Examples of such solutions include SEMMA and CRISP-DM [16, 6]. Thus defined, the second generation was related to the emergence of the concept of knowledge discovery systems that are expected to integrate diverse data mining models and support the pre-processing of data to be analyzed.

The current phase in the evolution of data mining does not simply seek to apply this or other kind support to the knowledge discovery process, but it also endeavors to provide support to business organizations, whereby the data mining process itself becomes an element of a broader business process. Within this approach – from the end user's perspective – data mining constitutes at once a tool and a process and becomes just one of the stages in the operation of the organizational information system. This being the case, the end user is not equipped with the knowledge of a systems analyst and, more often than not, is oriented on the outcome to be generated rather than on the knowledge discovery process. Yet, to be able to employ such an approach to the application of data mining within organizations, we need to have mechanisms to store and propagate data mining models across the system. The latest technology solutions make it possible to define data mining models as part of the organizational global metadata set, so that their structure and their outcomes can be accessed by different systems. The multi-agent system architecture using data mining to support customer service, discussed further in the chapter, represents an example of such a solution. Software agents, which can at the same time handle the knowledge discovery process and the delivery of pre-processed information to users, are thus capable of mediating between the end user and the knowledge discovery process.

When analyzing the current trends in the development of organizational knowledge discovery systems through the prism of multi-agent applications, two main streams of research relating to the Distributed Data Mining (DDM) concept can be identified ([19, 11, 18]). The former sees agent technology as an instance of intelligent systems designed to replace or assist humans in using information systems such as, in particular, ERP and Business Intelligence solutions. Within this class of systems, software agents may be components of an intelligent system infrastructure which is able to carry out routine activities and thus support the functioning work of the system as a whole ([24, 15, 17]). In this case, software agents can employ data mining in pursuing their objectives, such as e.g. performance monitoring or generating recommendations. Besides, software agents may by used as part of a data mining system, where software agents equipped with expert knowledge can control the knowledge discovery process by analyzing the output generated by the system. In this case, the software agent can be considered as a component of the data mining model. A deductive problem solving process within a software agent's knowledge system can be enhanced by the addition of an inductive knowledge discovery process. This makes it possible to build hybrid multi-agent systems that are capable of supporting various experts. Typical examples of such solutions are Bodhi, Padma, JAM, Papyrus [18] or DAMSA (DAta Mining System based multi-

Agent) [1]. Within architectures of this kind, software agents can assist in selecting data for analysis, choosing the data mining method, validating the results, and optimizing the data mining process.

The authors of this chapter intend to present a proposed multi-agent system architecture incorporating normative solutions as well as knowledge discovery processes, designed to support customer service at a local government office. The discussion will be structured as follows. The first sub-chapter describes sample multi-agent systems supporting knowledge discovery processes, along with the underlying concepts, pointing out the directions in which such systems currently evolve. The second sub-chapter outlines the theoretical foundations for developing normative solutions with a view to building normative multi-agent systems. The third sub-chapter discusses examples of architectures and implementations of systems that make use of norms to support system operation. The fourth sub-chapter puts forth a proposal for a multi-agent system architecture including knowledge discovery systems and supporting customer service at a local government office.

20.2 Theoretical issues – A Multi-Agent System vs. Norms

From the viewpoint of both data mining and software agent technology, a multi-agent system can be treated as a data source for the data mining process, particularly in the construction of simulation models, where software agents can generate a large amount of output. In such systems, the social behaviors of software agents can be analyzed with regard to interaction and collaboration as well as with respect to the dynamic structures that emerge during the simulation. Such solutions may be deployed in economics (Agent-Based Computational Economics) [1], [26] sociology [23], anthropology [2], etc. On the other hand, results generated by data mining systems may be part of a software agent's knowledge. In this case, the process of knowledge extraction can extend the software agent's learning process [27], [22], providing an opportunity to build self learning systems with adaptive capabilities.

Currently, business oriented knowledge discovery systems are closely integrated with business solutions such as SAS Data Miner 5.3, in which the management of data mining models is performed via metadata servers. The approach supports the use of multi-agent systems and offers new prospects for systems integration and management. The autonomy of software agents, being an essential factor in a system's operation, entails the need for the construction of mechanisms to support decision making by agents. Normative systems, which provide for the integrity and soundness of software agents' actions by allowing a possibility to pre-define these, could be seen as a viable response to this requirement. In society, norms will support the processes of task performance by members of the community through establishing standards for individual behaviors. In executing a specific task and interacting with the environment, we can rely on the accepted norms, rules and standards to help us predict what actions can be taken by other individuals as a result of our own behavior. Norms can be broken down into formal ones, which have been written down as laws and regulations, and informal ones, which do not have binding legal force but, nevertheless, hold within a specific society or social group. Social norms can also be divided into legal, religious, traditional, cultural, ethical, etc. However, in terms of their application in software agent theory, a broad distinction of norms is not required. [30] identifies three types of norms that can affect software agents:

1. norms that define the **meaning** of words, expressions, names, etc.
2. norms that define **actions** as well as their constituent activities, such as plans, rules, procedures, etc.
3. norms that define **obligations, permissions** and **restrictions**.

Another distinction which can be used in agent-based systems has been introduced in [2] and [5]:

1. **regulative norms**: obligations, permissions, prohibitions,
2. **constitutive norms**: counts-as conditionals in which are manifest the general norms defining relationships among the components of the system,
3. **procedural norms** defining procedures for the performance of specific tasks, e.g. the rules of administrative proceedings.

Within multi-agent systems, according to [3] and [29], norms are not addressed to any specific agents but to the roles they play. In case problems arise that are attributable to some agents' overlapping objectives, the relevant objectives can be altered, depending on their priorities. Normative reasoning is founded on the fact that [3] a norm is not identifiable with an objective or an obligation but it forms a set of criteria that allow an agent to choose a valid objective or adopt a suitable attitude. Besides norms that are specific to each autonomous agent and can be modified by the agent itself, there ought to exist system-wide norms that hold for all entities throughout the system. Within systems where autonomy represents a determining factor for an agent's behaviors, the application of knowledge discovery systems may enhance the process of learning and contribute to defining new norms. Prolonged examination of the behaviors of groups of agents may lead to discovering important relationships and optimizing the agents' behaviors by generating further internal norms.

In decentralized systems (i.e. those with networked, or distributed, architectures), where agents are fully autonomous, some authors [8] will employ the notion of social order, which is achieved through social control. To ensure control mechanisms in such environments [8], each agent in the system is assigned an additional role, which is to supervise another agent. This makes it possible to monitor and analyze its behavior. While a similar degree of control over the agents' behaviors can be exercised by introducing a superior agent, this latter approach will increase the number of software agents, which immediately poses issues relating to structural complexity. [13] indicates two paths for the implementation of norms into an agent system, where:

1. The former defines sets of norms as specifications of correct behaviors, which are implemented directly in a software agent's code. In this case, it is the software

engineer that decides what norms the agent will abide by. The writer of [8], however, insists that norms should not be implemented in an agent's code within systems of this sort. This is due to likely heterogeneity issues as well as to the fact that, within such an environment, agents should be able to analyze and modify their own norms.
2. The latter path is based on an assumption that norms are no more than definitions of correct behaviors which may be accepted by a software agent or not. In this case, it is the agent that decides what norms will be approved and which norms will be breached. Norms can thus become an element of the system's knowledge.

[21] argues that the approach involving constraints on actions relates to the notion of social laws which impact on the agent by determining the behaviors that are appropriate at a particular moment or in specific circumstances. In their present form, the applications of norms within multi-agent systems described in topical literature do not show a sufficient degree of formalization and do not provide convenient tools for the development of such systems [13]. In the case of software agent norms will perform functions that support its interactions with other agents, with humans, and with resources. As far as social order is concerned, norms can be perceived in terms of interaction between the normative system and its environment [4]. According to [4] and [2], a normative multi-agent system can be conceived of as a combination of a multi-agent system and a normative system where, on the one hand, agents can make decisions concerning the norms built in the system and their observance while, on the other hand, the normative system specifies how the norms can be altered by agents. What is remarkable about the approach under consideration is that a normative system may be perceived by an agent as a normative agent. If this is the case, a normative agent is equipped with mechanisms to enabling it to analyze norms. The concepts discussed in this section signify a possibility to build systems enriched with norms and supported by data mining processes. The following sub-chapter will present examples of normative multi-agent systems along with recommended system architectures.

20.3 A Normative Multi-Agent Enriched Data mining Architecture and Ontology Frameworks

[10] claims that ...*artificial societies are typically characterized by agents that interact with each other in accordance with common rules or norms. Similarly to a human society, members of the artificial society must be allowed to coexist in a shared environment and to follow their respective goals in the presence of others. Here, the application of norms serves an important purpose in that they govern the rules of participation and provide important measures to achieve the desired behavior in a society.*

These concepts are founded on the use of trust deriving from the observance of applicable norms. In subject literature, trust is typically defined in the context of specific goals, where agents resident within the system can form small groups and temporarily cooperate. The recent trends in the design of hybrid multi-agent systems are headed toward the development of **heterogeneous, open** and **dynamic** systems. This is illustrated in Fig. 20.1. From the perspective of a software agent, norms

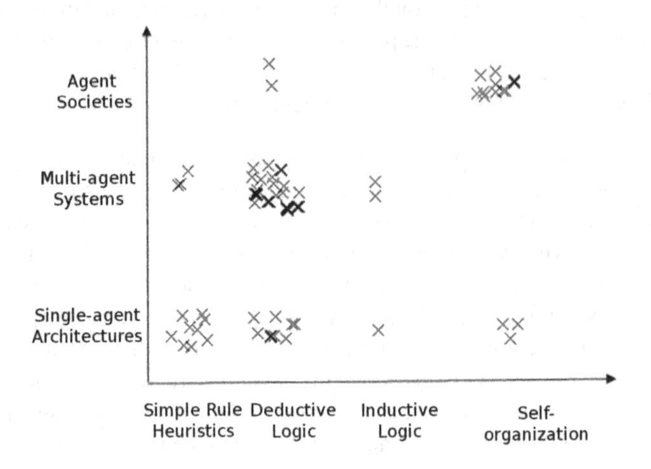

Fig. 20.1: A review of approaches to the construction of the reasoning component. Source: [24]

may determine the principles of its interaction with the user, while in a multi-agent system they will usually regulate an agent's operations within the system, and in artificial societies they will affect the rules of participation and thus support activities involving trust [12, 14] and reputation [12, 9]. Some authors view agent organizations as a space in which agent institutions can be defined. Agent organizations are then regarded as groups of agents that have been united for a cause. Such groups make up [12] social entities, have a definite structure as well as its own resources and authorities, and are created for the pursuit of emergent objectives.

Another term designating solutions of this kind, encountered in some literature, is "electronic institutions" (EI) [12]. Electronic agent institutions may be found functioning within an organization and determining social conventions [7] which are to be observed within this or other system. Social conventions adopted within a system will affect its general objectives that are translated into specific agents' goals. A system of this kind is exemplified by road traffic regulations which define the behaviors of drivers. Observance of the norms can streamline car traffic and help avoid accidents. Unless these general objectives are identified with agents' direct goals (an agent's goal may be e.g. to travel from *a* to *b*), they will support the performance of agents' goals by determining the rules for communication within the system. Software agents operating within open societies may cause a number of threats relating to the use of their knowledge and to the autonomy of their behaviors. Trust – specifically defined in the context of software agents and their interactions and communications — is therefore a concept that can largely contribute to reducing such threats in the construction of agent-based organizational support systems. In [12], the following two main types of trust are distinguished:

1. personal – which is subjective and relies on a person's beliefs, observations, reasoning, stereotypes, communication, experience; hence, this perception of trust is centered on the human;
2. impersonal – which arises from information and experience acquired from other persons, or from third-party subjects.

[12] points to three moments when agents' behaviors can be audited: **prior to task execution, in real-time, and on task completion**. In the first case, trust can decrease due to the agent's insufficient knowledge on the task being performed or on the user's intent. In the second instance, trust is built through constant monitoring of the agent's progress and performance. In the third variant, trust depends on a post factum validation of the effects of the agent's actions. The concepts being presented concerning the use of norms as an element defining an agent's role in the system and determining its behavior are closely linked to the notion of a system's knowledge which, given an adequate structure and extent, would inform an agent's actions.

The lack of standards for the model of agents' knowledge within a system employing norms to define the agents' desirable states has encouraged the authors of this chapter to propose their own model of agent' knowledge, both in terms of norms and domain knowledge, being an example of a normative knowledge meta-model in which knowledge on norms constitutes an element of domain knowledge representation. The considerations and recommendations on the design and development of multi-agent systems, knowledge discovery and the use of norms have all been incorporated to support the user's activities related to customer service at a local government office. Since it was the designers' ambition to address not just isolated problems handled by the office, but to provide citizens with the widest possible support, the application of a multi-agent solution was contemplated. It was particularly justified considering the need to adopt a comprehensive approach to problem solving and to stimulate the acquisition of experience by persons involved. The sample fragment of the multi-agent system architecture, otlined further in the chapter, emphasizes the idea of the user's involvement in defining the system's knowledge and assumes the user's participation in strengthening the multi-agent system's learning process. As a follow-up for research done, and taking account of all the considerations discussed above, the proposed multi-agent system's knowledge representation was created. Ongoing research efforts are aimed at using the OWL language to represent the knowledge of a multi-agent system being an extension of the functionality of the system presented in this chapter. Within the structure of the agents' knowledge 20 main classes of objects have been specified, including Address, Working Hours, Document, Worker, Citizen, Service, Coordinator, Contact, Payment, Legal Basis, Processing Time, Category, etc. [28].

The specification of the normative knowledge meta-model proposed in this chapter makes it possible for software agents to access knowledge on norms, rules and restrictions defined on the basis of the system's domain knowledge. To ensure that the meta-model specification is easily expandable, simple to interpret, and applicable across a range of information systems including multi-agent systems, an OWL ontology was used to build a normative meta-model of knowledge which was then employed within a hybrid multi-agent system. As a key element of the research project, it was assumed that regulatory norms would be used to determine what behaviors of agents are desirable in terms of system performance.

The proposed fragmentary structure of the normative knowledge meta-model defines the following classes of objects: Norm, Effect, Agent, Role, Recipient, Location, Restriction, Beliefs, Goals and Plans. The system knowledge models mentioned above have been used to specify the knowledge of the system described in the following sub-chapter.

20.4 Case Study – A Multi-Agent System Architecture

The introduction of electronic media to support public administration entails the use of tailor-made solutions offering intelligent guidance to at least match the level of competence that can be expected of a public officer. This challenge can be faced by systems equipped with interface agents and backed up by multi-agent approaches. A number of system modules have been defined within the multi-layer architecture of our multi-agent system to support the work of an office of public administration. The specific components (modules) of the proposed architecture, illustrated in Fig. 20.2, are described in subsequent paragraphs A through D.

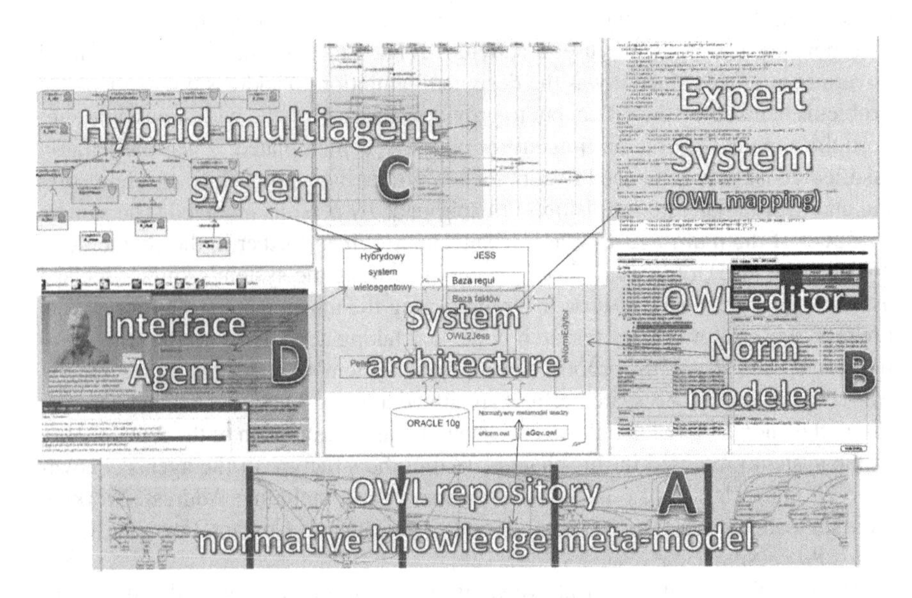

Fig. 20.2: Components of the hybrid support system for customer service at a local government office. Source: own

A. The knowledge representation and processing module

The apparent lack of insightful experiences in implementing the concepts discussed earlier in this chapter, i.e. using norms and domain knowledge to build hybrid multi-agent systems, has made it necessary to elaborate the theoretical foundations and develop a prototype solution for the management of the system's knowledge. In the proposed models, knowledge can be stored locally in OWL files or in an ORACLE 10g data base. A central element which served as the platform for the system was JENA – a semantic Web framework for Java[1]. To ensure consistency of the knowledge with the adopted model, JENA and Pellet[2] validation tools were used, both capable of reasoning based on an ontology provided.

The Jess[3] expert system was utilized to define rules, from which norms were subsequently derived. Although Jess cannot directly use knowledge saved in OWL files, a mechanism based on XSD templates was put in place to transform the knowledge into a fact base, which made inference possible and, as a result, permitted defining the rules.

B. The system knowledge definition module

Following is a list of key features required of the prototype normative knowledge meta-model editor. Namely, the module should:

1. allow storage, retrieval and validation of a model of norms and knowledge in the Oracle 10g database system in the form of OWL DL, so that the knowledge could be used by a multi-agent system;
2. allow storage, retrieval and validation of norms kept locally in the OWL DL format;
3. permit a graphical preview of the system's normative knowledge meta-model structure;
4. provide a possibility to create, edit and save norms in a normative knowledge meta-model, in keeping with the considerations discussed earlier in this paper;
5. make it possible to interactively define rules and restrictions being part of a given norm, using the knowledge stored within the normative knowledge meta-model;
6. permit operations on the system's knowledge using the SPARQL language;
7. allow a graphical preview of selected instances of any specific system norm;
8. provide a transformation mechanism to convert the system's knowledge represented in the OWL language into a fact base of the expert system;
9. enable inference using a fact base with rules and restrictions defined in an ontology.

[1] http://jena.sourceforge.net

[2] http://pellet.owldl.com

[3] http://herzberg.ca.sandia.gov/

C. The multi-agent system module

The first role defined in the system was that of **Agent-Protector** (agent A_obr), which was required as the entity providing access to the agent system and safeguarding observance of the norms governing the system. It is responsible for making, keeping and terminating connections with the interface agent, initiating the creation of an agent to perform the role of Agent-Representative, and supervising that agent's behaviors. Another role which was isolated was that of **Agent-Representative** (agent A_rep1). It is a hybrid agent implementation which is supported by a normative system in pursuing its goals. The role of **Agent-Legislator** (agent A_leg) is, in line with what is demanded of a normative system, supposed to analyze and update knowledge concerning norms. This sort of knowledge, defined in the system owing to a normative knowledge meta-model, is also used by an agent performing the role of **Agent-Normative** (agent A_nor), which handles queries from agents playing the role of Agent-Representative and responds to these in the context of their current tasks and goals. Other roles include **Agent-Manager** (agent A_two), which creates hybrid Agents-Representative and **Agent-Supplier** (agent A_dos), which acts as an intermediary between the interface agent and an agent performing the role of Agent-Representative. In addition, there are the roles: Agent-Chat (agent A_chat), Agent-Visualize (agent A_wiz), Agent-Speak (agent A_mow), which are responsible for automated communication with the user via different media, respectively, as well as a group of agents which handle access to external resources, e.g. a database access agent, a knowledge discovery agent, a met-data agent, etc.

For the sake of analysis of the relationships between the agents' behaviors, their knowledge, and the user support processes, the proposed system was integrated with the SAS system, specifically with the Enterprise Miner module that provides support for electronic document processing. The agents mentioned above will control the different system modules. Based on the considerations shaping agent roles within the system, elements of the architecture of the prototype hybrid multi-agent system were defined. The next step was to specify the hierarchical relationships resulting from dependencies among the types of roles which specific instances of agents perform within the system. For example a role of Representative is central, as this is the entity which controls the process of communication with the customer.

The overal structure of the system is visualized in Fig. 20.3.

D. The interface agent module

The module is supposed to separate the presentation functions of the hybrid system and to standardize the user interface so that it takes on a uniform appearance and reveals similar operating properties regardless of which of the diverse information systems handles the customer's business. The diversity of solutions applied to develop public information portals is the primary difficulty in the case of public administration offices. Problems crop up as users are forced to use several platforms at the same time, or where a system does not have an electronic interface customized

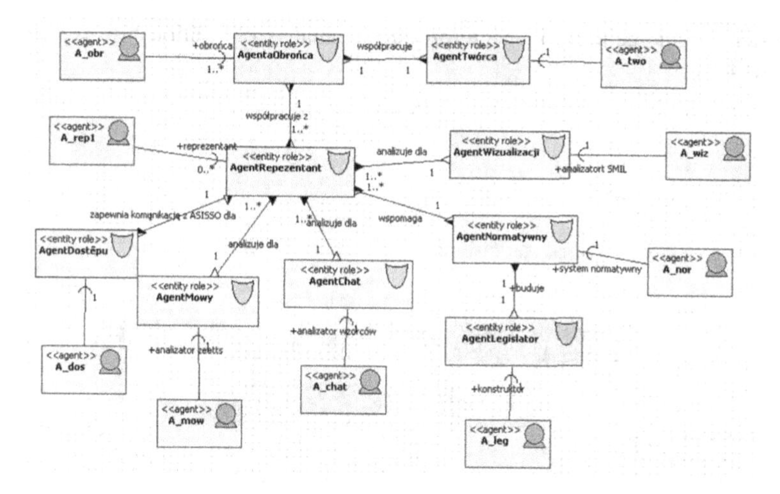

Fig. 20.3: Representation of the quantitative relationships and the types of agents operating in the system. Source: own

to cater for citizens. Within the system presented, the ASISSO agent is just an example of a "light-weight" agent which, if necessary, can be steered and controlled by a multi-agent system. The key system modules include: the advisory module which handles presentation of knowledge (base) on procedures and on frequently asked questions (viz. visualization, speech synthesis, etc.), the document processing module, the chat module, the e-mail module, the about-the-office module, the multimedia presentation module, and the update module.

The following example aims to highlight the application of a knowledge discovery module in the context of support for the process of document filling by users. To run the experiment we had prepared 135 different documents and 15 Agents-Representative. The objective was to discover relevant relationships between documents, and then work out and deliver clues for user. A simplified cycle was as follows:

1. A document selected by the user.
2. The parameters of the selected document sent to Agent-Representative along with the names of all documents available to the user.
3. A query transmitted to the multi-agent system.
4. A local matrix of relationships between fields in the user's documents created by each Agent Representative.
5. Relationship matrices collected by the Agent Representative querying the system; output from mistrusted agents rejected, all remaining matrices processed to build up a global relationship matrix.
6. Clues derived from the global matrix and presented by the interface agent.
7. One of the relationships strengthened by the user by choosing the most relevant clue.

A sample clue generated by interpreting the matrix of relationships is shown in Fig. 20.4.

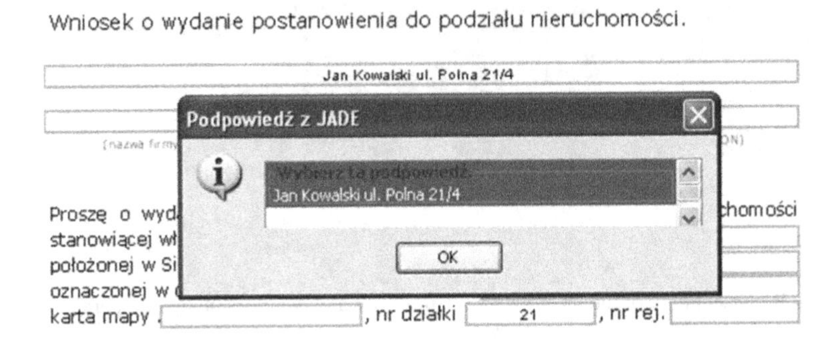

Fig. 20.4: A sample clue generated by system. Source: own

20.5 Concluding Remarks

The concepts involving the use of the OWL to represent a multi-agent system's knowledge are the starting point the first step in exploring the potential of hybrid multi-agent systems coupled with norms defining agents' desirable actions in supporting customer service at local government offices. The general structure of the knowledge model developed within this project, which has been discussed in greater detail and illustrated with a figure earlier in this paper, allows to:

1. support local citizens by supplying them with up-to-date information on what kind of services are available from a particular office;
2. support hybrid multi-agent systems by enforcing a standardized structure for knowledge storage, which facilitates interpretation of the knowledge by agents, enabling the agents to make use of it in pursuing their goals;
3. support officers by providing them with a possibility to store information on the cases processed independently and outside of the office's individual computer systems and create relationships between the stages of processing at the organizational level, i.e. in the context of the overall process of case processing.

Among others, this paper delivers findings and outcomes of research on the application of agent technologies to support citizens in filling official documents. A hybrid agent system was designed to assist the customers of a public office and to guide the process of discovering knowledge in analyzing the data fed by the customers. Once in place, the system was able to form recommendations concerning the creation of new official documents.

It is the author's belief that, in general, the paper has been successful in demonstrating that there might be a wide scope for integrating knowledge discovery methods, normative solutions and multi-agent systems.

References

1. Baazaoui Zghal, H., Faiz, S., Ben Ghezala, H.: A Framework for Data Mining Based Multi-Agent: An Application to Spatial Data. In: Proceedings of World Academy of Science, Engineering and Technology, Vol. 5 (2005)
2. Boella, G., van der Torre, L.W.N.: A Game-Theoretic Approach to Normative Multi-Agent Systems. In: Boella, G., van der Torre, L.W.N., Verhagen, H. (eds.) Normative Multi-agent Systems, 2007. Dagstuhl Seminar Proceedings 07122. Internationales Begegnungs- und Forschungszentrum fr Informatik (IBFI), Schloss Dagstuhl, Germany (2007)
3. Boella, G., van der Torre, L.W.N.: The ontological properties of social roles in multi-agent systems: Definitional dependence, powers and roles playing roles. Artificial Intelligence and Law 15, 201-221 (2007)
4. Boella, G., van der Torre, L.W.N., Verhagen, H.: Introduction to normative multiagent systems. Computational & Mathematical Organization Theory 12, 71-79 (2006)
5. Caire, P.: A normative multi-agent systems approach to the use of conviviality for digital cities. In Boella, G., van der Torre, L.W.N., Verhagen, H. (eds.) Normative Multi-agent Systems. Dagstuhl Seminar Proceedings 07122. Internationales Begegnungs- und Forschungszentrum fuer Informatik (IBFI), Germany (2007)
6. Chapman, P., Clinton, J., Kerber, R., Khabaza, T., Reinartz, T., Shearer, C., Wirth, R.: Cross Industry Standard Process for Data Mining (CRISP-DM) 1.0. Step-by-step data mining guide. SPSS Technical Report (2000). Available from http://www.crisp-dm.org. Cited 15 December 2008
7. Campos, J., Lpez-Snchez, M., Rodriguez-Aguilar, J.A., Esteva, M.: Formalising Situatedness and Adaptation in Electronic Institutions. In: Proceedings of The Fifth Workshop on Coordination, Organizations, Institutions, and Norms in Agent Systems COIN@AAMAS 2008, pp.103-117 (2008)
8. Castelfranchi, C.: Formalising the informal? Dynamic social order, bottom-up social control, and spontaneous normative relations. Journal of Applied Logic 1(1-2), 47-92 (2003)
9. da Silva, V.T., Hermoso R., Centeno, R.: A Hybrid Reputation Model Based on the Use of Organizations. In: Proceedings of The Fifth Workshop on Coordination, Organizations, Institutions, and Norms in Agent Systems COIN@AAMAS 2008 (2008)
10. Davidsson, P., Jacobsson, A.: Aligning Models of Normative Systems and Artificial Societies: Towards Norm-Governed Behavior in Virtual Enterprises, Dagstuhl Seminar Proceedings, `http://drops.dagstuhl.de/opus/volltexte/2007/908Cited1January2007`
11. Di Fatta, G., Fortino, G.: A Customizable Multi-Agent System for Distributed Data Mining. In: Cho, Y., Wainwright, R.L., Haddad, H., Shin, S.Y., Koo, Y.W. (eds.) Proceedings of the 2007 ACM Symposium on Applied Computing (SAC), pp. 42-47. ACM (2007)
12. Fasli, M.: Agent Technology for e-Commerce. John Wiley & Sons (2007)
13. Grizard, A., Vercouter, L., Stratulat, T., Muller, G.: A Peer-to-Peer Normative System to Achieve Social Order. In: Noriega, P., Vazquez-Salceda, J., Boella, G., Boissier, O., Dignum, V., Fornara, N., Matson, E. (eds.) Coordination, Organization, Institutions, and Norms in Agent Systems II, LNCS/LNAI 4386, Springer (2007)

14. Harbers, M., Verbrugge, R., Sierra, C., Debenham, J.: The Examination of an Information-Based Approach to Trust. Springer, Berlin Heidelberg (2008)
15. Kehagias, D., Symeonidis, A.L., Chatzidimitriou, K.C., Mitkas. P.A.: Information Agents Cooperating with Heterogeneous Data Sources for Customer-Order Management. In: Haddad, H., Omicini, A., Wainwright, R.L., Liebrock, L.M. (eds.) Proceedings of the 2004 ACM Symposium on Applied Computing (SAC), pp. 52-57. SAC (2004)
16. King, J., Linden, O.: Data Mining Isn't a Cookbook Activity. National Underwriter 106, 11-12 (2002)
17. Kirn, S., Herzog, O., Lockemann, P., Spaniol, O. (Eds.): Multiagent Engineering Theory and Applications in Enterprises. Springer (2006)
18. Klusch, M., Lodi, S., Moro, G.: Agent-Based Distributed Data Mining: The KDEC Scheme. In: Intelligent Information Agents. Springer (2003)
19. Klusch, M., Lodi, S., Moro, G.: Issues of Agent-Based Distributed Data Mining. In: Proceedings of the 2nd International Joint Conference on Autonomous Agents and Multiagent Systems, pp. 1034-1035. AAMAS (2003)
20. Kohler, T.A., Gumerman, G.J., Reynolds, R.G.: Simulating Ancient Societies: Computer Modeling is Helping to Unravel the Archaeological Mysteries of the American Southwest. Scientific American 293, 76-83 (2005)
21. Lopez y Lopez, F.: Social Power and Norms: Impact on Agent Behaviour. PhD thesis, University of Southampton (2003)
22. Praca, I., Viamonte, M.J., Vale, Z., Ramos. C.: Agent-Based Simulation of Electronic Marketplaces with Decision Support. In: Wainwright, R.L., Haddad, H. (eds.) Proceedings of the 2008 ACM Symposium on Applied Computing (SAC), pp. 3-7. ACM (2008)
23. Rennard, J.P.: Artificiality in Social Sciences. In: Rennard, J.P. (ed.) Handbook of Research on Nature Inspired Computing for Economics and Management, pp. 1-15. IGR, Hershey (2006)
24. Symeonidis, A.L., Mitkas, P.A.: Agent Intelligence through Data Mining. Springer Verlag, New York (2005)
25. Tesfatsion, L.: Agent-Based Computational Economics: Growing Economies from the Bottom Up, Artificial Life, Vol. 8, No. 1, 55–82 (2002)
26. Tesfatsion, L.: Agent-Based Computational Modeling and Macroeconomics. In: Colander, D. (ed.) Post Walrasian Macroeconomics: Beyond the Dynamic Stochastic General Equilibrium Model, pp. 175-202. Cambridge University Press, Cambridge (2006)
27. Tozicka, J., Rovatsos, M., Pechoucek, M.: A Framework for Agent-Based Distributed Machine Learning and Data Mining. In: Durfee, E.H., Yokoo, M., Huhns, M.N., Onn, S. (eds.): 6th International Joint Conference on Autonomous Agents and Multiagent Systems, AAMAS (2007)
28. Stanek, S., Pankowska, M., Soltysik, A., Zytniewski, M.: Agent system application for geoinformation management at municipal office. Annals of Geomatics, Polish Association for Spatial Information, Vol VI, Number 2,Warsaw, 81-87
29. Vazquez-Salceda, J.: The Role of Norms and Electronic Institutions in Multi-Agent Systems Applied to Complex Domains. The HARMONIA Framework. Springer-Verlag, New York (2004)
30. Vazquez-Salceda, J., Aldewereld, H., Dignum, F.: Norms In Multiagent Systems: From Theory to Practice. International Journal of Computer Systems Science & Engineering 20, 225-236 (2005)

Chapter 21
CV-Muzar - The Virtual Community Environment that Uses Multiagent Systems for Formation of Groups

Ana Carolina Bertoletti De Marchi and Márcia Cristina Moraes

Abstract The purpose of this chapter is to present two agents' societies responsible for group formation (sub-communities) in CV-Muzar (Augusto Ruschi Zoobotanical Museum Virtual Community of the University of Passo Fundo). These societies are integrated to execute a data mining classification process. The first society is a static society that intends preprocessing data, investigating the information about groups in the CV-Muzar. The second society is a dynamical society that will make a classification process by analyzing the existing groups and look for participants that have common subjects in order to constitute a sub-community. The formation of sub-communities is a new functionality within the CV-Muzar that intends to bring the participants together according to two scopes: interest similarity and knowledge complementarities.

21.1 Introduction

Over the last years, we were able to notice people's increasing interest in making use of the available resources on the Internet to improve their knowledge and interact with others. The virtual learning communities have proved to be suitable environments for this practice, because their participants are related to the construction of knowledge based on common goals. According to Pallof and Pratt [1] the virtual learning communities are dynamical components that emerge when a group of people shares certain practices, they are interdependent, make joint decisions, identify themselves with something larger than the total sum of their individual relationships and establish a long term commitment with the well being (theirs, others' and the

Prof. Dr. Ana Carolina Bertoletti De Marchi
Universidade de Passo Fundo, Faculty of Informatics, BR 285, Bairro So Jose, Zipcode: 99052-900, Passo Fundo, Brazil e-mail: carolina@upf.br

Prof. Dr. Marcia Cristina Moraes
Pontificia Universidade Catolica do Rio Grande do Sul, Faculty of Informatics, Av. Ipiranga 6681, Prdio 32, Zipcode: 90619-900, Porto Alegre, Brazil e-mail: marcia.moraes@pucrs.br

L. Cao (ed.), Data Mining and Multi-agent Integration, DOI: 10.1007/978-1-4419-0522-2_21,

group's in all their inter-relationships). The idea of groups formation inside a virtual learning community is interesting, because learning in groups aims to develop and to improve individual skills, to accept responsibilities for individual and group learning process, to develop abilities of reflecting about its own suppositions (expressing its ideas to the group) and to develop social and group abilities.

This chapter intends to present the groups formation within the CV-Muzar (Augusto Ruschi Zoobotanical Museum Virtual Community of the University of Passo Fundo). The groups are called sub-communities and are formed from two concepts: interest similarity and knowledge complementarities. In order to automate sub-communities construction we use predictive modeling, more specifically classification. Two agents' societies are used in this process. The first society is a static society that intends to investigate the information about groups in CV-Muzar. The second society is a dynamical society that will analyze the existing groups and look for participants that have common subjects in order to constitute a sub-community. So, the first society is responsible for preprocessing data using search algorithms to collect information for sub-communities establishment and the second society uses the preprocessing data to executes a classification technique that is based on the Dependence-Based Coalition Model, established on the Social Reasoning Mechanism and Contractual Network, based on Sichman's Economic Market Theory [2].

The chapter is organized in four sections. The second section presents the background related to museums and virtual learning communities and data mining as well as related works. The third section describes the group formation proposal for CV-Muzar. The fourth section presents some final considerations and future works.

21.2 Background and Related Work

21.2.1 Interactive Museums and Virtual Learning Communities

Among the proposals for the educational work in museums, Silva [3] suggests the use of new technologies in the study techniques of exhibitions, allowing new forms of interaction with experiments. The interactive museums are rich, attractive and motivating places that involve visitors in the process of scientific investigation. To improve this process, virtual communities are being used.

Rehingold [4] describes virtual communities as "social aggregations that emerge from the Net when enough people carry on public discussions long enough, with sufficient human feeling, to form webs of personal relationships in cyberspace". Nowadays there are several kinds of virtual communities, such as: learning communities, practice communities and entertainment communities [5]. In this chapter we are going to work on virtual learning communities, that are communities which intends to promote an environment that favors knowledge construction, where the members of the community are related to common learning objectives.

Virtual learning communities can promote learning in two different ways: a formal one and an informal one. The formal way of learning considers that the community is going to be formed based on a real physical structure, where we have a teacher as a mediator of learners [1]. The informal way of learning considers that a community is going to be formed based on the personal interest of their members. These interests are going to define a net of self-organized relationships which have common goals that lead the members of a community into a continuous and permanent process of learning.

We approach in our work the informal way of learning in virtual communities. We share the idea of Souza [6] that learning occurs even apart of formal programs. One way to improve the informal learning process is to allow the members of a community to form groups or sub-communities. These groups will focus theirs discussions on specific subjects that are considered particularly interesting for their members. According to Vygotsky's [7] work, the use of groups is relevant because through discussions it is possible to have a knowledge consolidation and the findings of new solutions.

21.2.2 Data Mining

According to Han and Kamber [8], data mining refers to extracting or "mining" knowledge from large amounts of data. Data mining is an integral part of knowledge discovery in databases (KDD), which is the overall process of converting data into useful information. This process consists of a series of transformations steps, from data preprocessing to postprocessing of data mining results [9], as showed in Fig. 21.1. The input data can be stored in a variety of forms and can be located in a

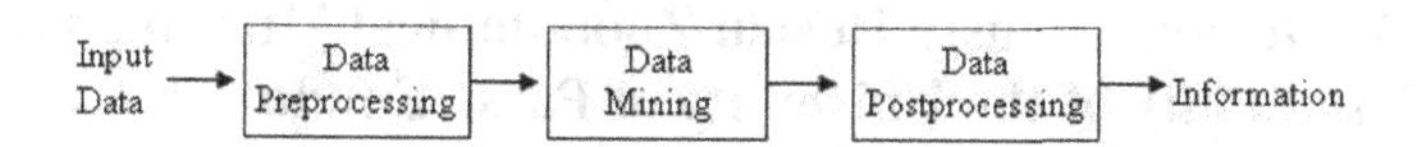

Fig. 21.1: The process of knowledge discovery in databases [9]

centralized repository or distributed in multiple sites. The objective of preprocessing is to transform the input data into an appropriate format to be used in data mining process. The postprocessing phase is responsible for integrating data mining results into a decision support system.

There are two kinds of data mining tasks: predictive and descriptive. The purpose of predictive task is to predict the value of a particular attribute based on the values of other attributes, one sample is predictive modeling. Predictive modeling intends to build a model for a target variable as a function of explanatory variables. There are two kinds of predictive modeling: classification (used for discrete target variables) and regression (used for continuous target variables). The purpose of the descriptive task is to derive patterns that summarize the underlying relationships in data [9], samples of this kind of task are: association analysis, cluster analysis and anomaly detection. Association analysis is used to discover patterns that describe associated features in the data. The discovered patterns are usually represented by implication rules. Cluster analysis aims to find groups that have closely related observations. Anomaly detection aims to identify observations whose characteristics are significantly different from the rest of the data.

In this work we use predictive modeling, more specifically classification. Classification is the task of learning a target function f that maps each attribute set x to one of the predefined class labels y. The target function is also known as a classification model. A classification technique (or classifier) is a systematic approach to building classification models from an input data set. Samples of classifiers include decision tree, rule-based classifiers and neural networks. Each technique employs a learning algorithm to identify a model that best fits the relationship between the attribute set and class label of the input data. Our classification technique is implemented through a multiagent system that uses an algorithm which implements Dependence-Based Coalition Model [2]. This algorithm is going to be presented in detail in section Sub-Communities Formations Assisted by a Multiagent System.

21.2.3 Related Works on Data Mining and Virtual Community

Silva et al [10] present distributed data minig algorithms focus on one class of distributed problem solving task executed by multiagent systems, analysis and modeling of distributed data. Gorodetskiy et al [11] describes an agent-mediated protocol based collaboration of distributed designers performing learning of agents of multi-agent distributed classification system. Lina and Hsuehb [12] propose a knowledge map management system (based on information retrieval and data mining) to facilitate knowledge management in virtual communities of practice.

21.3 CV-Muzar (Augusto Ruschi Zoobotanical Museum Virtual Community of the University of Passo Fundo)

The CV-Muzar (http://inf.upf.br/comunidade) was developed with the main purpose of involving the museum visitors more and make them part of the experience, putting an end to the passive receiver of the expositive speech that was established unilaterally [13]. For the environment development we made use of the virtual communities' concepts to promote the exchanges among the visitors and the Learning Objects (LOs) that comprise materials developed for the experiments, materials kept in the Museum and users' productions. In order to promote the exchanges among visitors and LOs, CV-Muzar is organized into two main modules: the repository manager of teaching resources (that comprises all the LOs available) and the learning environment support (that is responsible for organizing the didactic activities). Inside the learning environment support is the sub-module of sub-communities formation.

21.3.1 Sub-Communities Formations Assisted by a Multiagent System

The sub-community term represents the formation of small groups within the CV-Muzar. In these groups will take place discussions about subjects of common interest among the participants. A sub-community can be created by any participant previously registered in the environment and its formation (constitution of its components) occurs in light of two needs: interest similarity (group formed by participants that have similar profiles) and knowledge complementarities (group formed by participants, that are gathered to accomplish complex tasks which require the composition of abilities for problems solving).

The sub-communities formation is undertaken through a multiagent system. A multiagent systems (MAS) is a society formed by agents that coexist in the same environment and interact in order to accomplish a common goal [14]. In this work the multiagent system is composed by two societies: static (aims to investigate the information about groups), and dynamical (aims to analyze the existing groups and try to look for participants that have similar subjects to participate). Both societies are used in a data mining classification method. The static society, named Investigating Society of Sub-Community (SIS-C) is responsible for doing the preprocessing phase, preparing data do be used during the data mining process. The dynamical society, named Investigating Society of Participants (SIP) is responsible for data mining, executing a classification method that implements an algorithm based on the Dependence-Based Coalition Model, found on the Social Reasoning Mechanism and Contractual Network, based on Sichman's Economic Market Theory [2].

The next sections are going to explain how Investigating Society of Sub-Community (SIS-C) and Investigating Society of Participants (SIP) works.

21.3.1.1 Investigating Society of Sub-Community (SIS-C)

The Investigating Society of Sub-Community (SIS-C) is characterized as a kind of static organization, because the roles that each agent will play within the society are already pre-defined. The roles that an agent can execute inside this society are a service provider agent and a leader agent. The service provider agent is the one responsible to provide service to others. The leader agent is responsible to find out which services agents can fulfill the necessary requirements to execute a task. One of the society agents is defined as the leader agent, regardless of his knowledge. The first task that will be accomplished by the leader agent in SIS-C is the search for the group profile. This search is composed by the following steps: to verify all the information provided by the group coordinator (this information includes objectives, keywords, area of interest and communication tools used by the group); to verify if the proposed profile is not similar to any other existing profile and to verify the group's central theme, if it is in accordance with the environment central idea. If all the listed requirements above are in accordance, the other activities are carried out; otherwise, the leader agent sends a message to the group coordinator advising him about the items that must be reviewed. The other activities are the search within the concentration area for the sub-communities and the search within the interest area for the possible participants. In order to do these tasks, the leader agent has to distribute tasks among service provider agents. He does that establishing a communication with a list of services provider agents that could have the capability to fulfill the requirements demanded by a specific task.

The option to use a leader agent enables the interoperability among the heterogeneous agents that are part of the society. After the communication cycle among the SIS-C society agents is over, the obtained information is stored in a database for a possible migration of some static society agents to the dynamical society. Thus, if it is necessary to migration, the coalition formation can occur for the search of participants that have profile similar to the group.

With the first part of information about the stored groups in a database, it is necessary to obtain information about the participants in order to accomplish the formation of the sub-communities. This task is carried out by the Investigating Society of Participants (SIP).

21.3.1.2 Investigating Society of Participants (SIP)

The Investigating Society of Participants is characterized as the dynamical type, because in this kind of organization there is the need of social interaction. The option to use the Dependence-Based Coalition Model (DBC) for the dynamical organization is due to the fact that if the agent that integrates the society does not have the autonomy to carry out a certain activity, another member who can help is required. So, over the time the agents improve their knowledge about other agents.

The formation of coalitions on the DBC model occurs in the following way: (the steps of the model are written considering an example of a procedure that occurs in the society):

1. Choice of a goal: an agent Ag1 chooses a certain goal to be achieved. In case there is no longer a goal, agent Ag1 does not try to form coalitions anymore. The choice of a goal can be the search for participants that have interest in discussing issues about "Environment Pollution". The goal is always chosen based on the formed groups.
2. Choice of a plan: supposing that Ag1 chose the G1 goal, the next step is to choose a plan to accomplish it. As the agent can have more than one plan for the same goal, the choice of the plan is based on the notion of feasible plan. In case there are no more plans, step one is resumed. Based on the participants' profile, agent Ag1 can have several plans for this objective and this way he chooses one that can be used. If Ag1 finds the plan worthwhile, he executes the analysis of the plan actions.
3. Analysis of the plan actions: once a plan is chosen, Ag1 analyses its objective situations concerning G1, in case the situation is independent or dependent. If the situation is independent, Ag1 is considered independent to accomplish that

objective and this way does not need cooperation from any other agents. In the dependence situation, however, Ag1 cannot initiate the execution of his plan immediately, because he first needs to find an agent that accomplishes the action he does not know how to execute.

4. Choice of the partner: through the social resolution mechanism, Ag1 considers his relationships and dependence situations with the other agents related to G1 and through the pre-established criterion, Ag1 chooses the best possible partners. In case there are no possible partners for the actual action, Ag1 chooses a new plan to achieve G1 returning to step 2.
5. Coalition formation between the agents: once the best possible partner is chosen, here called Ag2, Ag1 will send it a coalition proposal, which can contain the following proposals:

    - Ag2 accepts the proposal and the coalition is formed. From this moment on, the works to solve G1 are started. At the end of this process, if the actions were accomplished correctly, G1 is considered concluded and an invitation is sent to the participants that have a profile similar to the group's to participate; and Ag1 can return to step 1;

    - Ag2 refuses the proposal and in this case Ag1 tries respectively to find another partner, returning to step 4. The proposal refusal by Ag2 can occur through the following factors:
        - Ag1 misunderstood Ag2, probably for having incorrect or incomplete information about Ag2. In this case, Ag2 informs such information to Ag1, and Ag1 can review his opinion about Ag2.
        - Ag2 did not find the proposal interesting for his goals.

The MAS uses the rules previously described to executes a classification method to look for sub-community participants.

Fig. 21.2 shows the algorithm that will calculate the total number of messages in a society with n agents to establish presentation communication and search for a partner, for a total number of cycles g. The algorithm is based on a previous analysis of the process within the CV-Muzar. At each cycle, all the communication between the agents takes place through the messages exchanges. The active agent sends messages of coalition proposal until he finds a partner or until there are no possible partners. The possible partner always responds to the coalition proposal sending a message of acceptance or review. When the active agent gets an acceptance message, he sends a coalition message establishing the agreement with the partner agent. If no partner is found the coalition message is not sent. Thus, considering a society with n agents, where m agents can accomplish the desired action, and coalition proposal messages are sent to k agents (means that k - 1 agents sent messages of refusal or review), the total number of sent messages in each cycle is: Scycle = 2m in case it didn't find any partner; Scycle = 2k + 1, where $0 > k \leq m$ and; Scycle = 0 if the agent is independent.

Now, considering a g cycles competition, the total number of exchanged messages between the agents after all the accomplished cycles (SDBC) is showed in

```
depint(n,g)
Sap = n* (n-1);        //calculates agents presentation
For i = 0 until i < g
   if Active autonomous then start a new cycle;
   else
     m = Active.searchPartner(plan);
     //search and calculates the total os possible partners
     k = 0;
     //initiates the amount of proposals carried through
     findPartner = false;
     while (k <= m) or (not findPatrner) do
          k = k + 1;
          If probably partner accepts proposal of collation
          formation then
             findPatrner = true;
             k = k + 1;
     end while
     Sc = Sc + k; //total of messages during coalition
                   //formation
     end else
     Stot = Sap + Sc; //total of exchanges messages
end for
```

Fig. 21.2: Algorithm that calculates the total number of messages in a society

Fig. 21.3. where: Scycle = 2m in case it didn't find any partner; Scycle = 2k + 1, where $0 > k \leq m$; Scycle = 0, if the agent is independent.

These societies were implemented using the platform JADE and they were integrated to CV-Muzar, which is implemented in PHP. After the societies SIS-C and SIP execute, the system sends an invitation e-mail to the community participants that were chosen by the society to participate in a sub-community. The participants can accept the invitation or not.

$$S_{DBC} = S_{presentation} + \sum_{i=1}^{g} S_{cycle} = n\,(n-1) + \sum_{i=1}^{g} S_{cycle},$$

Fig. 21.3: Total number of exchanged messages between the agents after all the accomplished cycles (SDBC)

21.3.2 Initial Evaluation

In order to evaluate the group formation functionality, we have conduced two kinds of evaluation, a user's evaluation and a usability evaluation. For the user's evaluation, an experiment in the formation module of sub-communities was carried out. Fifteen people related to the museum were invited. The participants were divided into two different groups. The first group received a small description of how they should fill out the individual profile and the subject nominations for the groups formation. The second group didn't have any help to filling out the individual profile as well as for creating the groups. Over the two weeks' tests, the participants were invited, through messages sent by e-mail, to take part in the sub-communities created by the two groups. In all simulations carried out, the MAS (considering the data mining process) nominated correctly the sub-communities related to the participant's profile. However, it will be necessary to optimize the processing time of the information exchange between agents on the Dependence-Based Coalition Model, because it took a long time for sending the invitations.

The tool used for the usability evaluation was Ergolist. This tool provides a way to evaluate the facility of use of interactive software considering ergonomics aspects [15]. The tool is composed of 18 checklists. Each one of these checklists comprises from three to twenty-seven subjects. The usability inspection has been completed for all environment. The data obtained through the inspection tests shows the percentage of conformity, or not, with certain characteristics in the system. Most of the applicable approaches obtained percentage of conformity higher than 50%. The best categories are: Explicit Actions (100%), Minimum Actions (100%), Consistency (90%), Density of Information (88,8%) and Grouping for Location (88,8%). In contrast, Users Experience (25%) and Flexibility (33,3%) are the worst categories. This occurs due to certain characteristics are not present in CV-Muzar such as: existence of dialogues with the user's abilities and various forms of presenting the same information to different types of user.

21.4 Final Considerations and Future Work

The main contribution of this work is to present a multiagent system that uses data mining classification implemented using dependence-based coalition model in order to automate the sub-communities construction. The integration of multiaget systems and Dependence-Based Coalition Model within a museum virtual learning community is a new idea.

So, in this chapter we presented the background related to the proposed sub-communities formation sub-module. We have described how data mining classification is executed by the two agents' society (SIS-C) and SIP executing an agent-mining interaction that involves users' preferences and human intelligence. In order to evaluate the proposal we made two initial evaluations and users' evaluation and a usability evaluation.

To examine carefully the data mining generated by the MAS, considering the Dependence-Based Coalition model (DBC) and to obtain a clear analysis of the exchange flow when the dependence-based coalition process occurs, studies are being carried out to analyze the time that the society takes to process these data and send the invitation message to the participants. In this way, we choose to work with the Exchange Values theme, considering that the nature of the social relations depends, mostly, on the proper representation of the norms and social conventions. We are analyzing the work proposed by Dimuro and Costa [16].

Thanks to: Conselho Nacional de Desenvolvimento Científico e Tecnológico for the support through the edictal MCT/CNPq 15/2007 - Universal.

References

1. R. M. Palloff, K. Pratt. Construindo Comunidades de Aprendizagem no Ciberespao: Estrategias eficientes para salas de aula on-line, Artmed, Porto Alegre, 2002.
2. J. S. Sichman. Du Raisonnement Social chez les agents: Une Approche fondee sur la thorie de la dependence. Thse de doctorat de l'NPG, Grenoble, Franca, 1995.
3. M. P. Silva. Ao Pedagógica: Uma questo a ser (re)pensada nos Museus de Arte. Porto Alegre: UFRGS, 2003. Master Thesis - Programa de Pos-Graduacão em Educacão, Universidade Federal do Rio Grande do Sul, Porto Alegre, 2003.
4. H. Rheingold. The Virtual Community: Homesteading at the Electronic Frontier, [1993]. http://www.rheingold.com/vc/book, Dec. 2008.
5. L. Palazzo et al. Comunidades virtuais de formaão tecnológica: fundamentaão pedagógica e metodologia de construo. In: XIII Simpósio Brasileiro de Informática na Educacão, 2002, UNISINOS.
6. R. R. Souza. Aprendizagem Colaborativa em Comunidades Virtuais, Florianópolis: UFSC, 2000. Master Thesis - Universidade Federal de Santa Catarina, Florianópolis, 2000.
7. S. L. Vygotsky. The Collected Works of L.S.Vygotsky. v. 1. New York: Plenum Press, 1987.
8. J. Han and M. Kamber. Data Mining: Concepts and Techniques. San Diego: Morgan Kaufmann Publishers, 2001.
9. P-N. Tan et al. Introduction to Data Mining. Boston: Addison Wesley, 2006.
10. J. C. Silva et al. Distributed data mining and agents. In: Engineering Applications of Artificial Intelligence Volume 18, Issue 7, October 2005, Pages 791-807.
11. V. Gorodetskiy et al. Infrastructural Issues for Agent-Based Distributed Learning. In: Proceedings of the 2006 IEEE/WIC/ACM international conference on Web Intelligence and Intelligent Agent Technology, 2006, Pages 3-6.
12. F. Lina and C. Hsuehb. Knowledge map creation and maintenance for virtual communities of practice.In: Information Processing & Management Volume 42, Issue 2, March 2006, Pages 551-568.
13. A. C. B., De Marchi. Um ambiente de suporte a comunidades virtuais baseados em repositório de objetos de aprendizagem informal em museus. 2005. Thesis, PGIE, UFRGS. Porto Alegre.
14. M. Wooldridge. Introduction to MuliAgent Systems. Wiley, 2002.
15. Ergolist Project (2000). Available at: http://www.labiutil.inf.ufsc.br/ergolist. Access in: 08/20/2008.
16. G. P. Dimuro, A. C. R. Costa. Exchange Values and Self-Regulation of Exchanges in Multi-Agent Systems: the provisory, centralized model. In: Engineering Self-Organising Systems: Revised Selected Papers of the Third International Workshop, ESOA 2005, Utrecht, The Netherland s, July 25, 2005 (Lecture Notes in Artificial Intellligence). Berlin: Springer, 2006, v. 3910, Pages 75-89.

Chapter 22
Agent based Video Contents Identification and Data Mining Using Watermark based Filtering

HeungKyu Lee[1]

Abstract This chapter describes the agent based video contents identification scheme using watermark based filtering technique. To prevent a user from uploading illegal video contents into the WEB storages, two strategies are employed. First stage is the upload blocking of illegal contents including copyright ownership information as a watermark when a user tries to upload the illegal video content. Second stage is to monitor illegal video contents that are already uploaded. For this stage, the monitoring agent obtains video content link information, and then extracts the watermark from corresponding content using the Open API. For two stage video identification strategies, two types of watermark extraction schemes are employed. Gathered data obtained from agents is analyzed using data mining method, and reporting process is done. To show the effectiveness of the described system, some experimental evaluation and test are conducted.

22.1 Introduction

The multimedia digital data and representation service has emerged something valuable and applied and spread to the commercial system. With this wide spread, the security issue is also emerged because it can be copied easily without loss of quality and illegally distributed on the high speed network. To resolve this issue, multimedia contents protection technique to prohibit the unauthorized copying and redistribution of multimedia contents has researched. It includes the digital rights management (DRM), digital watermarking [8], and the feature (or signature) based fingerprinting [1][2][9]. The DRM is means to protect multimedia contents based on the cryptography. It has a weakness that decrypted multimedia data can be redistributed easily. Meanwhile, digital watermarking provides secure contents protection scheme because the watermark is included into the content itself by modifying the pixel in-

Dept. of Electronics and Computer Engineering, Korea University, Seoul, Korea., E-Mail:hklee@ispl.korea.ac.kr

L. Cao (ed.), Data Mining and Multi-agent Integration, DOI: 10.1007/978-1-4419-0522-2_22,

tensity. Thus, it cannot be removed without destroying the value of the multimedia content. Therefore, embedding of unique user identification as a watermark into data can be used to identify illegal copies of multimedia contents. The identification number of a user buying content is embedded into the sold content. If an illegal copy is found, the malicious user's identification can be traced from the embedded identification. The feature based fingerprinting is means to find a highly similar content with a queried one in a stored feature database.

As a video content identification technique for protection of security and privacy, feature based fingerprinting technique is currently preferred choice. Recent works are done in [10][11][12]. However, the defect of the feature based fingerprinting technique is that the unique feature database of a large number of multimedia contents must be constructed [3]. In addition, feature based video fingerprinting system requires much time to extract video frame features and match them with feature vectors stored in a database. Thus, it is not sufficient to apply it to the practical and commercial products.

To cope with this issue, this chapter proposes the agent based multimedia contents identification scheme using watermark based filtering technique. Basically, watermark information embedded into multimedia data for enforcing copyrights must uniquely identify the data and must be difficult to remove, even after various media transformation process. To prevent a user from uploading video contents into the WEB server or the WEB hard disk, two strategies are employed. First stage is the upload blocking of illegal contents including copyright ownership using watermark by a contents blocking agent when a user uploads a video content. Second stage is to monitor and identify video contents that are already uploaded into the WEB storage. For this stage, a monitoring agent finds video contents that exists within a web server or in other locations, and extracts the watermark from them. To identify video contents that exist in other WEB server location, watermark extraction function using the Open API is provided.

22.2 Multagent Integration and Data Mining Concept

For agent based video contents identification, contents blocking agent and monitoring agent are used on a specific web site to provide data mining service for protection of security and privacy[18]. These agents can be located in every different web site [13][14]. The contents blocking agent only prohibits illegal contents by inspecting the presence of copyright ownership information using watermark extraction method. Meanwhile, the monitoring agent not only inspects the presence of copyright ownership information using more complex watermark extraction method, but also communicates with other site's monitoring agents to gather extra statistical information. This makes the multi-agent integration and data mining framework on distributed network environment [15][16][17]. This framework decreases unnecessary watermark extraction overloads that can be occurred repetitively by a monitoring agent.

Contents blocking agent is to prevent users from uploading copyrighted contents by simply extracting watermark information. Contents blocking agent has proactive role of preventing illegal contents distributions. This filtered some portion of illegal contents when users try to upload illegal contents. This decreases the server process overload of a monitoring agent to detect an illegal distribution of copyrighted contents using complex watermark extraction mode.

Monitoring agent extracts a watermark that represents copyright ownership information and traitor tracing information with complex extraction mode. This extracted information is stored in a database. Then, this information is queried to other monitoring agent located in other content providing server. If homogeneous content is stored on that server, the responded information is stored together to gather status information of the content with respect to the use of illegal distribution. This information provides statistical and relationship information that is not reliant on an existing database and has previously been discovered by monitoring agents of other specific sites without extra processing [18]. This decreases unnecessary watermark extraction overloads that can be occurred repetitively by a monitoring agent.

22.3 Watermark Embedding for Contents Identification

Two kinds of watermark information are embedded into the video contents as shown in Fig. 22.1. First watermark is embedded simultaneously when the contents are encoded for video on demand service. Meanwhile, second watermark is embedded when the demanded video is played. It requires real-time service. Thus, watermark pattern is generated at the display beginning time, and then represented on the graphics plane. The video frame is presented on the video plane of the STB device. To

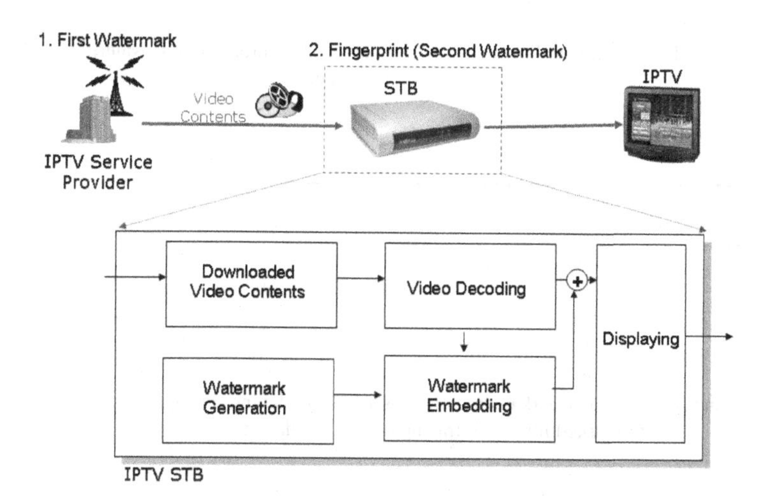

Fig. 22.1: Watermark embedding stage at the front-end and back-end system

reduce the computing time, the human visual system model is computed once per short time interval that we used 5 frame intervals experimentally. Finally, video and graphics plane are blended for displaying.

First watermark representing copyright ownership information is embedded at the back-end system. Two kinds of watermark signals are orthogonal between them because they have distinct secret key that is used for generation of random pseudo noise pattern. However, interference can be occurred between them if two signals are overlapped. Thus, first watermark is recursively embedded into specific time range positions TSi as shown in Fig. 22.2. The term, TSi represents the time range that first watermark can be embedded. It is assumed that an example video has 30 frames per second (FPS) as shown in Fig. 22.2. Second watermark representing buyer information is embedded at the front-end system that is set-top box device for IPTV. This watermark is called in fingerprints that are traitor tracing information. In the proposed system, the device ID and playing time information are embedded into specific time range positions TMi as shown in Fig. 22.3. The device ID can be associated with a user ID that was registered information in service application time. Thus, we can know his name and address if the embedded watermark is extracted. The playing time can be associated with a capturing time. From this information, we can know when and who captured the illegally uploaded video.

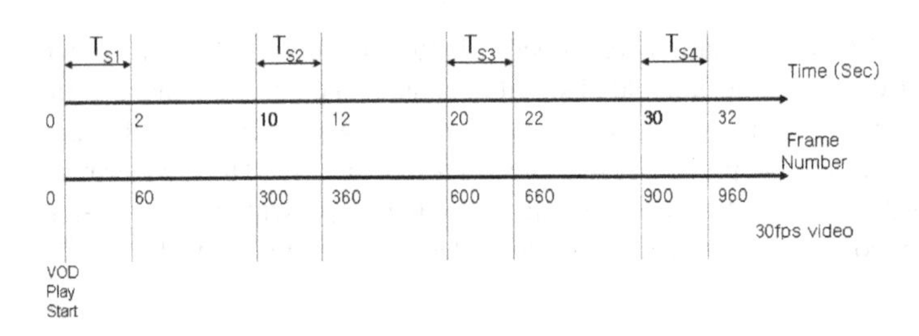

Fig. 22.2: First stage of watermark embedding: copyright ownership information is watermarked at the contents broadcasting side.

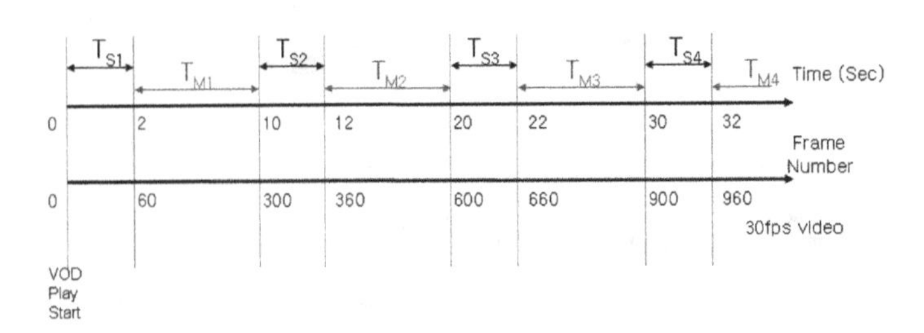

Fig. 22.3: Second stage of watermark embedding: User ID and playing time information is watermarked for traitor tracing at the STB device.

22.4 Agent based Content Blocking and Tracing

22.4.1 Contents Blocking Agent

The role of content blocking agent is to prohibit an illegal content from uploading into a Web server when a user tries to upload it. The one of the issues in video contents identification is fast computation time. Especially, file upload blocking technique requires nearly real-time computing speed because file upload speed is very fast. So, watermark extraction speed must be highly fast. To cope with this, the content blocking agent uses simple watermark extraction technique as shown in Fig. 22.4. Under consumption that the content is not attacked or only scale down attack is done, watermark extraction is performed. If the copyright ownership information embedded in Fig. 22.2 is extracted, the warning message is displayed on the user's screen and then uploading is prohibited. Generally, scaling down attack is frequently occurred one in case of a video intentionally or unintentionally. If we use a HD (High-Definition) video in watermark embedding time as shown in Fig. 22.1, we can predict that the content is transformed from a HD resolution to a SD one. So, direct geometrical recover is done, and then extracts a watermark. In addition, compression and digital filtering attacks do not change geometric parameters. Direct watermark extraction without considering attack and watermark extraction considering scaling down prediction are very useful strategy. Actually, 50% of illegal video contents can be prohibited in uploading time.

22.4.2 Monitoring Agent

The role of the monitoring agent is to examine already uploaded video contents once again by detailed watermark extraction method. In addition, it gather the illegal contents distribution information from other web site's monitoring agents.

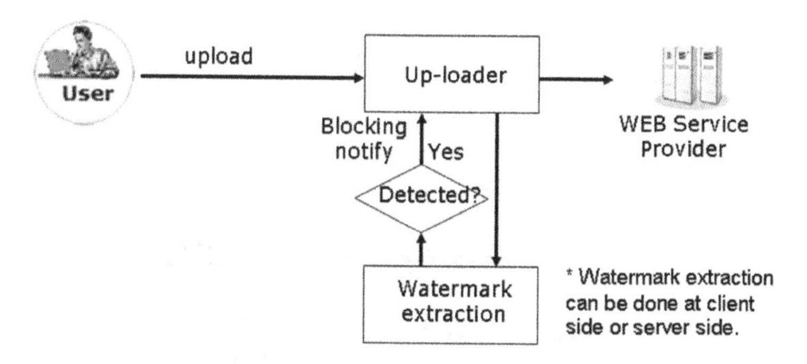

Fig. 22.4: Watermark based file upload blocking scheme using an agent. Watermark extraction can be done at client side or server side by the agent.

For traditional content searching, the web robot is generally used. It is often called as web wanderers, web crawlers, or web spiders. The web robot is a software based agent that automatically traverses the web's hypertext structure by retrieving a document and recursively all referenced documents. However, traversing the Internet and collecting contents requires tremendous time and storage. In addition, it can give the network and server overload. To reduce these faults, some effective method is proposed in [4]. For example, Digimarc's MarcSpider [5] is the traditional web robot based image tracking system that is the first trial to detect watermarked content on the web. MarcSpider uses hundreds of individual web spiders that look through the web to search for images that is watermarked one. However, it can not trace illegal distributor but prove only the copyright ownership information of that image. In addition, the traditional content searching method based on the web robot has still network and server overload even if the optimal searching is developed. It is a time consuming process. To resolve this issue, the monitoring agent provides the open API functions as shown in Fig. 22.5. The content search and watermark extraction functions are provided between the monitoring agents. For doing this, video content management database is constructed. The location of a video and service menu name is stored in a database. Thus, the monitoring agent at the site A requests contents link information to the one at the site B using the Open API function. Then the monitoring agent at the site A request the watermark extraction result of a specific content URL using a secret key to the one at the site B using the Open API function.

The contents link information is also stored in a database, and then its information is updated by periods. This content link information table includes the le-

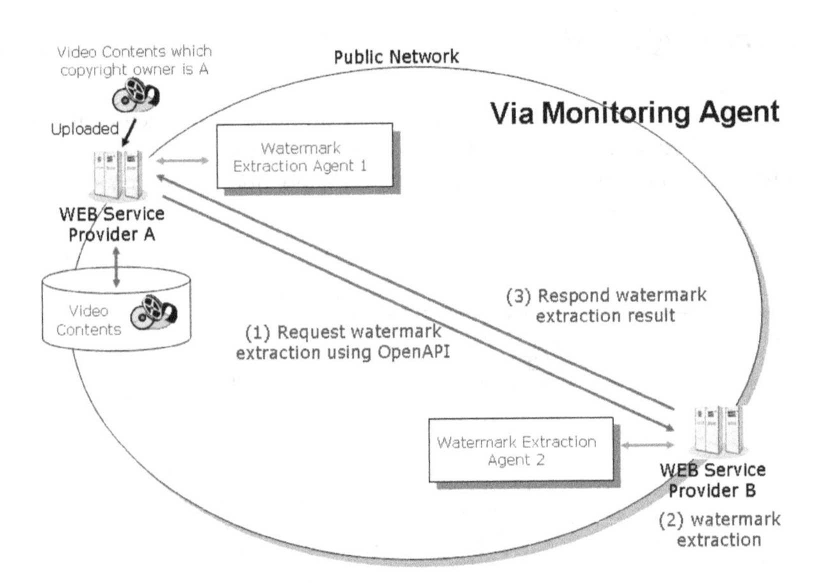

Fig. 22.5: Watermark based monitoring and identification scheme using an agent. Video contents can be identified irrespective of their location by using watermark extraction Open API function.

gal/illegal type field. This field is updated after watermark extraction. Thus, the monitoring agent has only to extract a watermark from a newly updated content. This process does not require network usage and burden for uploading and downloading contents. So, it provides the very practical content tracing function. The response of the Open API function is returned with the requested information. The monitoring agent saves only the updated link information by comparing it with the content link database. Then, the updated contents are requested for watermark extraction. This procedure can be repeated by periods.

22.4.3 Data Mining and Reporting Process

The watermark is extracted by the content blocking agent and the monitoring agent. These two agents are located in the specific site A. They have their own database to record a watermark extraction result and its behavior. The content blocking agent has a record that is content blocking result after watermark extraction. This information is only used as reference information because the illegal content is not uploaded. However, we can know how many contents the specific person tried to upload from the analysis. The monitoring agent has records that are copyright ownership information and its traitor tracing information of contents. This information can be directly provided to the content providers. In addition, it is notified to the specific site's administrator in order to delete the illegal content.

The gathered information by the content blocking agent and the monitoring agent is data mined to obtain the statistics and valuable information with respect to the uploaded contents, specific site, and illegal users. The result is represented by summary table and multiple level-of-abstraction rules to associate it with a meta content database [6][7]. This mining result exists in every web site. Thus, the copyright protection center such as governmental organization can gather the mined results in each web site for second data mining. After first mining phase of the process is finished, the system has produced the final set of illegal contents statistics. The final sets can be obtained from the web sites. Then we proceed to the post-processing step for data enrichment. Data originating from different parts of the websites are gathered and integrated with the data mining results. The goal of data enrichment is to take as input the illegal contents statistics, and correlate them with all the relevant fields of information. The enriched statistics information that was produced during the data enrichment phase can then be analyzed. Finally, a series of reports summarizes the results of the previous data analysis phases. According to the first watermark information, they can analyze which site or service has a defect.

22.5 Experimental Evaluations

The proposed method's prototype system has developed in order to detect and trace video contents in Windows XP environment. The database is constructed using SQL server, and the Aparch web server is used. The system use ODBC to manage the content link information and blocking information. The prototype system is tested on the designated web sites, which has watermarked video contents. In addition, illegal contents are tried to upload video contents. From the test results, we obtained 100% of retrieval rate. This retrieval rate is actually associated with the watermark extraction accuracy. The employed watermark extractor is developed against compression, digital filtering, scaling, and rotation attacks as shown in Fig. 22.6. Figure 11, (a) presents a robustness against frame rate change (30– >25, 30– >15), scanning mode change from interlaced to progressive, video format change from MPEG-2 to H.264/AVC, Flash, MPEG-4, and color conversion from color to gray. Figure 11, (b) presents a robustness against scaling down attack of a HD video from 1920*1080 resolution to 1280*720, 852*480, 640*360, 720*480, 320*240, and 400*300 one. Figure 11, (c) presents a robustness against scaling down attack of a SD video from 720*480 resolution to 360*240, 320*240, 400*300, and 500*400 one. Figure 11, (d) presents a robustness against rotation attack ranged between -5' and 5'.

Under the assumption of the content blocking agent, compressed and digital filtered watermarked videos are prohibited from uploading. In addition, known scaling attacked video is blocked. However, the watermark failure is occurred with respect to the compositely attacked videos. The content blocking agent is downloaded into the user's PC, and then is installed. The watermark extraction processing is done

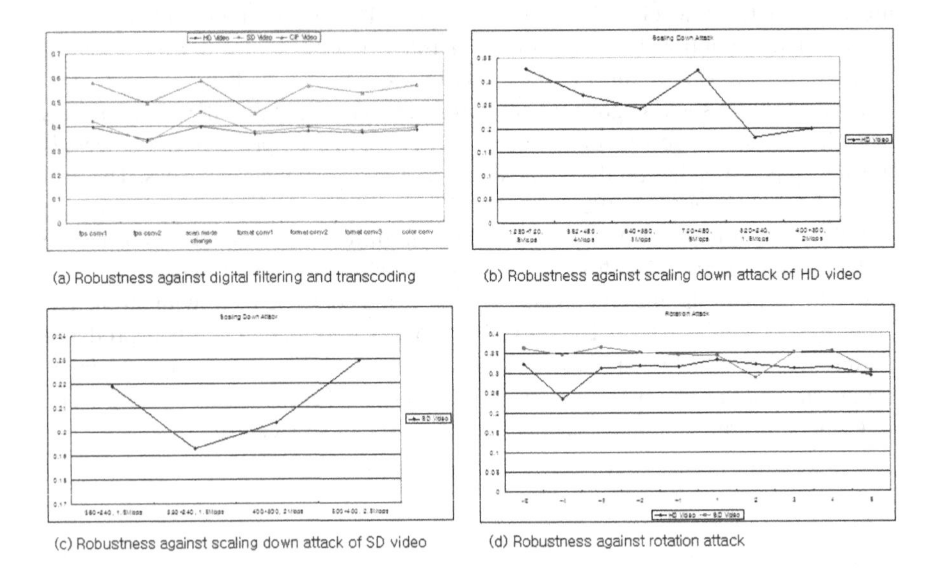

(a) Robustness against digital filtering and transcoding

(b) Robustness against scaling down attack of HD video

(c) Robustness against scaling down attack of SD video

(d) Robustness against rotation attack

Fig. 22.6: The watermark extraction robustness against attacks on user's machine. Thus, it does not give the processing overload to the web server. Meanwhile, 50% of CPU resource is required on the user's machine. The monitoring agent is installed on a PC server that is connected via the Internet with a web server because the watermark extraction requires high CPU and memory resource. It has access privilege of contents located in the web server. The monitoring agent obtains content link information using the Open API and can request watermark extraction. Meanwhile, the monitoring agent extracts a watermark from a specific content that is requested from the monitoring agent of the other site.

Fig. 22.7 represents comparison between agent-mining based and non-agent-mining based video content identification. Total 1,000 numbers of copyrighted video contents are used for experimental evaluations of site A and site B where 15% of redundant videos are included. In non-agent based video content identification, watermark detection rate was 95% and CPU overload was 80%. But, in agent based video content identification, watermark detection rate was same in non-agent based one while CPU overload due to watermark extraction was distributed. In addition, CPU overload is decreased into average 32.5% by removing unnecessary watermark extraction with communicating between the monitoring agents.

22.6 Conclusion

In this chapter, the agent based video contents identification scheme using watermark based filtering technique is described. To prevent a user from uploading illegal video contents into the WEB storages, watermark based filtering technique based on the agents are employed. First stage is the upload blocking of illegal contents including copyright ownership by extracting a watermark when a user try to upload illegal

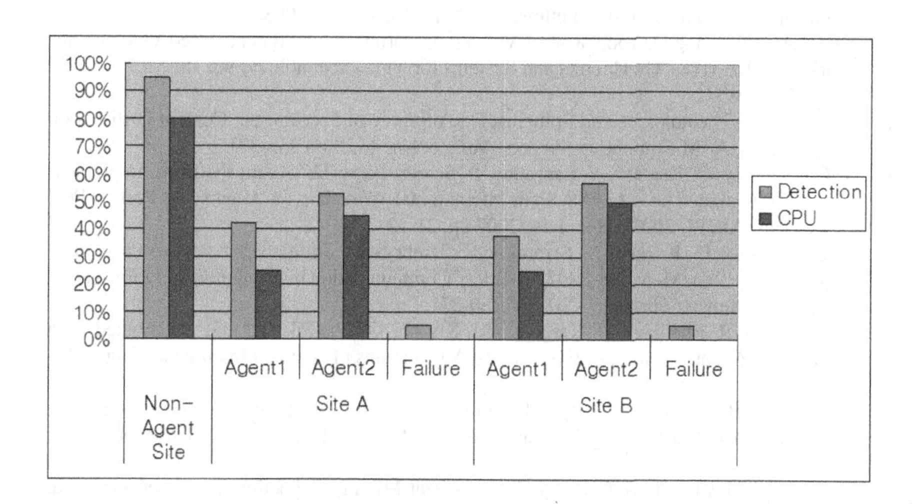

Fig. 22.7: The watermark extraction robustness against attacks

video content. Second stage is to monitor video contents that are already uploaded. For this stage, the monitoring agent obtains video content link information that exists within a web server, and then extracts the watermark from them using Open API. Gathered information obtained from agents is analyzed using data mining method, and reporting process is done.

References

1. J. Haitsma and T. Kalker, "A highly robust audio fingerprinting system," in Proc. Int. Conf. Music Information Retrieval, 2002.
2. P. Cano, E. Batlle, T. Kalker, and J. Haitsma, "A review of algorithms for audio fingerprinting ," in Proc. IEEE Workshop Multimedia Signal Processing, pp. 169-173, 2002.
3. B. Chor, A. Fiat, M. Naor, and B. Pinkas, "Tracing Traitors," IEEE Trans. Inf. Theory, Vol.46, pp.893-910, May 200
4. M. Koster, "WWW Robots, Wanderers and Spiders," URL:http://www.robotstxt.org/wc/robots.html
5. Digimarc Corporation, http://www.digimarc.com
6. J. Han, Y. Cai, and N. Cercone, "Data-Driven Discovery of Quantitative Rules in Relational Databases," IEEE Trans. on Knowledge and Data Eng., vol. 5, pp.29-40, 1993.
7. C. Bohm, S. Berchtold, and D. Keim, "Searching in high-dimensional spaces: Index structures for improving the performance of multimedia databases," ACM Comput. Surv. Vol. 33, no. 3, pp.322-373, 2001.
8. I. J. Cox, J. Killian, T. Leighton, and T. Shamoon, "Secure spread spectrum watermarking for multimedia," Tech. Rep. 95-10, NEC Research Institute, 1995.
9. J. Oostveen, T. Kalker, and J. Haitsma, "Feature extraction and a database strategy for video fingerprinting," in Proc. Int. Conf. Recent Adv. Vis. Inf. Syst., 2002, pp. 117-128.
10. Changick Kim and Bhaskaran Vasudev, "Spatiotemporal Sequence Matching for Efficient Video Copy Detection," IEEE Trans. On Circuits and Systems for Video Technology, Vol.15, NO.1, pp.127-132, Jan. 2005.
11. Li Chen, and F.W.M. Stentiford, "Video sequence matching based on temporal ordinal measurement," Pattern Recognition Letters, Vol.29, pp.1824-1831, 2008.
12. Sunil Lee and Chang D. Yoo, "Robust Video Fingerprinting for Content-Based Video Identification," IEEE Trans. On Circuits and Systems for Video Technology, Vol.18, No.7, pp.983-988, July 2008.
13. Androutsellis-Theotokis, S. and Spinellis, D.: A Survey of Peer-to-Peer Content Distribution Technologies, ACM Computing Surveys, Vol. 36, No. 4 (2004) 335-371.
14. V. Gorodetskiy, O. Karsaev, V. Samoilov, S. Serebryakov. P2P Agent Platform: Implementation and Testing. The AAMAS Sixth International Workshop on Agents and Peer-to-Peer Computing (AP2PC 2007), Honolulu, 2007 pp. 21-32.
15. V. Gorodetskiy, O. Karsaev, V. Samoilov, S. Serebryakov. Multi-Agent Peer-to-Peer Intrusion Detection. MMM-ACNS-2007. In series "Communication in Computer And Information Systems", volume 1, Springer 2007, pp. 260-271.
16. C. Giannella, R. Bhargava, H. Kargupta, M. Klusch, J.C. Da Silva (2005): Distributed Data Mining and Agents. Journal of Engineering Applications of Artifical Intelligence, 18(4), Elsevier Science
17. H Kargupta, B Park, D Hershberger, E Johnson, "Collective data mining: A new perspective toward distributed data mining," Advances in Distributed and Parallel Knowledge Discovery, 1999.
18. Longbing Cao, Chao Luo, Chengqi Zhang. Agent-Mining Interaction: An Emerging Area, AIS-ADM07, LNAI 4476, 60-73, Springer, 2007.

Index

Data Mining and Multi-agent Integration
DOI 10.1007/978-1-4419-0522-2

ERRATUM

Data Mining and Multi-agent Integration

L. Cao

Erratum to: Data Mining and Multi-agent Integration
Stanislaw A. B. Stane and Mariusz Zytniewsk

The original version of chapter 20, "Normative Multi-Agent Enriched Data Mining to Support E-Citizens," unfortunately contained a mistake. The correct spelling of the authors' names is as follows:

Stanislaw A. B. Stanek

Mariusz Zytniewski

The online version of the original article can be found under
DOI: 10.1007/978-1-4419-0522-2

L. Cao
University of Technology
Sydney, Australia